PESTICIDES IN THE ENVIRONMENT

Volume 1 (in Two Parts)
Part II

Edited by

ROBERT WHITE-STEVENS

CHAIRMAN, BUREAU OF CONSERVATION AND
ENVIRONMENTAL SCIENCE
COLLEGE OF AGRICULTURE AND
ENVIRONMENTAL SCIENCE
RUTGERS UNIVERSITY–THE STATE UNIVERSITY OF NEW JERSEY
NEW BRUNSWICK, NEW JERSEY

MARCEL DEKKER, INC., NEW YORK 1971

MARCEL DEKKER, INC. 95 Madison Avenue, New York, New York 10016

LIBRARY OF CONGRESS CATALOG CARD NUMBER 77-138499
ISBN# 0-8247-1784-8
PRINTED IN THE UNITED STATES OF AMERICA

CONTENTS OF VOLUME 1, PART I

CONTENTS

CONTRIBUTORS TO VOLUME 1, PART II

A. W. A. Brown, *Department of Zoology, University of Western Ontario, London, Canada*

Daniel MacDougall, *Executive Director of Research and Development, Chemagro Corporation, Kansas City, Missouri*

Bernard L. Oser, *Food & Drug Research Laboratories, Inc., Maspeth, New York*

C. H. Van Middelem, *University of Florida, Gainesville, Florida*

Chapter 4

THE ANALYSIS OF TECHNICAL PESTICIDES AND THEIR FORMULATIONS

Daniel MacDougall

EXECUTIVE DIRECTOR OF RESEARCH AND DEVELOPMENT
CHEMAGRO CORPORATION
KANSAS CITY, MISSOURI

4-1   INTRODUCTION

For the last ten to twenty years, the agricultural chemical literature has contained a great deal of information on the determination of pesticide residues in food and feed. Relatively less has been written on the methods that are used for the analysis of pesticide formulations. This is partly because most of the work on formulations is done in industrial laboratories and is never published, and partly because the interest in formulation analysis has been less than in residue analysis. It is not the purpose of this chapter to catalog the analytical methods for pesticides. To a large extent this has been done already (1–3). Rather, we discuss the analytical problems that face the pesticide manufacturer and the approaches that are used to solve these problems. As much of this work is unpublished, the author, of necessity, leans heavily upon the methods and approaches in use in his own laboratory.

In many respects, the control of the quality of a pesticide and its formulations is similar to the control of the quality of a pharmaceutical product. The manufacturer seeks to ensure the biological efficacy of his product together with minimum hazard to the biological systems (either plant or animal) that the product may come in contact with. To accomplish this the pesticide manufacturer carries out the following types of analyses.

(1) The product must be analyzed for its active ingredient content to ensure that it meets label claims, which is important not only to guarantee performance, but also to move legally in interstate commerce. Products are, therefore, continually checked by state analysts to ensure that the claimed levels for active ingredients are met.

(2) The physical properties of the formulation must be carefully controlled. The effectiveness of a formulation depends upon a satisfactory application, and therefore properties such as particle size distribution, flowability, wettability, and suspendability are determined with solid formulations, and emulsification and emulsion stability with liquid formulations.

(3) The nature of the impurities in the product is checked to ensure

uniformity of quality. This may change because of variations in manufacturing procedures, contamination of formulation ingredients, or use of formulation equipment that has not been thoroughly cleaned between runs of different materials. In addition, impurities, can materially alter the odor of a product, which is often a criterion used by the applicator for judging consistency and uniformity. It must be standardized.

(4) At periodic intervals, formulations are also checked for biological efficacy, phytotoxicity, and mammalian toxicity.

The analysis of pesticide formulations for active ingredient content is complicated by the presence of formulation ingredients, e.g., solvents, emulsifying agents, wetting agents, carriers, etc. In many cases it is necessary to separate the active ingredient from the formulation adjuvants before conducting the analysis.

In addition, changes often occur during storage, and a procedure that is used for formulations that have been stored must allow for interference from the additional impurities formed in this way. In some cases the method that is most conveniently used for manufacturing control is not suitable for storage stability purposes, and a new procedure must be developed.

It can be seen from the above description that the control of quality of a pesticide involves performance of a whole series of analyses rather than a single determination of active ingredient content only. It is obvious that several analytical methods may be required.

The Association of Official Agricultural Chemists plays a useful role in sponsoring collaborative studies to develop methods that eventually receive the Association stamp of approval as "official." This whole process requires a number of years, during which the methods are checked in a number of laboratories and are then initially recommended as "tentative procedures" and, finally, as "official methods." The length of time involved in making methods official is shown by the relatively small number of the newer pesticides for which "official" methods are described in the 1965 edition of the Book of Methods(3). Methods are described for only 17 or 18 of the modern synthetic organic pesticides.

In this chapter the basic problems in controlling the quality of pesticides are reviewed.

4-2 ANALYSES FOR ACTIVE INGREDIENT CONTENT

In deciding upon methods to use for pesticide quality control, the manufacturer is concerned primarily with specificity, speed, and simplicity.

Specificity is the most important consideration. It is obvious that a method must be specific for the active ingredient in a formulation if it is to be useful at all. This problem is a complex one and is discussed below. It is desirable, of course, to attain speed and simplicity, along with specificity. In the control of a manufacturing process, it is obvious that, economically, the most rapid method is the most efficient. In most cases the manufacturer supplies analytical methods to formulation plants or to government agencies. These groups usually do not have available the elaborate equipment present in industrial laboratories and, therefore, it is important that the methods be such that they can be employed in the ordinary analytical laboratory. To fulfill all these requirements, some compromises must be made and, in many cases, several alternative analytical methods are required.

When deciding upon the particular method to use for the pesticide analysis, the analytical chemist must first review the manufacturing process and the quality of the reactants employed to produce the final product. This will provide clues as to the possible contaminants in the technical material. Such contaminants may be unreacted raw materials, the products of side reactions, or the products of reaction with contaminants in the raw materials. The type of problem that may be encountered can be best illustrated by a specific example: the contaminants that were considered in developing analytical methods for fenthion.

Fenthion is manufactured by reaction of 0,0-dimethyl thiophosphoryl chloride with 3-methyl-4-(methylthio)phenol. The reaction is shown in Fig. 4-1.

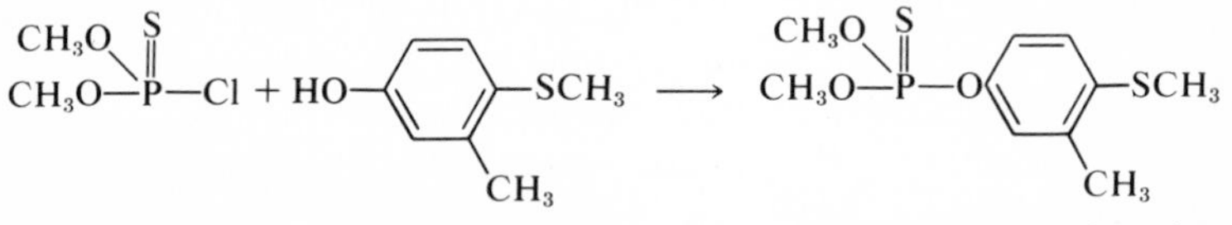

| O,O-dimethyl thiophosphoryl chloride | 3-methyl-4-(methyl-thio)phenol | fenthion (structural formula) |

Fig. 4-1

The 0,0-dimethyl thiophosphoryl chloride is formed by reaction of methanol with thiophosphoryl chloride (see Fig. 4-2).

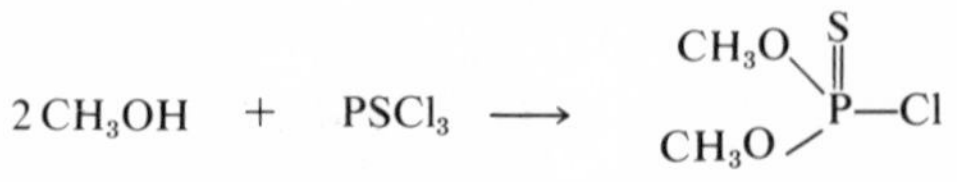

| Methanol | thiophosphoryl chloride | O,O-dimethyl thiophosphoryl chloride |

Fig. 4-2

It can be seen readily that the compounds CH_3OPSCl_2 and $(CH_3O)_3$ P=S could also be formed by this reaction. If trimethyl thiophosphate is produced, it will appear in the final product. If the monomethyl derivative is present, it will react with two molecules of 3-methyl-4-(methylthio)-phenol, as shown in Fig. 4-3. The derivative formed would appear in the final product.

3-methyl-4-(methylthio)phenol

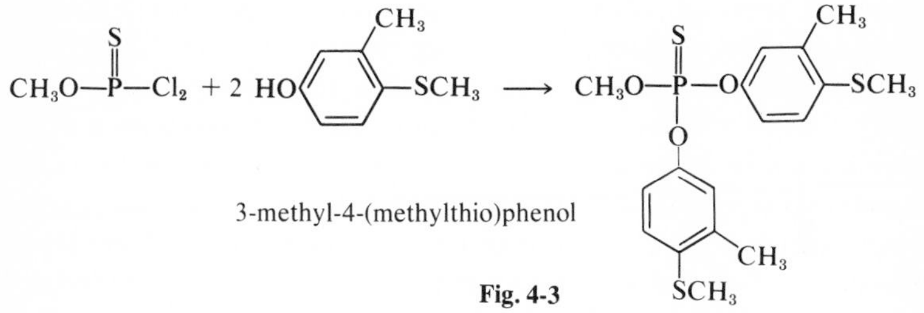

Fig. 4-3

Further, if there is any phosphoryl chloride ($POCl_3$) in the thiophos-phoryl chloride, this will result in formation of the corresponding phos-phate derivatives.

The 4-methylthiocresol is prepared from *m*-cresol. It is obvious that if any *o*-cresol or *p*-cresol is present, derivatives of these will be formed. It is also possible that, in preparing the methylthiocresol, some dimercapto-methyl derivatives may be produced.

In view of all these possibilities, it is apparent that as many as 20 or more related compounds could be present in technical fenthion. By purifying the reactants and controlling the coupling reaction carefully, however, these impurities can be held to relatively low levels. This exam-ple illustrates, however, the complexity of the problem which confronts the pesticide analytical chemist.

Analytical methods for pesticides can be based on a number of general procedures, and the scope to which different approaches have been utilized includes those discussed below.

4-3 VOLUMETRIC METHODS

A. Chlorine Determination

The determination of total chlorine is one of the most common methods used for the analysis of pesticides. This procedure is recommended, or can be used, for approximately one-quarter of the most commonly used pesticides described in Ref. (*1*). Among the pesticides for which this method is used are DDT(*3*, pp. 63–64), aldrin(*4*), dieldrin(*5*), chlordane (*3* [pp. 65–66],*6*), chlorobenzilate(*7*), ARAMITE(*8*), endrin(*4*), PERTHANE (*9*), methoxychlor(*10*), phenkapton(*11*), TDE(*12*), tetradifon(*13*), toxaphene(*14*), dichlorvos(*15*), CDAA(*16*), propanil(*17*), CDEC(*18*), phygon(*19*), and 2,4-D esters(*3*, p. 55).

The usual procedure involves reduction with sodium in isopropyl alcohol(*20*). There are many modifications of this method but usually the sample is refluxed with metallic sodium in 99% isopropyl alcohol. After refluxing for half an hour, the excess sodium is destroyed cautiously and the liberated chloride is titrated by the classical Volhard method. In cases where sulfides are present(*11*) it is necessary to oxidize with hydrogen peroxide before carrying out the chloride titration.

As an alternative to the sodium reduction procedure, a sodium biphenyl reduction has been used for many chlorine-containing pesticides (*3* [pp. 59–60, 63–64, 65–66], *4*, *5*, *13–15*). This procedure was originated by Pecherer et al.(*21*), after a study of various types of organosodium compounds for this purpose. Liggett(*22*) has described an improved method for preparation of the reagent. In the procedure the sample is extracted into a suitable organic solvent. Toluene has been used for aldrin(*4*) and dichlorvos(*15*). Benzene has been used for toxaphene(*14*). For tetradifon analysis(*13*), the sodium biphenyl is dissolved in ethylene glycol dimethyl ether and this reagent is added directly to either the technical pesticide or the formulation.

Many chlorinated pesticides are still analyzed for total chlorine content by the classical Parr bomb method(*2*, pp. 370–373). This method is applicable to most compounds.

With compounds containing readily hydrolyzable chlorine, increased specificity can be incorporated into analyses by exploiting this property rather than the determination of total chlorine. This method is applicable to compounds such as ARAMITE, BHC, DDT, TDE(*2*, pp. 347–350), and dicofol(*23*). The procedure involves refluxing with alcoholic potassium hydroxide. One or more molecules of hydrochloric acid are produced and this can be determined subsequently by the standard Volhard titration. Trichlorfon has been analyzed by a dehydrohalogenation in monoethanolamine(*24*). The sample is dissolved in absolute ethanol, monoethanolamine is added, and it is allowed to stand at room temperature for 1 hr. The liberated chloride is determined by the standard Volhard method. One must take great care that the reagents contain no moisture, or more than a single equivalent of chloride will be liberated from the trichlorfon. Even under very dry conditions there is a gradual increase in the amount of chloride liberated if the sample is allowed to stand for longer than the specified time. This latter criticism is also valid for the other pesticides for which dehydrohalogenation methods are used. The main reason that this method is not more widely used is the difficulty in achieving an exact stoichiometric reaction.

The official method for determination of heptachlor(3, p. 67) involves determination of labile chlorine only. This is accomplished by refluxing the sample with acetic acid in the presence of 0.1 N silver nitrate. The excess silver nitrate is determined by titration with standard sodium chloride.

The relatively labile chlorine in the herbicides simazine(25) and propazine(26) can be liberated by heating on a boiling water bath with morpholine. After dilution with distilled water, acidification with sulfuric acid, and cooling, the liberated chloride is titrated.

In general, the chlorine methods can be applied directly to pesticide formulations, provided that the other formulation ingredients contain no chlorine. The total chlorine methods are more widely applicable and are more precise than the dehydrohalogenation methods, and these are usually used even on compounds to which the latter procedures are applicable.

B. Hydrolytic Methods

Many pesticides are analyzed by some modification of a hydrolysis procedure. In some cases the amount of acid or base consumed in hydrolysis is determined. This type of procedure is subject to interference from any other hydrolyzable compounds which are present in the formulation. More specificity is obtained by analyzing for one of the products formed on hydrolysis.

The organophosphorus pesticides, demeton(27), disulfoton(28), and oxydemetonmethyl(29), have all been analyzed by hydrolyzing in a measured amount of standard potassium hydroxide. For disulfoton, 1.0 N sodium hydroxide in isopropyl alcohol is used. For demeton, 0.1 N of sodium hydroxide solution in isopropyl alcohol is used. For oxydemetonmethyl, 0.1 N aqueous sodium hydroxide is used. Disulfoton is hydrolyzed by boiling for 30 min while demeton is hydrolyzed by refluxing for 4 hr. Oxydemetonmethyl is hydrolyzed by standing at room temperature. After cooling, the excess base is determined by titration with standard acid.

Demeton is a mixture of two isomers: 0,0-diethyl-0-[2-(ethylthio)ethyl] phosphorothioate (I) and 0,0-diethyl-S-[2-(ethylthio)ethyl] phosphorothioate (II). As the biological efficacy of these isomers is somewhat different, a determination of the ratio of the isomers in the product is of importance. This can be accomplished by hydrolysis of the thiol isomer (II) in methanolic sodium hydroxide(27). The mixture is allowed to stand at room temperature for 5 min. The ethylthioethylmercaptan formed is titrated with standard iodine solution. The thiono isomer (I) is less than 2% hydrolyzed under these conditions.

Groves(30) has described a procedure in which the ratio of the isomers in demeton is determined by titration with bromine before and after alkaline hydrolysis. The thiono isomer reacts with 10 bromine equivalents both before and after hydrolysis. The thiol isomer requires 2 bromine equivalents before hydrolysis and 8 bromine equivalents after hydrolysis.

A variety of techniques has been employed in which hydrolysis is followed by determination of one of the decomposition products formed. The four dimethyl carbamate insecticides, dimetan, dimetilan, isolan, and PYROLAN, are analyzed by hydrolysis in an acid solution (1 N sulfuric) by refluxing for 15 min(31). The mixture is then made strongly alkaline and the liberated dimethylamine is distilled into an aqueous solution of boric acid. The dimethylamine is trapped in the boric acid and is titrated directly with standard 0.1 N sulfuric acid. The method is sensitive and reproducible. Ammonia, ammonium salts, or other volatile amines will interfere.

Carbaryl and its formulations can be analyzed by a similar procedure (32). In this case the technical compound or formulation is placed in a Kjeldahl distillation apparatus. The sample is hydrolyzed with alkali and the liberated methylamine is distilled into boric acid. It is then titrated with standard 0.1 N hydrochloric acid. The dimethylurea herbicides (monuron, diuron, and neburon) can be assayed by a procedure that is essentially identical to that described above for carbaryl, with the difference that the compounds determined with these products are dimethylamine and n-butyl methylamine(33). The organophosphorus pesticide dimethoate is analyzed(34) by acid hydrolysis to form methylamine, which is distilled into a known amount of standard acid. The excess acid is titrated with standard sodium hydroxide.

The carbamate herbicides CIPC(35) and IPC(36) can be analyzed by hydrolysis in acid and trapping of the liberated carbon dioxide in a measured excess of standardized sodium hydroxide. The hydrolysis is carried out by boiling the sample for 1 hr with a concentrated sulfuric acid–phosphoric acid mixture (1 : 10).

The dithiocarbamate fungicides, ferbam, ziram, thiram, maneb, zineb, nabam, and metham(3 [pp. 73–74], 37,38), can all be determined by hydrolysis with 4.5 N sulfuric acid. The formulation is boiled in an aqueous solution of EDTA and hot acid is added slowly to the boiling reaction mixture. The liberated carbon disulfide is absorbed in alcoholic potassium hydroxide in an absorption tower. On completion of the reaction, the contents are washed from the absorption tower with distilled water and the excess potassium hydroxide is neutralized with 30% acetic acid to the phenolphthalein end point. The neutralized xanthate is immediately titrated with standard 0.1 N iodine solution.

The triazine herbicides ametryne and prometryne(39) can be hydrolyzed by boiling with dilute sulfuric acid to produce methyl mercaptan. The liberated mercaptan is collected in 0.1 N iodine solution contained in a series of gas absorbers. The excess iodine is determined by titration with standard sodium thiosulfate. Any other compounds that liberate mercaptans or that form volatile oxidizable substances under the conditions of the determination will interfere.

Methyldemeton [0,0-dimethyl S-2-(ethylthio)ethyl phosphorothioate] can be analyzed by hydrolysis in ethanolic sodium hydroxide containing lead acetate(40). The mixture is heated for 2 hr at 60°C. The ethylthioethyl mercaptan formed is precipitated as the lead mercaptide. The lead mercaptide is extracted into chloroform in which it is soluble. In the final determination, a portion of the chloroform solution is acidified, and the liberated mercaptan is titrated with standard iodine solution.

One of the possible methods for analysis of mevinphos involves hydrolysis in 15% sulfuric acid(41). Acetone is produced and this is reacted with hydroxylamine hydrochloride to give one equivalent of hydrochloric acid. The hydrochloric acid formed is titrated with standard alkali. Any other compound that can be hydrolyzed under these conditions to form ketones will interfere. Of course, the presence of other acids, bases, or buffering salts in the hydrolysis mixture will necessitate quantitative separation of the acetone before it is added to the hydroxylamine reagent.

A hydrolysis method has been used as a general method for the analysis of endosulfan dusts, wettable powders, and emulsifiable concentrates(42). The endosulfan is hydrolyzed with methanolic sodium hydroxide by refluxing for 2 hr. After neutralizing the excess base, the inorganic sulfite produced is titrated with 0.5 N iodine solution.

C. Nonhydrolytic Methods

1. Perchloric Acid Titration. The direct titration of basic nitrogens with perchloric acid in glacial acetic acid has been used for analysis of a number of pesticides. The method was first described by Suter et al.(31) for diazinon. In the method described, 1-naphtholbenzein is used as an indicator. The method is applicable only if other compounds titratable with perchloric acid are absent. In the case of technical diazinon and diazinon formulations, other substances that react with perchloric acid are removed by washing a petroleum ether solution of the product with 3 N sulfuric acid. Most of the impurities in diazinon formulations are more basic than the active ingredient and are, therefore, separated by the sulfuric acid wash. The petroleum ether is then evaporated and the residue is dissolved in glacial acetic acid and titrated with 0.1 N perchloric acid in glacial acetic acid.

The triazine herbicides ametryne, atratone, prometryne, and prometone, can be determined by essentially the same method(31) as described for diazinon. Formulations of these compounds are extracted into ethyl ether. After evaporation of the solvent, the residue is dissolved in glacial acetic acid and titrated with 0.1 N perchloric acid in glacial acetic acid. By this method, one of the alkyl amino groups is titrated.

Dodine(43) can also be analyzed by essentially the same method. Dodine formulations are extracted by swirling for 3 to 5 min with a mixture (1:9) of glacial acetic acid and acetic anhydride. After filtering, the solution is titrated to methyl violet end point with standardized 0.05 N perchloric acid in glacial acetic acid. Urea, amides, and primary and secondary amines do not interfere with the dodine determination. Tertiary amines and organic or inorganic salts will interfere, however, as they are soluble in acetic acid and titratable in the same pH range.

2. Direct Titration. Several pesticides have sufficiently basic nitrogens to allow for a direct volumetric determination.

Glyodin(44) is first dissolved by heating in anhydrous isopropyl alcohol. After cooling to 5°C a solution of phenylisothiocyanate in isopropyl alcohol is added. The mixture is held at 5–10°C for 30 min. This decomposes the acetate and frees acetic acid. It is then titrated to thymol blue end point with 0.1 N hydrochloric acid in isopropyl alcohol.

3-Amino-s-triazole(45) can be analyzed by extracting with dimethylformamide. An aliquot is diluted with water, an excess of standard acid is added, and the solution is back titrated with standard 0.5 N sodium hydroxide. In the titration, base is added rapidly until a pH of 1.9 is reached, and then in 0.5-ml increments until pH 3.5 is reached. Base is again added rapidly until pH 6.5 is reached, then dropwise to pH 7.5. The equivalence point for the first inflection is determined by plotting pH against volume of base consumed. Next, the volume of base required to titrate from this point to the second equivalence point (pH 7.5) is determined. This volume of base is a measure of the 3-amino-s-triazole content of the sample.

Maleic hydrazide is formulated as a solution of the diethanolamine salt of maleic hydrazide in water (MH-30). The maleic hydrazide content of this solution can be determined(46) by dissolving it in absolute ethanol and then titrating with 0.1 N alcoholic sodium hydroxide to the thymolphthalein end point.

3. Titrations Following Reduction. Several pesticides contain groups that can be readily reduced. This property allows these compounds to be determined by either a direct or an indirect volumetric method.

Oxydemetonmethyl (0,0-dimethyl S-[2-(ethylsulfinyl)ethyl] phosphoro-thioate) contains a sulfoxide group that can be readily reduced with titanous chloride(29). The sulfoxide is reduced by treating oxyde-metonmethyl in concentrated hydrochloric acid with a measured excess of standard titanous chloride solution in the presence of a catalytic amount of ammonium thiocyanate. After refluxing for 30 min under nitrogen, the excess titanous chloride is determined by titration with standard ferric ammonium sulfate.

Chloranil (2,3,5,6-tetrachloro-1-4-benzoquinone) can be reduced to tetrachlorohydroquinone by potassium iodide in acetic acid. The liberated iodine is titrated with standard sodium thiosulfate(47). The parathion content of a formulation can be determined by extraction of the dust, the wettable powder, or the emulsifiable concentrate with ethyl ether. The ether solution is then reduced with zinc in acetic acid–hydrochloric acid. After heating gently for about 45 min on a steam bath, the reduced para-thion is titrated with standard sodium nitrite solution(3 [pp. 75–76], 49).

4. Oxidation. The pre-emergence herbicide 2,4-DEP is determined by dissolving the sample in alcoholic benzene-pyridine solution(48). The solution is then titrated rapidly with standard 0.1 N alcoholic iodine. The reaction is rapid.

$$(RO)_3P + I_2 \longrightarrow (RO)_3PI_2$$

The per cent of bis-2,4-dichlorophenoxyethyl phosphite can be determined by adding an excess of standard iodine to another portion of the solution and allowing it to stand for 15 min before titrating the unreacted iodine with standard alcoholic sodium thiosulfate solution. The bis compound reacts much more slowly with iodine than does the 2,4-DEP.

D. Determination of Kjeldahl Nitrogen

The secondary amine herbicide naptalam (ALANAP)(50) and the tertiary amine herbicides EPTC(51) and pebulate(51) can all be analyzed by means of the classical Kjeldahl nitrogen determination. In the case of the latter two compounds, care must be taken to prevent losses during the acid digestion. Dinocap(52) can also be determined by the Kjeldahl method provided that the salicyclic acid–zinc reduction procedure is followed prior to the acid digestion.

4-4 SPECTROPHOTOMETRIC METHODS

Spectrophotometric methods are not used so extensively as volumetric procedures for the analysis of pesticides and formulations. The reason for this is that the precision is generally not as great as with the classical macro chemical procedures. In many instances, however, the spectrophotometric procedures, especially in the infrared portion of the spectrum, are more specific than classical volumetric methods. Infrared measurements are generally more specific than are colorimetric or ultraviolet absorption methods, although this is not invariably so.

A. Infrared Methods

Infrared analysis methods have been described for a large number of pesticides. Methods are available for the chlorinated hydrocarbons: aldrin(3 [pp. 60–62], 4, 53), dieldrin(3 [pp. 60–62], 5, 53, 54), DDT(3 [pp. 64–65], 54–56), BHC(3 [pp. 66–67], 56), endrin(3 [pp. 60–62], 53), methoxychlor(10), toxaphene(14, 55); for the organophosphorus pesticides fenthion(57), naled(58), disulfoton(28, 59), azinphosmethyl (60, 70), malathion(3 [pp. 72–73], 61), mevinphos(41, 62), methyl parathion(62), phorate(3 [pp. 76–77], 63, 64), dichlorvos(3 [pp. 66–67], 15, 65), and DEF(66); for the dithiocarbamate fungicides(67); and for rotenone(3 [pp. 49–50], 68, 69). The measurements are most commonly made in carbon disulfide, chloroform, or carbon tetrachloride. Cyclohexane(63, 64), isooctane(28), and dimethoxyethane(70) have also been used as solvents.

Care must be taken with infrared methods to remove any interfering absorption. In the procedure for methoxychlor(10), five possible wavelengths for measurement are proposed – the one of choice being the one at which there is the least interference. The amount and nature of the interference vary with the formulation. With DEF emulsifiable concentrate, the emulsifying agents interfere with the P=O absorption band at $8.33\,\mu(66)$. These can be removed by passing the formulation through a column of acid-washed alumina and eluting the DEF with carbon tetrachloride. If the interference is not too great, a calibration curve can be developed in the presence of the interfering substances. This type of procedure has also been described for naled(58).

Mevinphos(41) is separated from solid formulations by placing the sample on top of a column of HY-FLOW SUPER-CEL and eluting with chloroform. Any interfering substances are absorbed on the clay. For liquid formulations, a magnesium silicate column is used with hexane that contains a little chloroform as the eluting solvent.

With toxaphene emulsifiable concentrate formulations, interfering emulsifiers are removed by passage through an alumina column. Benzene is used as a solvent(*14*).

B. Ultraviolet Methods

Ultraviolet spectrophotometric methods are not as widely used for analysis of pesticides and formulations as infrared procedures. The reason for this is that they are somewhat less specific. In common with other spectrophotometric methods, the precision is not as great as with volumetric or gravimetric methods.

A variety of solvents is used for these methods. Parathion and dicofol are measured in ethanol(*49,71*), carbophenothion, and METHYL TRITHION in isooctane(*72*), coumaphos in dioxane(*73*), barban in a mixture of ethylene dichloride and *n*-hexane (2 : 1)(*74*), warfarin in dilute aqueous alkali(*75*) or petroleum ether(*3*, p. 54), and CDEC in methanol(*18*). 2,4-D and 2,4,5-T esters have also been determined by ultraviolet absorption after clean-up on a FLORISIL column(*76*). By the difference in absorbance, mixtures of these two can be analyzed.

Barrette and Payfer(*77*) have described an ultraviolet spectrophotometric method for rotenone in formulations of low concentration. The final measurement is made in 95% ethanol.

C. Colorimetric Methods

In general, it is possible to obtain greater specificity with colorimetric methods than with ultraviolet spectrophotometric methods. The reason for this is that the colorimetric determinations are usually carried out on a derivative of the pesticide, or on a hydrolysis product. The two- or three-step procedure confers greater specificity than is obtained with a direct absorptiometric procedure.

Colorimetric methods have been employed more widely for residue, however, than for formulation analyses.

A colorimetric method applicable to aldrin(*78*), dieldrin(*5*), and endrin (*79*) involves reaction of these materials with phenyl azide and heating to form phenyldihydrotriazole(*80*) derivatives. The triazole is dissolved in ethanol and treated with hydrochloric acid. The secondary amine which results is coupled with diazotized dinitroaniline to form an intense red color. The method is very sensitive and, therefore, a very high dilution factor is necessary. This reduces the precision of the method, but it is reported that, with care, an accuracy of about 3% can be obtained with formulations.

The colorimetric method originated by Polen and Silverman(*80*), is recommended as an alternate method for analyzing heptachlor formulations. By this method, a benzene or hexane solution of the active ingredient is allowed to react with a solution of monoethanolamine and potassium hydroxide in butyl cellosolve to produce a reddish color. The transmittance of the solution is measured at 564 mμ. Other common insecticides do not interfere. The mean relative error is reported by the authors to be ± 1.5 to 2.0%.

The phosphorodithioate pesticides malathion, azinphosmethyl, and ethion can all be analyzed by the method first described by Norris et al. (60, 81, 82). In this method, the pesticide is decomposed by alkali, and the sodium 0,0-dialkylphosphorothioate is converted to a copper chelate that is soluble in carbon tetrachloride (60, 81) or cyclohexane (82) and is intensely yellow in color. The color is measured at 418 mμ. This method has the advantage of excellent specificity, as only phosphordithioates will form the colored complex. Upham (61) modified the original procedure by replacing the carbon tetrachloride with cyclohexane, the colored complex being more stable in the latter solvent. Ware (83, 84) further modified the method by eliminating the standard curve and carrying through a comparison standard with each sample. Orloski (85) reports a thorough evaluation of the variables such as temperature, malathion concentration, volume of alkali, shaking time, etc.

A colorimetric method has been described for parathion (49) in which an alcoholic extract of dusts or wettable powders is hydrolyzed with potassium hydroxide and the resulting potassium p-nitrophenol is determined colorimetrically at 400 mμ. The official method (3, p. 76) calls for measurement at 405 mμ. Helrich (86) reported that the absorbance of the solution of potassium p-nitrophenol must be measured within 90 sec after dilution to volume to obtain consistent results.

In 1948 Averill and Norris (87) described a method for determination of parathion that was based on reduction of the nitro group with hydrochloric acid and zinc dust. The amino compound which is formed is then diazotized and coupled with N-1-naphthylethylenediamine to produce a red colored complex which can be measured at 560 mμ. The same principle has been applied to the analysis of azinphosmethyl formulations (88). Azinphosmethyl is readily hydrolyzed by 0.5 N potassium hydroxide in absolute isopropyl alcohol at room temperature. The anthanilic acid formed by decomposition of the benzazimide group can be diazotized and coupled as in the method described for parathion (87). The resulting red colored complex is measured at 555 mμ.

The fungicide DEXON is water soluble and forms an intensely yellow colored solution (89). This property is utilized for quantitative estimation of the compound in formulations. The compound and the color are rapidly decomposed by light. However, the addition of sodium sulfite stabilizes the color and allows for a satisfactory quantitative determination.

Yuen(90) has described a colorimetric method for the determination of carbaryl in formulations. The carbaryl is coupled with either diazotized sulfanilic acid or sulfanilamide at room temperature in an alkaline solution. The maximum color develops in 10 min and is stable for another 10 min. The color is read at 520 mμ. It has been shown that 1-naphthol in amounts up to 10% of the formulation (equivalent to 20% of the carbaryl content of 50% wettable powder) does not interfere.

Colorimetric methods have been described for the analysis of pyrethrins (91), cyclethrin(92), and piperonyl butoxide(93). The method for cyclethrin(92) is based on the fact that heating with reagents containing orthophosphoric acid produces a red color with maximum absorbance at 545–550 mμ. In the determination, the compound is extracted into petroleum ether. After evaporation of the solvent, the residue is heated with 85% orthophosphoric acid for 40 min in a boiling water bath. The color is suppressed by piperonyl butoxide. However, this can be separated by chromatography on silicic acid. Pyrethrins also give the color reaction. Allen, Beckman, and Fudge(94) used partition chromatography to separate pyrethrins from piperonyl butoxide. The pyrethrins are determined colorimetrically by heating with a mixture of ethyl acetate and 85% orthophosphoric acid (4:1) as described by Williams et al.(91). Piperonyl butoxide is determined in solutions of the compound by heating with purified tannic acid in a mixture of phosphoric and glacial acetic acids(93, 95). A stable blue color is formed. The reaction appears to be specific for piperonyl butoxide and closely related compounds. The color is not obtained with many other compounds containing the methylene dioxyphenyl group but it is produced by 3,4-methylenedioxy-6-propylbenzyl alcohol.

4-5 GAS–LIQUID CHROMATOGRAPHIC METHODS

The gas–liquid chromatographic (GLC) method has recently gained favor in the analysis of pesticide formulations. It is rapid, simple, and highly specific. It is extremely sensitive, however, and it is difficult, therefore, to obtain the precision that is possible with the standard volumetric or gravimetric procedures. A great many pesticides can be detected and determined quantitatively by this procedure.

Burke(96) has recently given an excellent description of the use of programmed temperature gas chromatography for the separation of a large number of chlorinated hydrocarbon pesticides. Barrette and

Payfer (97) have described column conditions for the chromatographing of pesticides of all types. In this study, a 6-foot column filled with 30/60 mesh Chromosorb W, coated with 20% (w/w) Dow-Corning silicone grease, and the hydrogen flame detector were used. Bonelli et al. (98) have reported similar information. In the latter study the electron-capture detector was used. This detector is too high in sensitivity for analysis of pesticide formulations.

A few instances of specific cases in which gas–liquid chromatography has been used for analysis of pesticide formulations include the following.

Formulations of the chlorinated hydrocarbon insecticides aldrin (4) and dieldrin (5) have been analyzed by this method. For these determinations, columns packed with silicone oil on firebrick are used. At a column temperature of 240°C and a flow rate of 100 ml/min, the retention times for aldrin and dieldrin are 14 min and 35 min, respectively. These compounds are extracted from solid formulations with n-hexane, which can be applied directly to the chromatographic column.

A gas–liquid chromatographic method for endrin has been described by Phillips et al. (99). The column used consisted of 710 silicone oil on GC-22 Supersupport. At 230°C two peaks were obtained, neither of which was due to endrin itself. These compounds are formed by isomerization of endrin on the column. These results emphasize the need for caution in interpreting results from high-temperature gas chromatograms.

Wesselman and Koons (100) have reported a GLC method for heptachlor using lindane as an internal standard. The 6-ft column was composed of silicone gum (5% w/w) on Chromosorb W. The detection was by means of a radium sulfate ionization detector. The precision obtained for analysis of heptachlor on fertilizer (approximately 3%) was ±3% of the amount present. Koons and Wesselman (101) have described a similar procedure for determining lindane and DDT in mixtures. Lindane is used as an internal standard. The procedure is reported to give a relative standard deviation of ±3.62% for lindane and ±1.83% for DDT.

Zweig and Archer (102) have described a procedure for gas–liquid chromatographic determination of endosulfan emulsifiable concentrate, wettable powder, and dust formulations. The method has the advantage over other available methods in that the endosulfan isomers are separated and can be estimated separately. The endosulfan is extracted into benzene and chromatographed on a silicone-Chromosorb column.

Gas chromatography is recommended for the determination of tetradifon (13) in technical material, wettable powders, and dusts. The tetradifon is extracted into benzene and passed through a silicone-Chromosorb column. Good precision is claimed for this method.

Gas chromatography is the recommended procedure for analysis of formulations of the herbicides EPTC(*51*) and pebulate(*103*). With these compounds, liquid emulsifiable formulations can be introduced directly into the chromatograph. Solid formulations are extracted with a mixture of anhydrous methanol, water, and benzene (15:1:25); more water is then added, and a portion of the benzene phase is removed, dried with anhydrous sodium sulfate, and chromatographed. Chloroform saturated with water is also used to extract EPTC from granular formulations. At a column temperature of 160°C the retention time of EPTC is about 8 min. Pebulate is retained on the column for a longer time than EPTC. By raising the column temperature to 190°C however, the retention time of pebulate can be reduced to about 6 min. The results from 18 injections were reported to show a relative standard deviation to be ±1.2%. The authors do not make it clear, however, whether this deviation is for the entire determination or just for the chromatographic step.

Gas–liquid chromatography is the obvious method of choice for volatile halogenated grain insecticides, nematocides, and soil fumigants such as dichloropropene/dichloropropane, ethylene dibromide, methyl bromide, and 1,2-dibromo-3-chloropropane. Archer et al.(*104*) have described a procedure for the last mentioned compound.

4-6 POLAROGRAPHIC METHODS

Polarography is not used widely for the analysis of pesticide formulations. Relatively few pesticides contain the necessary readily oxidizable or reducible groups.

The fungicide DEXON can be reduced at the dropping mercury electrode (*105*). The solution is stabilized by the addition of dilute sodium sulfite. A small amount of 0.2% TRITON x-100 is added as a maximum suppressor. The polarogram is from 0 to −1.5 V. The method has a standard deviation equivalent to about 4% of the active ingredient content.

Giang and Caswell(*106*) have described a procedure for the determination of trichlorfon in formulations. The compound is reduced at the dropping mercury electrode at 25°C in an aqueous solution containing 0.2 N potassium chloride as a supporting electrolyte and 0.002% gelatin as a maximum suppressor. An accuracy of about 2% is claimed for the method.

Gajan and Link(*107*) have reported on the use of the oscillopolarography for the analysis of DDT and a number of other chlorinated hydrocarbon pesticides. It was found that under certain conditions only those compounds that contain the trichloroethane group are reduced at the dropping mercury electrode. Methoxychlor, *p,p*-DDT, *o,p*-DDT, and dicofol give well-defined waves, while TDE, PERTHANE, and DDE produce no response. Methods are described for the analysis of both technical DDT and formulations.

It has been reported(13) that the Philips-Duphar Company uses a polarographic titration method for the routine analyses of tetradifon.

4-7 MISCELLANEOUS METHODS

When formulations are free of interfering materials, elemental analyses can often be used. These procedures are relatively nonspecific, of course. Elemental analyses are particularly useful for manufacturing control when what is required is actually a determination of the degree of dilution of a technical product with various formulation ingredients.

With this procedure the technical material is usually analyzed by a more specific method, in addition to the element that is subsequently assayed in the diluted formulation. Chlorine assay in the analysis of chlorinated hydrocarbon pesticides and Kjeldahl nitrogen determinations for certain nitrogenous materials are examples of element analytical procedures already described.

LETHANE(3, pp. 74–75) is analyzed by a method that is based upon decomposition of the organic thiocyanates by reaction with alkaline polysulfide to form soluble potassium thiocyanate. The thiocyanate is precipitated as cuprous thiocyanate, which is separated by filtration and analyzed by the Kjeldahl method. The separation procedures eliminate interference by other nitrogen-containing compounds.

The determination of phosphorus has been used for analysis of formulations of several organophosphorus pesticides. In the method described for carbophenothion(72), the active ingredient is extracted from dry formulations with isooctane. It is then oxidized with a mixture of concentrated nitric and perchloric acids. The inorganic phosphate formed is determined by precipitation as ammonium phosphomolybdate followed by filtration, solution in an excess of standard alkali, and titration of the excess base with standard acid. It is reported that mevinphos(41) and dichlorvos(15) formulations are also analyzed by determination of total phosphorus.

The thiono sulfur content of phenkapton is used to analyze formulations of this compound(108). The phenkapton is oxidized with elemental bromine. The sulfur linked to the aromatic ring is converted to an arylsulfonic acid. The procedure oxidizes aliphatic and thiono sulfur atoms to sulfate which is determined gravimetrically by precipitation as barium sulfate.

The official method for the determination of pyrethrins in formulations is a mercury reduction procedure(3, pp. 50–52). For the determination of pyrethrin I, the formulation (liquid or solid) is extracted with petroleum ether. After evaporation of the solvent, the residue is hydrolyzed with alcoholic sodium hydroxide to form chrysanthemum monocarboxylic acid. This compound reduces the mercury in Deniges reagent. The mercurous mercury is precipitated as Hg_2Cl_2 and filtered. The precipitate is suspended in hydrochloric acid and quantitatively oxidized by titration with standard potassium iodate. Pyrethrin II is hydrolyzed to chrysanthemum dicarboxylic acid, which is then titrated with standard sodium hydroxide.

The Association of Official Agricultural Chemists mercury reduction method(3, pp. 50–52) has not proved satisfactory for the analysis of allethrin. Several methods have been developed for its analysis. Hogsett et al.(109) describe a procedure in which the allethrin is converted to the amine salt of chrysanthemum monocarboxylic acid by reaction with ethylenediamine. This is titrated with sodium methylate in a pyridine solution.

A unique procedure has been used for the determination of the naled content of technical material(58). This involves the determination of the set point of the material. The sample is melted completely by heating gently (not above 35°C). The sample is supercooled to about 10°C and seeded with a crystal of naled. As the sample solidifies, the rise in temperature is recorded. The highest temperature noted is taken as the setting point. The percentage of naled is determined from a previously prepared graph in which set point is plotted against percentage purity. There is a linear relationship from 88% purity to 100%. The set point rises to about 26°C for pure material. This is only suitable for technical material.

Dieldrin(5) in formulations (liquid formulations, wettable powders, and dusts) can be determined by reaction of the epoxide group with hydrogen bromide. The sample is extracted with a mixture of carbon disulfide and acetone, and aliquots are reacted with an excess of anhydrous hydrogen bromide at room temperature. This forms the bromhydrin of the active ingredient. The excess hydrogen bromide is titrated with standard alcoholic sodium hydroxide to a thymol blue end point. Aldrin interferes to some extent with this determination. Any other organic materials that react with hydrogen bromide will also interfere. This would include olefinic or acetylenic materials, some emulsifiers, and other epoxy compounds.

Best and Hersch(110) have described a cryoscopic method for the analysis of the organophosphorus pesticides malathion, dimethoate, 0,0-diethyl 0-2-pyrazinyl phosphorothioate, and phorate. The cryoscopic method was selected because it does not require a standard, is not subject to interference from other compounds, and yields per cent purity on a molar rather than a weight basis. The method involves measurement of the equilibrium temperature of a solid–liquid system as the compound being analyzed is slowly frozen or melted. The method is especially useful in the analysis of a compound in which the purity approaches 100%. For each compound the time–temperature melting curve was analyzed by the Taylor–Rossini(111) or Witschonke(112) methods to determine the melting point of the pure compound. The cryoscopic constants, K, were established by deliberately adding impurities and measuring the new melting points. The analysis of samples is made by comparison with these data.

4-8   EXTRACTIONS AND SEPARATIONS

A. Wettable Powders, Dusts, and Granules

For analysis of solid formulations (wettable powders, dusts, or granules), it is usually necessary to separate the pesticide from its carrier before assay of the active ingredient. This can be done in a number of ways. The usual extraction methods are by exhaustive extraction in a Soxhlet extractor or by shaking for a short period with a suitable solvent. The Soxhlet method is employed where it is difficult to remove the active ingredient from the carrier. Soxhlet extraction has the disadvantage that the pesticide is heated for an extended period of time in the extraction solvent. This is not a problem if the compound is stable. With certain pesticides, however, some decomposition may occur during this period. In general, the time of extraction in a Soxhlet apparatus varies from 3 hr for aldrin(4) to 16 hr for ARAMITE(8). The longer time is probably not actually required but is convenient for overnight extraction. The Bailey–Walker extraction has also been used for a number of solid pesticide formulations(28). In general, its performance is equivalent to that of the Soxhlet.

Much time is saved if the pesticide can be removed from the solid carrier by simply shaking with a suitable solvent. This procedure is used with a large number of pesticides; the shaking times varying from a few minutes to a maximum of about 60 min. Although extractions by this method are probably not quite so complete as by the hot continuous extraction procedure, the differences are small and, where applicable, the direct cold extraction is preferable.

In some cases($4, 5$) the pesticide is extracted from a formulation by placing it over an absorbent in a chromatographic column and leaching with a suitable solvent. This method is somewhat slower than the direct extraction method but has the advantage that interfering formulation ingredients, such as wetting agents, may be separated from the pesticide during the extraction.

The choice of extraction solvent depends upon the solubility characteristics of the pesticide and the method intended for the final determination. When such methods as gas–liquid chromatography or infrared absorption are to be employed in the final determination, it is advantageous to use an extraction solvent in which the final reading can be made.

A wide variety of solvents are used for extraction purposes. For convenient reference, the solvents used with different pesticides are tabulated in Table 4-1. Information as to time of extraction is also shown, where available.

Of particular interest are the mixed solvent systems. In some cases it is possible to obtain the desired polarity properties by combining miscible solvents. Aldrin can be separated from solid formulations on a clay column using pentane containing 5% of acetone(4). Carbophenothion is extracted by shaking with a solvent containing equal amounts of methanol and benzene(72). EPTC(51) is extracted from 5% granules by shaking for 1 min with a solvent system composed of anhydrous methanol, water, and benzene in the proportion 15:1:25.

In many instances, extraction of the active ingredient from solid formulations is not complete. This may be due to irreversible absorption or to decomposition of a small amount of the pesticide on active sites on the absorbent. The problem is relatively more serious the lower the concentration of the formulation.

B. Liquid Formulations

In general, the analysis of liquid formulations does not present the extraction and separation problems encountered with solid formulations. In cases where the determinations are based on elemental analyses (chlorine, sulfur, phosphorus, or nitrogen), the entire formulation can be decomposed easily. When hydrolysis or other reactions are involved, the problem is only to find a solvent in which the formulation and reagent are mutually soluble.

1. Separation of Interferences. In many cases it is possible to choose an analytical method for pesticide formulations in which there is little or no interference from impurities. This is by no means, however, always the case. A few examples of methods used to separate interfering impurities are listed below.

With methods that involve hydrolysis and determination of one of the hydrolysis products, prevention of interference can sometimes be

TABLE 4-1

SOLVENT AND EXTRACTION METHODS FOR SOLID PESTICIDE FORMULATIONS

Solvent system	Soxhlet	Shaking	Column	Exhaustive[a]
Acetone	aldrin(4) (3 hr).	captan(113) (3 min)		simazine(25)
Acetonitrile		malathion(61) (2–3 min)	phorate[b](64)	
Acetic acid glacial		chloranil(47) (hot)		
Acetic acid = acetic anhydride (1:9)		dodine(43) (3–5 min)		
Benzene				tetradifon(13) endosulfan(42)
Carbon tetrachloride	DEF(66) (16 hr)			
Chloroform		carbaryl(32) botran(115) (30 min) azinphosmethyl(60) (60 min) coumaphos(73) (30 min) amino triazole(45) (2–3 min) warfarin(75) (30 min)	mevinphos(41) (Hyflo Super-Cel)	EPTC(51)
Dimethyl-oxyethane				
Dioxane				
Dimethyl-formamide				
Ethyl ether	chlorobenzilate(7) (6 hr) dimetan(114) (8 hr)			

292

Solvent				
Isopropyl alcohol	heptachlor(*116*) phenkapton(*11*) ARAMITE(*8*) (16 hr) disulfoton(*28*) (16 hr)			PCNB(*115*) glyodin(*44*)
Isooctane	disulfoton(*28*) (16 hr)			
Methanol	toxaphene(*14*)	endosulfan(*42*) (15 min) carbophenothion(*72*) (15 min)		
Methanol–benzene (1:1)				
Methanol–benzene–water (15:25:1)		EPTC(*51*) (1 min)		
Methylene chloride	CIPC(*35a*)			
Pentane–acetone (19:1)			aldrin(*4*) (clay)	
Petroleum ether	Heptachlor(*116*) Pyrethrum(*3a*, p. 50) (7 hr)		dieldrin(*5*) (clay)	
Toluene			dichlorvos(*15*) (filter acid)	

[a] Actual method of extraction not described.
[b] Leaching method. No adsorbent.

accomplished by a study of the relative hydrolysis rates of the active ingredient and the interfering materials. In the case of azinphosmethyl, one method of determination is alkaline hydrolysis to anthranilic acid, which can then be estimated colorimetrically by diazotization and coupling(88). Hydrolysis of azinphosmethyl during storage could result in formation of small amounts of benzazimide or hydroxymethyl benzazimide. Hot alkaline hydrolysis is required to form anthranilic acid from the two latter compounds, while azinphosmethyl is hydrolyzed rapidly at room temperature. Thus, it is a simple matter to prevent interference from the two compounds mentioned in this method for analyzing for azinphosmethyl(88).

The same principle is applied in the analyses of captan(113) and folpet(117) formulations. These compounds are hydrolyzed to form inorganic chloride, which is subsequently titrated. It has been reported that perchloromethyl mercaptan and impurities derived from it are hydrolyzed very rapidly in acetone–methanol to yield chloride ion. Under these conditions, captan and folpet are not hydrolyzed. The quantity of chloride ion derived from these materials is determined separately and subtracted from the total chlorine found in captan and folpet formulations.

DYRENE furnishes another example of the same type. Cyanuric chloride is a reactant used in the synthesis of DYRENE. This would constitute an interference in the chloride method for DYRENE if it were not for the fact that cyanuric chloride is hydrolyzed in water, while DYRENE requires sodium hydroxide. Correction for cyanuric chloride interference is based on this fact(118).

In spectrophotometric methods for pesticide formulations, emulsifiers, wetting agents, etc., often interfere. Barban(74) emulsifiable concentrate is analyzed by ultraviolet spectrophotometry after separation of interfering materials on a column of silica gel. In this method the formulation is diluted with ethylene dichloride. A portion of this solution is applied to a silica gel column. Part of the interfering materials is removed from the column by elution with a 1 : 2 mixture of ethylene dichloride and n-hexane. Then, by changing the solvent to a 2 : 1 mixture of ethylene dichloride and n-hexane, the active ingredient is eluted from the column.

Analysis of DEF emulsifiable concentrate is another example of the same type(66). In this case the formulation is placed on a chromatographic column of standardized, acid-washed alumina. The defoliant is washed through the column with carbon tetrachloride, the interfering emulsifier being retained. In chromatographic separations of this type, great care must be taken to standardize the column conditions very carefully, otherwise small losses of the active ingredient may occur.

Many ultraviolet absorptiometric methods are based on the absorption characteristics of a phenol constituent in the molecule. The uncoupled phenolic component constitutes a potential interference. This can sometimes be avoided by utilizing the bathochromic shift that occurs when the phenol is treated with alkali. This shift only occurs with the free phenol. This principle can be utilized for p-nitrophenol in parathion(49) and chlorferone in coumaphos(73).

Naptalam(50) is analyzed by the Kjeldahl nitrogen determination. 1-Naphthylamine is a potential interfering impurity. By extraction of a separate portion of the extract with ethyl ether, the free 1-naphthylamine is separated from the product that can then be measured by ultraviolet absorption.

The herbicide 2-4-DEP(48) is analyzed by titration with 0.1 N iodine in a solution of benzene, pyridine, and methanol. Bis-(2,4-dichlorophenoxyethyl) phosphite is a possible contaminant of 2-4-DEP and, while it does not interfere in the above determination, it is of value to determine its concentration. This can be accomplished by adding a potassium acetate–potassium iodide solution and an excess of 0.1 N iodine to the solution remaining from the above analysis. After standing for 15 min, the excess iodine is titrated with 0.1 N iodine. This determination is possible because the impurity reacts much less rapidly with iodine than does 2-4-DEP.

Dodine(43) is analyzed by dissolving it in an acetic acid–acetic anhydride reagent to form a solution that is basic enough to titrate with perchloric acid at a relatively low potential. Under the conditions of this titration, urea, amides, and primary and secondary amines do not interfere.

Where formulations of mixtures of pesticides require analysis, the ingredients to be analyzed can often be separated by column chromatography. Barrette and Payfer(120) have recorded the relative rates of movement of a large number of pesticides through silicic acid columns.

2. Contamination. The problem of contamination of pesticide formulations with other pesticides is a serious one against which all pesticide manufacturers must guard. Most pesticide manufacturers prepare several pesticides in the same manufacturing equipment. Care must be taken to avoid cross contamination on this account. While it is relatively simple to ensure absence of contamination in one's own plant, it is more difficult where formulations are prepared by custom formulators. In the past, many such formulators have not given adequate attention to cleaning of equipment between runs with different formulations. This problem has become more acute during recent years since the USDA Pesticide Regulation Division has set up a laboratory for detection of cross contamination of pesticides – cf. Grady Commission Report. During the past

few years many pesticide formulations have been seized and the manufacturer charged with mislabeling as a result of traces of contaminating
pesticides in a formulation. It should be emphasized that in the great
majority of the cases these cross contaminations have been very low. The
development of good gas–liquid chromatographic methods however, has
allowed the government analysts to screen pesticide formulations for
contamination to a degree that was not possible previously. The USDA
chemists have reported to the author that they are striving to achieve a
sensitivity of 0.01% in the detection of contamination of pesticide formulations. It has also been reported that seizures have been made in which
the contaminant is present at a concentration of less than 0.1% of the final
formulation. It is important to remember that in many cases where
seizures have been made because of very low-level contamination, no
conceivable harm could have occurred to the crops or the animals on
which the formulations would have been used. This problem is often a
legal rather than a practical one.

   Because of this situation, it is important for a manufacturer to devise
sensitive detection methods to ensure the absence of contaminating pesticides in formulations. One procedure is to wash equipment with suitable
solvents and to analyze the washings. A final check must often be made
however, on the formulation itself. Generally speaking, methods that will
show the absence of contaminant in the 0.1 to 0.01% range are satisfactory.

   The most satisfactory method in use in the author's laboratory for
detecting traces of impurities is thin-layer chromatography. In general,
silica gel G plates are used. The mobile solvents vary, depending upon the
compounds to be tested. The most commonly used detection method for
organophosphorus compounds is spraying the plates with 2,6-dibromo-N-
chloro-p-quinoneimine. This reagent may be used either with or without
a bromine oxidation. The methods used for detection of traces of impurities in rinse solvents are summarized in Table 4-2. The final estimations
are made on a semiquantitative basis by preparing standards containing
known amounts of the impurity and developing these along with the
unknown. Visual comparison of the relative density of the spots will give
an approximate estimate of the concentration of the contaminant in the
rinse solvent. By this means, it can be determined whether or not the
equipment is clean.

   For determination of contaminants in formulations, similar methods
can be used. The solvent systems are varied, depending upon the nature
of the compounds separated. To illustrate the number of methods that
must be checked out for a single compound, the thin-layer chromatographic methods used for detection of a number of possible contaminating
impurities in azinphosmethyl formulations are summarized in Table 4-3.

TABLE 4-2

THE DETECTION OF PESTICIDES IN RINSE SOLVENT BY THIN-LAYER
CHROMATOGRAPHY

Compound	Solid phase	Mobile solvent	Detection method
coumaphos	Silica gel G	Heptane–acetone (3:1)	Fluorescent spots under uv
demeton	Silica gel G	Benzene	DCQ[a] Bromine oxidation
DEF	Silica gel G	Chloroform	DCQ[a] Bromine oxidation
azinphosmethyl	Silica gel G	Acetone	NED[b]
oxydemeton-methyl	Uranyl acetate-treated silica gel G	Acetone	Ultraviolet quenching
aprocarb	Silica gel G	Heptane–acetone (3:1)	Ultraviolet quenching
fenthion	Silica gel G	Heptane–acetone (7:1)	DCQ[a] heat at 110°C
disulfoton	Silica gel G	Heptane–acetone (7:1)	DCQ[a] heat at 110°C

[a]2,6-Dibromo-N-chloro-p-quinoneimine 1% in acetone.
[b]N(1-Naphthyl) ethylenediamine dihydrochloride 1% in 1 N hydrochloric acid.

With all of these methods, suitable aliquots of the formulations are weighed out, and acetone is used as the diluting or extracting solvent. Silica gel G plates are used with the solvent systems shown in the tabulation. The detection system most commonly used is a 1% solution of 2,6-dibromo-N-chloro-p-quinoneimine in acetone. In some cases it has been found that exposure of the treated plate to bromine vapor or to heat enhances the amount of color developed, while, in other instances, the bromine treatment has been found to reduce the amount of color. With these procedures, it is easily possible to detect contaminants in the concentration range of 0.1 to 0.01%. A similar chart of methods could be prepared for each of the products manufactured by the author's company.

4-9 STORAGE STABILITY TESTING

One very important consideration with any pesticide formulation is the time over which it can be kept in storage without deterioration of either its biological activity or its physical properties. To this end a manufacturer must conduct both accelerated and field storage tests.

TABLE 4-3

THIN-LAYER CHROMATOGRAPHIC METHODS FOR THE DETECTION OF IMPURITIES
IN VARIOUS AZINPHOSMETHYL FORMULATIONS

Azinphosmethyl formulation	Contaminant	Mobile solvent	Detection method
25% Wettable powder	coumaphos	Chloroform	DCQ[a] Heat at 110°C
	DYRENE	Benzene	Ultraviolet quenching
	DDT	Chloroform	Silver nitrate
	carbaryl	Chloroform–benzene–acetone (2:2:1)	Sodium hydroxide-fluorescence under ultraviolet
Emulsifiable concentrate	fenthion	Heptane–acetone (7:1)	DCQ[a] – Bromine oxidation
	oxydemeton-methyl	Acetone–methanol (1:1)	DCQ[a] – bromine oxidation
	azinphosethyl	Heptane–acetone (3:1)	DCQ[a] – heat at 110°C bromine oxidation
	Parathion	Heptane–acetone (3:1)	DCQ[a] – bromine oxidation
	DEF	Heptane–acetone (3:1)	DCQ[a] – bromine oxidation; heat at 110°C
	disulfoton	Heptane–acetone (7:1)	DCQ[a] – heat at 110°C
	coumaphos	Chloroform	Fluorescence under ultraviolet

[a]2,6-Dibromo-N-chloro-p-quinoneimine 1% in acetone.

The decomposition of a pesticide during storage often results in the formation of decomposition products that may interfere with the determination of the active ingredient. Thus, methods that may be quite adequate for routine production control are often unsatisfactory for storage stability investigations. In these cases other procedures must be used for storage stability studies.

This type of problem is illustrated in the case of azinphosmethyl. There are several methods available for analysis of azinphosmethyl formulations. These are: an infrared method(60) in which the band at 15.25 μ is measured; a colorimetric method involving hydrolysis in alkali to form anthranilic acid, which can subsequently be diazotized and coupled(88); and a colorimetric method involving acid hydrolysis to form 0,0-dimethyl-phosphorodithioic acid, which forms a yellow copper salt(60). The methods all give identical results on highly purified material. Also, the results are essentially equivalent on newly prepared technical material.

The anthranilic acid method is not affected by benzazimide or methyl benzazimide. Some benzazimide derivatives are formed in storage, however, that interfere with this procedure. In addition, phosphorus-containing hydrolysis products interfere with the infrared method. Therefore, only the copper method has sufficient specificity to be completely reliable for storage stability tests. However, this method is more time consuming and less precise than the infrared procedure and, therefore, the latter is used for routine control purposes.

4-10   EXAMINATION OF PHYSICAL PROPERTIES

In order to ensure the quality of a pesticide formulation, it is not sufficient just to determine that the concentration of the active ingredient agrees with the label claims. In addition to this, a number of physical properties must also be examined. Some of the commonly determined physical characteristics are discussed below:

A. Appearance

This is usually done by making a visual comparison with a laboratory standard. The analyst is instructed to watch for any unusual characteristics in the formulation being tested and, if such are noted, the formulation is rejected.

Greater variations in appearance can be tolerated with some formulations than with others. For instance, it is more important to maintain an exact color standard for "Pest Control Operator" or "Home Use" formulations than for agricultural formulations.

Marked changes in color, however, are an indication of changes in quality of some of the raw materials or of contamination, and a strict appearance specification will reveal this, when other testing methods might not. If appearance varies appreciably from the standard specified, the analyst must determine the cause of the change.

B. Sediment

This test is applied to liquid formulations. It is obvious that the presence of insoluble sediment in liquid spray formulations can cause serious nozzle clogging. Sediment is determined by shaking the formulation thoroughly and measuring 10 ml into a conical, graduated centrifuge tube. After the sediment is centrifuged at 2500 rpm for 15 min, its volume is determined. The tube is recentrifuged for 5–8 minutes. If sediment volume does not increase, it is recorded; if an increase occurs, however, centrifuging is repeated until a constant volume results. In general, it is difficult to get rid of sediment completely, and a specification of a maximum of 0.1% on a volume/volume basis is usually satisfactory.

C. Sieve Analysis

The biological activity of a pesticide may be greatly influenced by particle size distribution, which can also have a marked effect upon phytotoxicity. Therefore, sieve analyses are carried out on wettable powder formulations to ensure the proper distribution of particle size. Similar assays are also made on granular formulations to ensure the absence of excessive amounts of "fines." Dry sieve analyses are used to measure particle size on wettable powders that are sufficiently dense and free-flowing to pass through sieve mesh openings comparable to the particle diameter. Ordinarily, a series of U.S. Standard Sieves (8 in. diam) are used on a shaker such as the Rotap. A typical dry sieve specification for trichlorfon 50% soluble powder requires that the maximum amounts of the formulation retained on 40-mesh, 20-mesh and 10-mesh screens be 10%, 2.0%, and 0.5%, respectively.

With wet sieve analyses, the material is stirred in water and passed through a specified sieve with the aid of water washing and brushing. Much material that would be retained on a screen by the dry sieve method will be forced through by the wet procedure. Friable agglomerates are broken down by the wet sieve method. For the trichlorfon 50% soluble powder formulation mentioned previously, the specification for a wet sieve analysis is a maximum of 0.05% on a 100-mesh screen.

Where pertinent, the hazards of application of solid toxic formulations can be minimized by using granular rather than dust formulations. In the case of granular formulations, the manufacturer must be concerned with the dust (fines) in the product. The dust may contain appreciable concentrations of the toxicant and could present a health hazard when inhaled. For this reason, disulfoton 10% granules carry the specification that the maximum amount of the formulation which may pass a 60-mesh screen is 5%.

Particle size distribution can be determined more accurately with the Coulter Counter than with sieve analysis. This is particularly useful for very fine preparations. To obtain adequate data, however, one must be sure that the entire system is measured. To accomplish this, it may be necessary to use several apertures, which renders the procedure very tedious. Therefore, the Coulter Counter is not used for routine product control of pesticide formulations in the author's laboratory. It is used occasionally, however, as an audit test to check problem formulations.

D. Emulsification

Emulsification properties are extremely important in the evaluation of liquid emulsifiable formulations. Minor impurities or water hardness can have a marked effect both on the formation of emulsions and upon their stability. The emulsification properties of a formulation often change in storage. The objective is to obtain as fine an emulsion as possible with as great a stability as possible.

Emulsification is estimated by emulsifying a standard amount of the formulation in hard water in a cylinder and determining the degree of separation of the phases upon standing. Standard (1400 ppm) hard water is used for this test. Five milliliters of the sample are pipeted into 95 ml of water in a 100 ml glass-stoppered, graduated cylinder. Emulsification must be complete on inverting the cylinder three times. A second sample is inverted 25 times and allowed to stand for a specified length of time. If the separation of the phases is more rapid than allowed by the specification, the sample fails. For azinphosmethyl liquid concentrate formulation, the maximum amount of separation allowable by the above test is 6 ml after 2 hr of standing.

E. Wettability

This property is important in determining the ease with which a wettable powder formulation will mix with water. Formulations that do not wet easily are difficult to prepare for use in the field.

Like the emulsification test, this procedure is carried out with standard hard water. A 2-g sample is poured into 95 ml of water in a 250-ml beaker. The time necessary to wet the sample completely is determined. The suspension is stirred gently to determine whether agglomerates disperse readily. With azinphosmethyl 25% wettable powder, the maximum time allowed for wetting under the above conditions is 1 min.

F. Suspendability

This property is important with wettable powder formulations where settling is a problem. If the spray equipment does not provide constant agitation, appreciable amounts of the solid will settle to the bottom of the container and, consequently, induce variations in the amount of material applied. Methods for determination of suspensibility of water dispersible wettable powders are discussed by Gooden and Ringel(121). In these methods, the concentration of the active pesticide is assayed at differing levels in a suspension after various periods of standing. They suggest that the common objective of all of the methods that they studied is to ensure that, under rigorous conditions, little of the material in a given powder shall have settling velocities of more than 1 cm/min.

4-11 CHEMICAL TESTS FOR OTHER THAN ACTIVE INGREDIENTS

A. Solvent

In many instances, technical materials contain traces of the solvents used in their manufacture. In some cases these have important effects on the properties of the finished formulations. These solvents can usually be determined by gas–liquid chromatography. An example of this type of specification is that of 5% ethylene dichloride in technical azinphosmethyl.

B. Water

The water content of formulations may be important. In some cases, the presence of more than a trace of water will increase the rate of decomposition of a compound during storage. Small amounts of water may also cause caking of some wettable powder formulations. Moisture is usually determined by the Karl Fischer method or by drying to constant weight at 105°C. In a product such as trichlorfon sugar bait, the moisture content is limited to a maximum of 0.2%.

C. Acidity

The presence of free acid in a technical product or a formulation may have an effect on the stability of the product. Because of this, the amount of titratable acidity in a product is often determined routinely.

The pH of a finished formulation may be of significance as far as container corrosion is concerned and it is, therefore, measured routinely where warranted.

4-12 BIOLOGICAL TESTS

The biological tests that are usually conducted on pesticide formulations are designed to measure biological efficacy, phytotoxicity, and mammalian toxicity.

Samples seldom fail such tests, but monitoring data of this type in a manufacturer's files is invaluable to contest claims for lack of efficacy from disappointed customers.

A. Efficacy

When a new product is first marketed, it is advisable to run periodic checks for its biological activity on the finished formulation. A suitable indicator species and technique must be established, therefore, for each product and formulation. For insecticides, tests on flies, cockroaches, mites, or aphids can usually be run in the laboratory. For fungicides, one of the agar plate laboratory testing methods can generally be used. For herbicides, suitable test plants can be grown in the greenhouse. In each case, the samples to be tested must be compared with a standard of known activity.

After the product has been marketed successfully for a period of time, the frequency of these tests can be reduced or they may even be discontinued.

B. Phytotoxicity

Phytotoxicity testing is usually done by applying the formulation to a sensitive test plant species grown in the greenhouse and, as with the tests for biological efficacy, comparative standards must be employed in every case. In such tests, it is advisable to make a side-by-side comparison of samples and their standards over a series of dose levels. Phytotoxicity testing is usually checked more regularly than biological efficacy testing. The frequency can be decreased, however, as a history of field usage of the product accumulates. Phytotoxicity testing is usually conducted on a regular, although essentially infrequent, basis in the case of products that have been marketed over a long period of time, i.e., several years.

C. Mammalian Toxicity

Mammalian toxicity testing is designed to check formulations for their possible hazard to the field operator. This test is more critical, of course, for those products which are marketed for use on animals, for "Pest Control Operator" formulations, and for "Home and Garden" products than are those tests for agricultural products. With the latter formulations, an occasional check should suffice, e.g., once every six months. With the first three types of products, mammalian toxicity should be checked on a number of batches, selected at random, from every production run.

In general, it is not necessary to determine an LD_{50} value to check mammalian toxicity. A screening dose can be established such that one or two animals out of five are killed, with a confidence of 95%. Then those samples which kill more than two animals of five are rechecked and, if need be, the actual LD_{50} is determined. As intraperitoneal LD_{50} values are more reproducible than are oral or dermal values, this route can be employed for these routine tests, even though the hazard from the use of the compound is actually from oral or dermal exposure. The precision and confidence of the method can be improved substantially by using eight or ten test animals for each test. The albino rat is the species most commonly employed in this type of toxicity testing, while the rabbit is the species of choice for dermal toxicity tests.

4-13   CONCLUSION

It was not the purpose of this chapter to review all the analytical methods used for pesticide formulations. Indeed, the official methods for such inorganic pesticides as those based upon arsenic and copper are omitted, simply because these methods are well established and there are few new developments in the analyses of these compounds.

The purpose here, then, is twofold. First, to serve as a guide to development of formulation analysis methods for new compounds; and, second, to give those unfamiliar with this field an appreciation of the complex problems involved in assay method development for pesticide formulations and of the multiple and diverse analytical techniques required to control their quality adequately.

REFERENCES

1. *Analytical Methods for Pesticides, Plant Growth Regulators, and Food Additives* (G. Zweig, ed.), Vols. I, II, III, IV, Academic, New York, 1964.
2. F. A. Gunther and R. C. Blinn, *Analysis of Insecticides and Acaricides*, Wiley (Interscience), New York, 1955.
3. *Official Methods of Analysis of the Association of Official Agricultural Chemists*, Chap. 4, Association of Official Agricultural Chemists, Washington, D.C., 1965.
4. P. E. Porter, *Analytical Methods for Pesticides, Plant Growth Regulators, and Food Additives* (G. Zweig, ed.), Vol. II, Chap. 1, Academic, New York, 1964.
5. P. E. Porter, *Analytical Methods for Pesticides, Plant Growth Regulators, and Food Additives*, (G. Zweig, ed.), Vol. II, Chap. 12, Academic, New York, 1964.
6. T. G. Bowery, *Analytical Methods for Pesticides, Plant Growth Regulators, and Food Additives*, (G. Zweig, ed.), Vol. II, Chap. 5, Academic, New York, 1964.
7. A. Margot, and K. Stammbach, *Analytical Methods for Pesticides, Plant Growth Regulators, and Food Additives*, (G. Zweig, ed.), Vol. II, Chap. 6, Academic, New York, 1964.
8. J. R. Lane, *Analytical Methods for Pesticides, Plant Growth Regulators, and Food Additives*, (G. Zweig, ed.), Vol. II, Chap. 3, Academic, New York, 1964.
9. C. F. Gordon, *Analytical Methods for Pesticides, Plant Growth Regulators, and Food Additives*, (G. Zweig, ed.), Vol. II, Chap. 30, Academic, New York, 1964.
10. W. K. Lowen, M. L. Cluett and H. L. Pease, *Analytical Methods for Pesticides, Plant Growth Regulators, and Food Additives*, (G. Zweig, ed.), Vol. II, Chap. 27, Academic, New York, 1964.

11. K. Stammbach, *Analytical Methods for Pesticides, Plant Growth Regulators, and Food Additives*, (G. Zweig, ed.), Vol. II, Chap. 31, Academic, New York, 1964.
12. C. F. Gordon, *Analytical Methods for Pesticides, Plant Growth Regulators, and Food Additives*, (G. Zweig, ed.), Vol. II, Chap. 37, Academic, New York, 1964.
13. C. C. Cassil and J. Yaffe, *Analytical Methods for Pesticides, Plant Growth Regulators, and Food Additives*, (G. Zweig, ed.), Vol. II, Chap. 41, Academic, New York, 1964.
14. C. L. Dunn, *Analytical Methods for Pesticides, Plant Growth Regulators, and Food Additives*, (G. Zweig, ed.), Vol. II, Chap. 44, Academic, New York, 1964.
15. P. E. Porter, *Analytical Methods for Pesticides, Plant Growth Regulators, and Food Additives*, (G. Zweig, ed.), Vol. II, Chap. 46, Academic, New York, 1964.
16. R. A. Conkin and L. S. Gleason, *Analytical Methods for Pesticides, Plant Growth Regulators, and Food Additives*, (G. Zweig, ed.), Vol. IV, Chap. 20, Academic, New York, 1964.
17. C. F. Gordon, A. L. Wolfe and L. D. Haines, *Analytical Methods for Pesticides, Plant Growth Regulators, and Food Additives*, (G. Zweig, ed.), Vol. IV, Chap. 24, Academic, New York, 1964.
18. R. A. Conkin and L. S. Gleason, *Analytical Methods for Pesticides, Plant Growth Regulators, and Food Additives*, (G. Zweig, ed.), Vol. IV, Chap. 27, Academic, New York, 1964.
19. J. R. Lane, *Analytical Methods for Pesticides, Plant Growth Regulators, and Food Additives*, (G. Zweig, ed.), Vol. III, Chap. 14, Academic, New York, 1964.
20. A. Stepanow, *Ber.*, **39**, 4056–4057 (1906).
21. B. Pecherer, C. M. Gambrill and G. W. Wilcox, *Anal. Chem.* **22**, 311–315 (1950).
22. L. M. Liggett, *Anal. Chem.*, **26**, 748–750 (1954).
23. C. F. Gordon and R. J. Schuckert, *Analytical Methods for Pesticides, Plant Growth Regulators, and Food Additives*, (G. Zweig, ed.), Vol. II, Chap. 23, Academic, New York, 1964.
24. D. MacDougall, *Analytical Methods for Pesticides, Plant Growth Regulators, and Food Additives*, (G. Zweig, ed.), Vol. II, Chap. 17, Academic, New York, 1964.
25. R. Suter (Analytical Laboritories, J. R. Geigy S. A., Basel, Switzerland), unpublished work described in *Analytical Methods for Pesticides, Plant Growth Regulators, and Food Additives*, (G. Zweig, ed.), Vol. IV, Chap. 23, Academic, New York, 1964.
26. R. Suter (Analytical Laboritories, J. R. Geigy S. A., Basel, Switzerland), unpublished work described in *Analytical Methods for Pesticides, Plant Growth Regulators, and Food Additives*, (G. Zweig, ed.), Vol. IV, Chap. 19, Academic, New York, 1964.
27. D. MacDougall, T. E. Archer and W. L. Winterlin, *Analytical Methods for Pesticides, Plant Growth Regulators, and Food Additives*, (G. Zweig, ed.), Vol. II, Chap. 40, Academic, New York, 1964.
28. D. MacDougall and T. E. Archer, *Analytical Methods for Pesticides, Plant Growth Regulators, and Food Additives*, (G. Zweig, ed.), Vol. II, Chap. 16, Academic, New York, 1964.
29. D. MacDougall, *Analytical Methods for Pesticides, Plant Growth Regulators, and Food Additives*, (G. Zweig, ed.), Vol. II, Chap. 26, Academic, New York, 1964.
30. K. Groves, *J. Agr. Food Chem.*, **6**, 30–31 (1958).
31. R. Suter, R. Delley, and R. Meyer, *Z. Anal. Chem.*, **147**, 173–184 (1955).
32. W. E. Whitehurst and J. B. Johnson (1958) described by H. A., Jr. Stansbury, and R. Miskus, *Analytical Methods for Pesticides, Plant Growth Regulators, and Food Additives*, (G. Zweig, ed.), Vol. II, Chap. 39, Academic, New York, 1964.
33. W. K. Lowen, W. E. Bleidner, J. J. Kirkland and H. L. Pease, *Analytical Methods for Pesticides, Plant Growth Regulators, and Food Additives*, (G. Zweig, ed.), Vol. IV, Chap. 16, Academic, New York, 1964.

34. G. L. Sutherland, *Analytical Methods for Pesticides, Plant Growth Regulators, and Food Additives*, (G. Zweig, ed.), Vol. II, Chap. 14, Academic, New York, 1964.
35. L. N. Gard and C. E. Ferguson, Jr., *Analytical Methods for Pesticides, Plant Growth Regulators, and Food Additives*, (G. Zweig, ed.), Vol. IV, Chap. 7, Academic, New York, 1964.
36. L. N. Gard, *Anal. Chem.*, **23**, 1685–1686 (1951).
37. D. G. Clarke, H. Baum, E. L. Stanley and W. F. Hester, *Anal. Chem.*, **23**, 1842 (1951).
38. M. Levitsky and W. K. Lowen, *J. Assoc. Offic. Agr. Chemists*, **37**, 555 (1954).
39. R. Suter (Analytical Laboratory, J. R. Geigy, Basel, Switzerland), described in *Analytical Methods for Pesticides, Plant Growth Regulators, and Food Additives*, (G. Zweig, ed.), Vol. IV, Chap. 18, Academic, New York, 1964.
40. W. Z. Pilz, *Anal. Chem.*, **164**, 241–246 (1958).
41. P. E. Porter, Yun-Pei Sun, and T. E. Archer, *Analytical Methods for Pesticides, Plant Growth Regulators, and Food Additives*, (G. Zweig, ed.), Vol. II, Chap. 32, Academic, New York, 1964.
42. J. R. Graham, J. Yaffe, T. E. Archer, and A. Bevenue, *Analytical Methods for Pesticides, Plant Growth Regulators, and Food Additives*, (G. Zweig, ed.), Vol. II, Chap. 43, Academic, New York, 1964.
43. G. L. Sutherland, *Analtyical Methods for Pesticides, Plant Growth Regulators, and Food Additives*, (G. Zweig, ed.), Vol. III, Chap. 4, Academic, New York, 1964.
44. J. N. Hogsett (Union Carbide Chemicals Co., South Charleston, West Virginia), described by H. A. Stansbury, Jr., in *Analtyical Methods for Pesticides, Plant Growth Regulators, and Food Additives*, (G. Zweig, ed.), Vol. III, Chap. 9, Academic, New York, 1964.
45. G. L. Sutherland, *Analytical Methods for Pesticides, Plant Growth Regulators, and Food Additives*, (G. Zweig, ed.), Vol. IV, Chap. 3, Academic, New York, 1964.
46. J. R. Lane, *Analytical Methods for Pesticides, Plant Growth Regulators, and Food Additives*, (G. Zweig, ed.), Vol. IV, Chap. 15, Academic, New York, 1964.
47. J. R. Lane, H. P. Burchfield, and E. E. Storrs, *Analytical Methods for Pesticides, Plant Growth Regulators, and Food Additives*, (G. Zweig, ed.), Vol. III, Chap. 3, Academic, New York, 1964.
48. G. M. Stone, described by J. R. Lane, *Analytical Methods for Pesticides, Plant Growth Regulators, and Food Additives*, (G. Zweig, ed.), Vol. IV, Chap. 13, Academic, New York, 1964.
49. G. L. Sutherland and R. Miskus, *Analytical Methods for Pesticides, Plant Growth Regulators, and Food Additives*, (G. Zweig, ed.), Vol. II, Chap. 29, Academic, New York, 1964.
50. J. R. Lane, *Analytical Methods for Pesticides, Plant Growth Regulators, and Food Additives*, (G. Zweig, ed.), Vol. IV, Chap. 1, Academic, New York, 1964.
51. G. G. Patchett, G. H. Batchelder, and J. J. Menn, *Analytical Methods for Pesticides, Plant Growth Regulators, and Food Additives*, (G. Zweig, ed.), Vol. IV, Chap. 12, Academic, New York, 1964.
52. W. W. Kilgore, *Analytical Methods for Pesticides, Plant Growth Regulators, and Food Additives*, (G. Zweig, ed.), Vol. III, Chap. 10, Academic, New York, 1964.
53. L. Lykken, *J. Assoc. Offic. Agr. Chemists*, **44**, 595–605 (1961).
54. G. E. Pollard, W. M. Saltman, and P. Yin, *J. Assoc. Offic. Agr. Chemists*, **38**, 478–482 (1955).
55. W. H. Clark, *J. Agr. Food Chem.*, **10**, 214–216 (1962).
56. W. B. Bunger and R. W. Richburg, *J. Agr. Food Chem.*, **5**, 127–130 (1957).
57. D. MacDougall, *Analytical Methods for Pesticides, Plant Growth Regulators, and Food Additives*, (G. Zweig, ed.), Vol. II, Chap. 4, Academic, New York, 1964.

58. D. E. Pack, J. N. Ospenson, and G. K. Kohn, *Analytical Methods for Pesticides, Plant Growth Regulators, and Food Additives*, (G. Zweig, ed.), Vol. II, Chap. 11, Academic, New York, 1964.
59. E. B. Corbett, *J. Assoc. Offic. Agr. Chemists*, **47**, 257–258 (1964).
60. D. MacDougall, *Analytical Methods for Pesticides, Plant Growth Regulators, and Food Additives*, (G. Zweig, ed.), Vol. II, Chap. 20, Academic, New York, 1964.
61. S. D. Upham, *J. Assoc. Offic. Agr. Chemists*, **43**, 360–364 (1960).
62. T. T. White and G. G. McKinley, *J. Assoc. Offic. Agr. Chemists*, **44**, 589–591 (1961).
63. N. R. Pasarela, *J. Assoc. Offic. Agr. Chemists*, **47**, 245–247 (1964).
64. G. L. Sutherland, P. A. Giang, and T. E. Archer, *Analtyical Methods for Pesticides, Plant Growth Regulators, and Food Additives*, (G. Zweig, ed.), Vol. II, Chap. 42, Academic, New York, 1964.
65. L. E. Mitchell, *J. Assoc. Offic. Agr. Chemists*, **47**, 268–271 (1964).
66. D. MacDougall, *Analytical Methods for Pesticides, Plant Growth Regulators, and Food Additives*, (G. Zweig, ed.), Vol. IV, Chap. 10, Academic, New York, 1964.
67. D. Firestone and P. J. Vollmer, *J. Assoc. Offic. Agr. Chemists*, **39**, 866–872 (1956).
68. R. G. Knoerlein, *J. Assoc. Offic. Agr. Chemists*, **44**, 577–579 (1961).
69. B. L. Samuel, *J. Assoc. Offic. Agr. Chemists*, **44**, 580–581 (1961).
70. Chemagro Corporation, unpublished test methods C-5.7, C-5.8, C-5.9, infrared analysis of GUTHION in technical material and formulations.
71. F. A. Gunther and R. C. Blinn, *J. Agr. Food Chem.*, **5**, 517–519 (1957).
72. J. J. Menn, G. G. Patchett, and G. H. Batchelder, *Analytical Methods for Pesticides, Plant Growth Regulators, and Food Additives*, (G. Zweig, ed.), Vol. II, Chap. 45, Academic, New York, 1964.
73. P. F. Kane, C. J. Cohen, W. R. Betker, and D. MacDougall, *J. Agr. Food Chem.*, **8**, 26–29 (1960).
74. K. J. Bombaugh and W. C. Bull, *J. Agr. Food Chem.*, **9**, 386–390 (1961).
75. C. H. Schroeder and J. N. Eble, *Analytical Methods for Pesticides, Plant Growth Regulators, and Food Additives*, (G. Zweig, ed.), Vol. III, Chap. 21, Academic, New York, 1964.
76. M. P. Milner, F. J. Holzer and J. B. Leary, *J. Assoc. Offic. Agr. Chemists*, **46**, 655–659 (1963).
77. J. P. Barrette and R. Payfer, *J. Assoc. Offic. Agr. Chemists*, **44**, 581–584 (1961).
78. A. A. Danish and R. E. Lidov, *Anal. Chem.*, **22**, 702–706 (1950).
79. L. C. Terriere, *Analytical Methods for Pesticides, Plant Growth Regulators, and Food Additives*, (G. Zweig, ed.), Vol. II, Chap. 18, Academic, New York, 1964.
80. P. B. Polen and P. Silverman, *Anal. Chem.*, **24**, 733–735 (1952).
81. M. V. Norris, W. A. Vail and P. R. Averell, *J. Agr. Food Chem.*, **2**, 570 (1954).
82. J. R. Graham and E. F. Orwoll, *J. Agr. Food Chem.*, **11**, 67–69 (1963).
83. J. H. Ware, *J. Assoc. Offic. Agr. Chemists*, **44**, 608–610 (1961).
84. J. H. Ware, *J. Assoc. Offic. Agr. Chemists*, **45**, 529–530 (1962).
85. E. J. Orloski, *J. Assoc. Offic. Agr. Chemists*, **47**, 248–252 (1964).
86. K. Helrich, *J. Assoc. Offic. Agr. Chemists*, **47**, 242–244 (1964).
87. P. R. Averell and M. V. Norris, *Anal. Chem.*, **20**, 753–756 (1948).
88. W. R. Meagher, J. M. Adams, C. A. Anderson and D. MacDougall, *J. Agr. Food Chem.*, **8**, 282–286 (1960).
89. D. MacDougall, *Analytical Methods for Pesticides, Plant Growth Regulators, and Food Additives*, (G. Zweig, ed.), Vol. III, Chap. 5, Academic, New York, 1964.
90. S. H. Yuen, *Analyst*, **90**, 569–571 (1965).
91. H. L. Williams, W. E. Dale and J. P. Sweeney, *J. Assoc. Offic. Agr. Chemists*, **39**, 872–879 (1956).

92. J. P. Sweeney and H. L. Williams, *J. Agr. Food Chem*, 5, 670–672 (1957).
93. H. A. Jones, H. J. Ackermann, and M. E. Webster, *J. Assoc. Offic. Agr. Chemists*, 35, 771–780 (1952).
94. P. T. Allen, H. F. Beckman, and J. F. Fudge, *J. Agr. Food Chem.*, 10, 248–251 (1962).
95. J. J. Velenovsky, *J. Assoc. Offic. Agr. Chemists*, 43, 350–352 (1960).
96. J. Burke, *J. Assoc. Offic. Agr. Chemists*, 46, 198–204 (1963).
97. J. P. Barrette and R. Payfer, *J. Assoc. Offic. Agr. Chemists*, 47, 259–263 (1964).
98. E. J. Bonelli, H. Hartmann, and K. P. Dimick, *J. Agr. Food Chem.*, 12, 333–336 (1964).
99. D. D. Phillips, G. E. Pollard, and S. B. Soloway, *J. Agr. Food Chem.*, 10, 217–221 (1962).
100. H. J. Wesselman and J. R. Koons, *J. Agr. Food Chem.*, 11, 173–174 (1963).
101. J. R. Koons and H. J. Wesselman, *J. Agr. Food Chem.*, 12, 550 (1964).
102. G. Zweig and T. E. Archer, *J. Agr. Food Chem.*, 8, 190–192 (1960).
103. G. G. Patchett, G. H. Batchelder, and J. J. Menn, *Analytical Methods for Pesticides, Plant Growth Regulators, and Food Additives*, (G. Zweig, ed.), Vol. IV, Chap. 25, Academic, New York, 1964.
104. T. E. Archer, A. Bevenue, and G. Zweig, *J. Chromatog.*, 6, 457–460 (1961).
105. Farbenfabriken Bayer Internal Report described by D. MacDougall in *Analytical Methods for Pesticides, Plant Growth Regulators, and Food Additives*, (G. Zweig, ed.), Vol. III, Chap. 5, Academic, New York (1964).
106. P. A. Giang and R. L. Caswell, *J. Agr. Food Chem.*, 5, 753–754 (1957).
107. R. J. Gajan and J. Link, *J. Assoc. Offic. Agr. Chemists*, 47, 1119–1123 (1964).
108. K. Stammbach, R. Delley, R. Suter, and G. Szekely, *Z. Anal. Chem.*, 196, 332–348 (1963).
109. J. N. Hogsett, H. W. Kacy, and J. B. Johnson, *Anal. Chem.*, 25, 1207–1211 (1953).
110. R. J. Best and N. R. Hersch, *J. Agr. Food Chem.*, 12, 546–549 (1964).
111. W. J. Taylor and F. D. Rossini, *J. Res. Natl. Bur. Std.*, 32, RP1585, 197 (1944).
112. C. R. Witschonke, *Anal. Chem.*, 24, 350 (1952).
113. J. N. Ospenson, D. E. Pack, G. K. Kohn, H. P. Burchfield, and E. E. Storrs, *Analytical Methods for Pesticides, Plant Growth Regulators, and Food Additives*, (G. Zweig, ed.), Vol. III, Chap. 2, Academic, New York, 1964.
114. A. Margot and K. Stammbach, *Analytical Methods for Pesticides, Plant Growth Regulators, and Food Additives*, (G. Zweig, ed.), Vol. II, Chap. 13, Academic, New York, 1964.
115. M. J. Kolbezen, *Analytical Methods for Pesticides, Plant Growth Regulators, and Food Additives*, (G. Zweig, ed.), Vol. III, Chap. 12, Academic, New York, 1964.
116. T. G. Bowery, *Analytical Methods for Pesticides, Plant Growth Regulators, and Food Additives*, (G. Zweig, ed.), Vol. II, Chap. 21, Academic, New York, 1964.
117. J. N. Ospenson, D. E. Pack, and G. K. Kohn, *Analytical Methods for Pesticides, Plant Growth Regulators, and Food Additives*, (G. Zweig, ed.), Vol. III, Chap. 13, Academic, New York, 1964.
118. P. F. Kane and K. G. Gillespie, *J. Agr. Food Chem*, 8, 29–32 (1960).
119. W. W. Kilgore, *Analytical Methods for Pesticides, Plant Growth Regulators, and Food Additives*, (G. Zweig, ed.), Vol. III, Chap. 6, Academic, New York, 1964.
120. J. P. Barrette and R. Payfer, *J. Assoc. Offic. Agr. Chemists*, 44, 606–608 (1961).
121. E. L. Gooden and S. J. Ringel, *J. Agr. Food Chem.*, 4, 244–248 (1956).

Chapter 5

ASSAY PROCEDURES FOR PESTICIDE RESIDUES

C. H. Van Middelem

UNIVERSITY OF FLORIDA
GAINESVILLE, FLORIDA

5-1 INTRODUCTION

Prior to 1946, residue assays were largely restricted to the arsenicals and several other nonpenetrating inorganic insecticides and therefore, the residue chemist of this period needed only to be familiar with analytical methods for arsenic, lead, fluoride, pyrethrins, rotenone, and a few other compounds. This rather serene analytical situation was suddenly changed by the commercial introduction in 1946 of the synthetic organic pesticides. During the past two decades, the analytical chemist in industry, government, and university laboratories has been faced with the task of analyzing a bewildering array of commercially available synthetic organic pesticides. Since 1954, the Food and Drug Administration of the U.S. Department of Health, Education, and Welfare has established more than 2000 tolerances for residues of more than 225 pesticide chemicals in, or on, a great variety of raw agricultural commodities, according to Gunther(1).

Until rather recently, because of the rapid commercial development of organic pesticides, methodology as an art was usually not adequate for the mushrooming needs of the dynamic pesticide field. Now, after almost a decade of intensive and forced evolution, this new field of chemical investigation is referred to as "residue chemistry." Gunther(2) has defined "residue chemistry" as a body of techniques and knowledge — biological, chemical, and physical — used to determine the pesticide residue composition and content of any foodstuff, where a "residue" is an original or any derived residuum from a chemical added to the foodstuff during any stage of its production. This relatively virgin field is largely empirical in outlook and pragmatic in philosophy, and too often theory is lagging far behind enthusiastic practice.

Twenty years ago, a few very basic instruments plus classical chemical manipulations were the primary tools with which the residue chemist could attack a problem. However, during the past few years, rapid strides in analytical techniques have even surpassed the impressive technicological advancements of the food chemical industry. As pointed out by Fischbach(3) advances in electronics, designed to reveal fundamental physico-chemical properties, furnish the informed chemist of today with an imposing array of selective and sensitive tools.

A. Responsibilities of Industry, Federal, and State Agencies

The basic impetus to this greatly increased emphasis on more sensitive and selective tools for measuring pesticide residues was the Pesticide Amendment of July 1954 and the Food Additives Amendment of

September 1958. These two amendments to the Food, Drug and Cosmetic Act by government statute specifically demanded adequate methodology for monitoring foods prior to the marketing of such products.

1. Industry. These statutes informed the affected industries that, in planning their expenditures for research and development of a pesticide, adequate means of detecting the pesticide must be furnished before the commodity could be sold to the consumer. The analytical method must be sensitive enough to detect residues at the levels to be found on the commodity as marketed and also be specific for the particular pesticide in question. Moreover, the method should be relatively simple, practical, and capable of use by field laboratories.

At the time of submitting a petition for the public use of a pesticide, the manufacturer, in addition to providing an analytical method, proof of the compound's usefulness, toxicity, and safety, must also submit sufficient residue data obtained under field conditions. The residue data should illustrate a disappearance rate falling below proposed tolerance levels following recommended pesticide applications.

2. Federal Agencies. After the U.S. Department of Agriculture certifies to the Food and Drug Administration (FDA) that the pesticide is useful and renders an opinion that the tolerance proposed is realistic for the dosage and crop suggested by the company, the FDA then directs its attention to the request. The FDA evaluates the adequacy of pharmacological data to prove safety at the requested level, pesticide chemical characterization, adequacy of the residue method submitted, and related effects such as residues in meat and milk resulting from contaminated feeds, etc. After the FDA has approved a residue method submitted by industry with the petition for pesticide clearance, this method can then be used by various state and federal agencies involved with regulatory and/or research responsibilities for this pesticide at the state level.

3. State Agencies. The states' laboratories are given, therefore, (1) a reliable, practical, and sensitive method, (2) a tolerance, plus (3) a suggested waiting interval on a given crop in order to meet the allowed tolerance. The state or federal laboratories may or may not choose to use the method proposed by industry. With the recent proliferation in instrumentation, an alternate method of detection is often resorted to, particularly if the petition method is somewhat cumbersome. One precaution that should be followed, if an alternate method of residue quantitation is used, is that a few comparative analyses on duplicate samples should be run by the approved (petition) method. Therefore, if the alternate method is as sensitive and reliable as the approved method, subsequent data will not be questioned. Another factor that the individual states or areas can alter according to their specific climatic and cultural practices is the recommended interval between pesticide application and harvest. The recommended interval on the label should not be changed, of course, without adequate local residue information. The only factor that cannot be altered is the tolerance set by the FDA. The method of detection and the interval are flexible at the local level, but the official tolerance established on the product as it crosses into interstate commerce cannot be altered. In Florida, as in many other states, there are three basic agencies involved in regulation, research, and dissemination of information pertaining to pesticide residues.

a. REGULATORY AND ENFORCEMENT. As an example, Florida has vested authority in the State Chemist in the State Department of Agriculture to regulate pesticide residues on crops destined for intrastate as well as interstate commerce. The Residue Laboratory at Tallahassee and satellite laboratories situated near the large production areas examine various crops and food products for pesticide residues. All federal tolerances are respected by the Florida regulatory agencies involved with pesticide residues. State field inspectors work closely with and cooperate fully with federal field inspectors within the state, and there is a healthy interplay of information and cooperation between the two agencies. The primary difference between the two enforcement agencies is that the state agency is not only looking for violations but often takes samples in the field to warn a particular grower of impending trouble from excessive residues. Often, because of this prior information, a grower can allow more time for the pesticide residue to "weather" or breakdown in his crop before it is marketed. Occasionally, growers may suspect excessive residues on their crops and will request that samples be taken in the field before initiating harvesting operations. If excessive residues are discovered, it is often possible to avoid costly "stop sales" and possible condemnation and confiscation of the crop once it reaches marketing channels.

b. RESEARCH. The Florida Agricultural Experiment Station has the responsibility of developing disappearance curves as well as conducting research programs on residue methodology and the fate of pesticides in plants, animals, and soils. The disappearance curves are developed in various areas of the state using the specific dosages and formulations recommended by the Experiment Station on the particular crop of interest. This phase is vital since insect and disease problems vary from region to region and even within the same state. A dosage level or interval suggested on a national label may be unrealistic for a particular state or region because of vastly different climatic conditions which, of course, affect insect populations and the prevalence of diseases as well as the pesticide persistence on the commodity. State experiment stations, based on their own research with a specific pesticide formulation and dosage, may often recommend to the growers a particular interval to wait between their last pesticide application and harvest.

c. DISSEMINATION OF INFORMATION. It is the responsibility of the agricultural extension service to disseminate this information to the growers and other interested state agencies. The Extension Services have the responsibility for the continuing education of growers in the precautions necessary for the correct use of pesticides, including both applicator and harvester safety as well as production of marketable crop and animal produce that contain levels of pesticide residues at or below the established tolerance.

d. INTERAGENCY COOPERATION. The two state agencies involved in actual pesticide residue detection cooperate closely in newly introduced analytical procedures. This interchange of information includes field sampling, sample storage, extraction, and cleanup as well as the final determinative procedures. An annual workshop is held in Florida in which the "bench" chemists of the various state agencies involved in pesticide residue determinations can compare notes on their intricate techniques. A semi-annual interagency meeting is also held, composed of all administrators within the state involved in pesticide residue regulation and research. The conferences have proven very useful in keeping all agencies informed of a rapidly expanding field, and include such diverse state agencies as the Department of Agriculture, Board of Health, Forest Service, Game and Fresh Water Fish Commission, Agricultural Experiment Station, and Agricultural Extension Service. In addition, federal agencies working within Florida such as U.S. Public Health Service, U.S. Fish and Wildlife Service, and FDA (Atlanta District) send representatives and report on their activities involving pesticides within the state since the previous interagency meeting.

B. Formulation versus Residue Analysis

There are several basic differences between pesticide residue and pesticide formulation methods of analyses. The most important difference is that most formulation chemists are generally analyzing for percentages ranging from 0.5 to 100%, whereas the residue chemist is analyzing in the parts per million or lower range. The formulation chemist usually employs macro methods known to have high reproducibility and relatively low sensitivity. Residue chemists, on the other hand, use micro methods that usually require special processing procedures including exhaustive extractions and rigorous cleanup. Since residue analysis is usually at such a low level of detection, large samples must usually be employed and often result in serious interference problems from co-extracted contaminants. Normally the formulation chemist will not be faced with as such serious difficulties with contaminant interferences, but will usually employ techniques capable of high specificity, similar to those used by the residue chemist. The formulation chemist often uses volumetric, gravimentric, colorimetric, and infrared techniques for quantitation. At the present time, the residue chemist primarily employs detection techniques that are both recent introductions and highly sensitive, particularly gas and thin-layer chromatography. Relatively little use is made of infrared spectrophotometry, primarily because of the customarily small quantity of pesticide in the sample.

C. *Pesticide Penetration, Migration, and Transformation*

In the past few years, the FDA has directed increasing attention to pesticide metabolic intermediates and end products. Concurrently, in the entire pesticide residue field, there has been increased emphasis on more sophisticated research to determine the "what, how, and where" involved when a pesticide is introduced to our environment. This type of research is often very complicated and time-consuming compared to that of a few years ago which involved simply obtaining parent compound residue data. Where a new compound forms metabolites in nature, methods must be devised to detect, identify, and measure such derivatives in the various extracts of plants, animals, soil, water, and air, and at levels ranging down to from 1 to 1000 ppb. This is indeed a formidable task and only through the combined talents of chemists in industry, government, and universities, using the latest techniques and instrumentation, can this intricate task be successfully undertaken for any given pesticide and all its possible metabolites in nature.

Generally, the nonsystemic organic pesticides dissolve in fats, waxes, and oils, whereas the systemic materials are usually deliberately designed to permeate aqueous media. Therefore, both types of pesticides may exist for short periods external to the cuticle or skin of the treated plant or animal, and then eventually migrate into the cuticular and subcuticular regions. The nature and extent of this migration depends on many factors, including structure and stability of the toxicant and the detailed structure and composition of the substrate surface. Crafts and Foy(4) have summarized and discussed the many modes of entry of both polar and nonpolar compounds into and through the plant cuticle, and Ebeling

(5) has reviewed the influences of pesticide formulation and physical environment upon these migrations. Many pesticides are quite transient on their substrate due to hydrolysis, spectral radiation, thermal instability, and physical dislodgement by wind, rain, etc [Gunther(6)]. Those pesticides that survive to permeate tissues below the epidermal surface are often subject to further alteration by enzymic attack, hydrolysis or other transformations as they migrate from cell to cell or from tissue to tissue. Both externally or internally produced alterations may be quite harmless toxicologically to mammals, or, on the other hand, they may be as toxic or more toxic than the originally applied pesticide. These products and the unaltered pesticide molecules on and in plant and animal tissues constitute "pesticide residues."

The chemical pathways through which pesticides pass in plant and animal systems or in the soil are of interest not only in understanding their mode of action but also in assaying the toxicity of their residues in edible tissues. Because of the recent development of rapid isolation techniques and the revolution in modern electronics and instrumentation, metabolite studies are not as formidable as in the past. As pointed out by Lisk(7), in addition to the well-known chromatographic (column, paper, and thin-layer) and radiotracer methods, gas chromatography now appears to be accepted as a primary analytical tool for the separation, identification, and quantitation of the parent pesticide and its metabolites.

D. Evolution of Residue Methodology

From 1946 to 1955, specific residue assay was largely restricted to colorimetric chemical methods as well as some nonspecific procedures such as total chlorine, cholinesterase inhibition, and bioassay. These techniques were often quite cumbersome and their sensitivity was not always adequate. Until 1955, most analytical methods were sensitive to about 0.05 ppm and would respond with reasonable reproducibility to as low as a few micrograms of pesticide in 30 to 100 g of food sample. Any residue in a food sample containing less than 0.05 ppm was usually reported, in all good faith, as zero or nondetectable. The advent of thin-layer and gas chromatography techniques, with a 1000-fold more sensitivity than the methods prior to the 1960, has placed the residue chemist in industry, government, and state regulatory and enforcement agencies in somewhat of a dilemma. The basic problem is that many registrations and recommendations were previously based on less sensitive methodology. The inevitable consequence is that a commodity, which 10 years ago might have been assigned a "no residue" registration, today might very well result in an excessive residue violation.

The art of pesticide residue analyses has undergone a revolution in the past few years with the introduction of gas–liquid chromatography as a detection and quantitation tool. The first breakthrough in pesticide residue analysis was the introduction by Coulson and Cavanagh(8) of a microcoulometric titration cell coupled to a gas chromatograph equipped with a high-temperature combustion chamber. Here, for the first time, was an analytical tool that could simultaneously separate and quantitate halogen or sulfur-containing compounds with a sensitivity in the milligram range. Shortly thereafter, a series of ionization detectors was introduced by Lovelock and Lipsky(9). Currently, the most widely employed ionization detector used in the pesticide residue field is the electron-capture detector, which is routinely coupled to a conventional gas chromatograph. With this detector, the residue chemist can now detect pesticides in suitably prepared biological extracts in the nanogram $(1 \times 10^{-9}\,\text{g})$ and the picogram $(1 \times 10^{-12}\,\text{g})$ range.

Other new tools and modifications of older concepts in instrumentation continue to increase the number of detection and quantitation devices available to the residue chemist. Thin-layer chromatography has more or less replaced paper chromatography as a basic screening tool for the residue chemist. This technique offers a level of sensitivity between that of gas–liquid chromatography and the relatively insensitive paper chromatographic techniques. In the past decade, applications of infrared spectroscopy, ion-exchange resins, polarography, fluorescence, phosphorescence, neutron activation, emission spectroscopy, etc., have been made for the cleanup, characterization, and quantitation of pesticide residue samples. New instrumentation, due to its greatly increased sensitivity, has placed increased demands on the isolation and preparation of the pesticide introduced for analysis. Contrary to popular belief, considerable time and effort on extraction and cleanup are usually required prior to the determinative step by these new and highly sensitive analytical tools.

E. Current Residue Research Areas

Research in the pesticide residue field currently falls into three main categories: (a) methodology, (b) metabolism studies, and (c) dissipation relationships. In the past five years, there has been a tremendous upsurge in research on all phases of pesticide methodology including extraction, cleanup, detection, identification, and quantitation. This emphasis on improved methodology will undoubtedly continue in the future since pesticide metabolism and dissipation studies are wholly dependent on sound methodology. This chapter is restricted primarily to the methodology involved in assaying residues from various biological extracts.

The three basic phases involved in pesticide residue methodology are as follows: (a) sample preparation, (b) extraction and cleanup, and (c) detection and quantitation. Greatest emphasis in the subsequent chapter is placed on phase (b) and (c) because the bulk of the reported research has been concentrated in these vital areas. Despite the relative paucity of published data pertaining to the sample preparation phase, it should nevertheless not be considered as less important. One cardinal rule that everyone involved in pesticide residue research should be aware of is that no analytical result can be better than the original sample assessed. An excellent review of current research in the field of methodology for residue analysis from 1964 through October, 1966, has been made by Williams and Cook(*10*). This review covers published research concerned with sample preparation, extraction, cleanup, detection, and quantitation.

5-2   SAMPLE PREPARATION

The importance of careful, unbiased, and representative sampling in the field by trained personnel cannot be overemphasized. The general subject of biometry has been well covered by Garber(*11*) and will not be elaborated on further. However, with valid sampling in mind there are other important considerations, such as sample collection, preparation, and storage, which deserve some elaboration because of their direct association with the eventual pesticide assay. As has been pointed out repeatedly by Poos et al.(*12*), Fahey(*13*), Gunther and Blinn(*14*), Van Middelem et al.(*15*), Huddleston et al.(*16*), Gunther(*17*), and Lykken (*18*), a selected residue sample must be representative to be valid.

A.   Field Sampling and Sub-sampling

1. Experimental Plot Sampling.   Most of the workers in this field are cognizant of the need for adequate sampling but may not be sufficiently experienced or close enough to the actual field sample preparation phase to avoid gross errors. In most cases, it is not practical or feasible to sample all of the crop from a given test plot. Therefore, it is necessary to devise a sampling scheme in selecting the "gross" field sample that will be as representative of the entire plot as possible. This field sample should be prepared in similar manner to the commercial crop. In other words, the usual trimming, washing, and other commercial processes which aid in reducing the total crop residues should also be carried out on the experimental samples. The field sample collected is usually many times the size of the ultimate laboratory sample. It is imperative, therefore, that the gross sample be chopped and well mixed, prior to the selection of the 500- or 1000-g laboratory sample. It is generally recognized that the larger the original field sample, if representative, the more valid will be the subsample. The field sample must be selected in a manner that takes into consideration the crop variations in the plot. Plot selections should be made to minimize error due to overlapping applications, drift contamination from adjacent plots, skips in coverage, etc. Adequate "check" or untreated commodity must be taken so that the residue analyst will have ample residue-free material with which to check his entire residue analyses, with and without fortification, to authenticate his skill in analyzing for the particular pesticide in the presence of this particular crop.

2. **Regulatory Sampling.** Although the regulatory inspector may often be presented with somewhat different sampling procedures, the same basic rules of valid and representative sampling must always be followed. In taking an "official" sample, the inspector should be sure that the sample is (1) sufficiently representative to reveal the violation suspected, (2) large enough to permit proper laboratory processing, and (3) correctly handled and identified so as to maintain its integrity as court evidence. An "investigational" sample is not used as a basis for specific legal action, but it serves to provide information which can be applied in many ways to the producer, processor, or the regulatory agency. The procedures for collecting investigational samples are usually not so rigid as those required for the collection of official samples. Some of these samples, however, are used to substantiate official samples and their integrity must be well established.

The practical regulatory inspector is usually faced with many types of sampling situations. For example, he might be confronted with a lot of 200 crates of produce, stacked four crates deep, ten long and five wide, from which a stratified statistical sampling should be made. This could be accomplished by dividing the lot into sections and sampling equally from each section at random. Each section should be so subdivided that the increments of the subsample will be representative of that subsection. This sampling procedure is suggested when there is little or no information on the sample or when the inspector feels that the lot is fairly uniform throughout.

B. Sample Preparation, Packaging, and Storage

The laboratory sample should be prepared from the field sample as rapidly after harvesting as possible in order to minimize residue loss, water content, etc. The representative field sample may be placed in a clean polyethylene bag or other type of container, labeled, and placed in a freezer as soon as possible. Some residue laboratories will at some later date place the frozen sample in a foodcutter and mechanically reduce the size of particles to a minimum without unnecessary loss of essential liquids. At this time a portion or all of the sample could be extracted or returned to the freezer for future extraction. Some laboratories prefer to store the crop extracts, whereas others prefer to store the finely chopped laboratory sample until a subsample is taken for extraction and analysis. There are pros and cons as to which is the better procedure. There is some evidence, however, that there may be a significant difference in extraction efficiency of internal residues depending on whether the commodity is extracted fresh or has been allowed to freeze prior to extraction. It is postulated that the freezing and subsequent thawing rupture the internal cells to such a degree that greater recoveries of internal residue are achieved with a given extraction procedure. If extracts or unextracted crop samples are to be stored any length of time it is best to keep them at $-20°F$. There are definite indications that certain pesticides, particularly the organophosphates, may break down slowly in 0°F storage. Therefore, it is important that the analytical chemist consider the chemical characteristics of the pesticide in question when considering the type and length of sample storage. It is highly important that all crops or extracts be tightly sealed to prevent moisture evaporation in storage. Extracts should be stored at 0°C or lower in glass-stoppered bottles or screw-cap bottles with aluminum-lined caps. Waxed paper or plastic liners should be avoided since the solvent will often dissolve minute quantities of liner impurities, resulting in serious contamination later during GLC analysis. Thornburg(*19*) cites instances where carbaryl was rapidly lost from chloroform extracts even at low-temperature storage. However, a small volume of ethanol added to the chloroform aids in preservation of the carbaryl residues. In general, it is the best policy to analyze the samples as soon as possible after extraction unless there is available experimental evidence to indicate that the pesticide is relatively stable in the solvent. To be on the safe side, the analyst should fortify a portion of the "check" extract at the beginning of the storage period. The level of fortification should be in the general range of residue level expected in the treated samples to be analyzed and well within the sensitivity range of the method.

5-3 EXTRACTION AND CLEANUP

A. General Considerations Involving Extraction

After the field sample has been composited, subsampled, and a suitable aliquot weighed out, the next step is to extract the pesticide residue from the sample into a suitable solvent. It is essential that the residue chemist consider the method of analysis that will be employed prior to initiating the sample extraction. The method of extraction and the type of solvent or solvent combinations will be dependent on the chemical and physical properties of the pesticide to be extracted, the type of substrate from which it will have to be quantitatively removed, and the final method of analyses.

1. Purification of Solvents. If the final determinative step will involve gas chromatography by electron capture, emission spectrometry, etc., it is essential that all solvents be redistilled in an all-glass apparatus. An excellent review of procedures involved in the purification of solvents for pesticide residue analyses has been presented by Thornburg(22). For some laboratories it may be best to purchase a high grade of solvents, specially redistilled for use with gas-chromatographic analysis. If commercial solvents are to be redistilled on the premises, all-glass systems with efficient columns should be operated at high reflux ratios. Great care must be taken with any solvent to avoid contact with rubber or any plastic, except TEFLON, before injection into the electron-capture or microcoulometric detector. Even detergent-contaminated glassware can negate hours of careful effort in the cleanup of a solvent. The use of reactive chemicals such as sodium followed by distillation has been a favored technique in the past for removing impurities from aliphatic hydrocarbons. Chlorine-containing compounds are removed and the solvent is dried by this procedure. Where large quantities of solvent are not required, the use of commercial, specially distilled solvents for pesticide residue analyses might be the least troublesome to most laboratories. Most of the chlorinated solvents should be redistilled just prior to use, since these solvents will form phosgene on standing any length of time. Ethers form peroxides upon storage and should be removed prior to redistillation.

2. Selection of Solvents. The choice of the extracting solvent, polar, nonpolar, or a mixture of both, will depend on the nature of the sample and the polarities of the pesticides to be extracted. After extraction and evaporation, the sample obtained from 1 liter of water or 100 g of a non-oily crop may weigh from 1 mg to 1 g. This reduction in mass is made possible by the fact that many samples contain 90% or more water, most of which will not partition into the organic solvent. Moreover, many cellular constituents such as amino acids, proteins, carbohydrates, and cellulose are only sparingly soluble in most organic solvents. Since most pesticides dissolve in polar and/or nonpolar organic solvents, they are separated from the bulk of the sample containing interfering materials.

No one solvent, or even a solvent mixture, will extract every pesticide or their metabolities from a weathered sample. If the pesticide history of a sample is unknown, Thornburg(23) has suggested that three extraction procedures be followed: first, a dual solvent-blending extraction, using either hexane–isopropanol or benzene–isopropanol; second, an acetonitrile extraction, correcting for the water content of the sample; third, a redistilled methylene chloride extraction is performed. This solvent, being heavier than water, is useful in separatory funnel separations. It should be remembered, however, that methylene chloride should not be injected directly into an electron-capture detector since it will overload the system. If used for extraction, the methylene chloride should be completely evaporated and the residue picked up in a nonchlorinated solvent before injection.

3. Evaporation of Solvent Solutions. Most analytical methods for the determination of any pesticide residue require that the solution containing the pesticide(s) be considerably reduced in volume, often to the point of dryness. Current methods of separation, identification, and quantitation by thin-layer or gas chromatography frequently require that the final volume of cleaned-up solution be less than 0.5 ml. Danish–Kuderna evaporative concentrators can effectively reduce the volume to 5 ml without loss of the pesticide. Further evaporation is usually done by directing a stream of clean, dry air onto the pesticide solution. This is a highly critical step, and considerable loss of the pesticide can occur if the solvent is allowed to go to absolute dryness.

Burke et al.(20) have reported on some of these evaporation losses of various organochlorine pesticides and have suggested a technique to protect against such losses. It was noted that losses increased as the volume approached dryness, and percentage losses were greater when smaller amounts of pesticide were present.

Smaller pesticide losses were noted when the solvent was carefully boiled off at steam bath temperatures than when an air stream was used at room temperatures. However, when a small amount of oil was added to the solution, evaporation losses were decreased. Excellent recoveries were achieved for five chlorinated hydrocarbons when 7 ml of pesticide solvent was evaporated to the smallest possible volume (0.3 ml) in a Danish–Kuderna concentrating tube. A two-bubble micro Snyder condensor, used with the concentrator on the steam bath, was found to be more satisfactory than either the one- or three-bubble type.

Burchfield and Johnson(21) stress the precautions to follow when removing solvents from extracted pesticide residue samples by removal with air currents, distillation at atmospheric pressures using Danish–Kuderna concentrators or evaporation under vacuum using rotary evaporators. These precautions are as follows:

(a) Usually the organic solvents containing the pesticide should be dried over anhydrous sodium sulfate.

(b) Solvent temperature during evaporation should not exceed 50°C in most cases.

(c) The sample should be watched carefully during the final stages of solvent removal. The beaker holding the residue should be removed from the heater just as the last few drops evaporate. Samples should not be heated or subjected to vacuum after all the solvent has evaporated.

(d) Solvents should never be completely removed from purified extracts or pesticide losses will occur. It is safer to evaporate crude extracts to dryness since the nonvolatile impurities in the sample retard sublimation of the pesticides. Sometimes it is advantageous to add a few milliliters of a high-boiling compound "keepers" to the solution of pesticide before evaporation of the solvent.

(e) Care must be taken to prevent condensation of moisture in the pressure lines and consequent contamination of the sample during evaporation.

In any analysis for pesticide residues there is always an interrelation between the method of extraction, the cleanup of the extract, and the technique used for the actual determination of the residues. In addition, the type of biological material or soil being extracted and the pesticide being investigated must also be considered.

4. Pesticide Recovery or Extraction Efficiency. The first step in pesticide residue analysis is to separate a very small amount of pesticide quantitatively from a relatively large proportion of biological material on or in which it is found. The quantity of pesticide removed by the particular extraction method is not a true measure of the actual original residue unless the extraction is complete. This degree of residue removal is usually referred to as the extraction efficiency. As pointed out by Gunther and Blinn(14) and Bann(24), extraction efficiency depends upon a high solubility of the pesticide in the solvents used, an intimate mixture of the solvent and pesticide, and the establishment of an equilibrium between the toxicant concentration in the recovered solvent and that remaining in the crop residue.

When residue analysts report extraction efficiency, it is usually expressed as a percentage recovery of a pesticide from a fortified or "spiked" sample. At best this procedure consists of adding known amounts of pesticides to an untreated tissue sample just prior to extraction. A fortification procedure sometimes followed, particularly when residues are suspected to be low, is to add a known quantity of pesticide to a treated sample and then determine the percentage recovery of the added compound by difference. The least valid procedure, but unfortunately one quite commonly employed, is to add the pesticide to the extract following the cleanup immediately prior to the analysis.

The primary difficulty in determining the efficiency of an extraction procedure is to ascertain the amount of pesticide originally present on the sample as it occurred in natural conditions in the environment. Because of this inherent problem, very few authentic extraction efficiency studies have been conducted in the pesticide residue field. The few studies that have been reported involved the removal of organochlorides from plant materials. Klein(25) has suggested establishing the amount of pesticide present on a given sample by using an exhaustive extraction procedure. This might be too laborious for routine work but could be used for comparison with a shorter extraction procedure carried out on a duplicate sample of the same commodity. Of course, it has to be assumed that the exhaustive extraction procedure quantitatively removes all of the pesticide.

Thornburg(23) in an attempt to avoid the obvious deficiencies in the fortification method for evaluation extraction efficiencies, has outlined a procedure called a "weathered residue study" in which three separate field-treated plant subsamples are extracted by three different extraction procedures. The first is extracted by the standard procedures, the second subsample is extracted three times with the standard procedure, and the third subsample is extracted in the standard manner except three times the normal volume of solvent is used. If all three extraction procedures yield approximately the same per cent recovery, then the first standard extraction procedure can be assumed to be satisfactorily quantitative according to Thornburg(23). However, there is a possibility that the last two of the above extraction techniques may not be entirely effective in removing all of the pesticide residue from the field-treated sample. If this were the case, somewhat erroneous per cent recoveries would be reported. Nevertheless, the "weathered residue study" procedure is certainly far superior to the majority of "percent recovery" information that has been reported in the past.

5. General Extraction Procedures. The primary extraction procedures that have been evaluated on plant tissue fall into three general categories: surface rinsing, macerating or blending with a solvent or solvent pair, and exhaustive extraction.

Klein(25) demonstrated the necessity for a "blending" type extraction by showing the effectiveness of an isopropanol–benzene extraction system. Following this study, Klein et al.(26) investigated the removal of chlorinated hydrocarbons residues on leafy vegetables by the three basic extraction procedures: Soxhlet, tumbling (surface rinsing), and blending, and he concluded that the procedure of choice was to blend the sample first with isopropanol and then with benzene. Moddes and Cook(27) reported the successful use of acetonitrile for the extraction of parathion and diazinon from lettuce. Webster and McKinley(28) investigated the relative extraction efficiencies of a simple surface rinsing versus the Klein blending procedure for use with plant materials that tend to exude water during processing. At least for the removal of DDT residues from apples, these investigators found that the simple stripping procedure was as efficient as the longer blending procedure. They found, however, by using ^{14}C-labeled DDT, that complete extraction was not obtained by either of the two procedures tested. If another crop such as a leafy vegetable had been tested in this study, the simple surface extraction technique would undoubtedly have been less efficient than the blending procedure. Van Middelem et al.(29) found that three times as much field-weathered parathion was extracted from celery by blending in benzene for 5 min plus end-over-end tumbling for 30 min compared with the tumbling only. The use of a cosolvent (isopropanol–benzene) blending was found to be superior to single solvent blending for the removal of parathion residues from mustard greens.

Hardin and Sarten(30) compared five extraction procedures for the recovery of DDT residues in field-treated collards and concluded that successive blendings with isopropanol followed by hexane offered the most satisfactory extraction procedure. A comparison of alcohol–hexane with acetonitrile blending for the removal of DDT from cabbage and lettuce was reported by Johnson(31). The results of the acetonitrile extraction compared favorably with those from the mixed solvent blending. Bertuzzi et al.(32) found that acetonitrile extraction was not effective for the complete removal of chlorinated hydrocarbons from dehydrated crops containing about 10% moisture. A 35% water–acetonitrile solvent mixture was found to be satisfactory for the extraction of dieldrin residues from dried beet pulp. Burke and Porter(33) compared nine blending type extractions' procedures for effectiveness in removal of field-applied p,p' TDE from kale. Extraction efficiences were based on the amount removed by an exhaustive extraction. The acetonitrile extraction procedure of Mills et al.(34) produced the highest residue recovery, i.e., 99% of that removed by the exhaustive extraction. This method was found to be very effective for the extraction of parathion and diazinon from field-sprayed kale.

A review of three basic extraction procedures has been made by Wheeler and Frear(35). Wheeler et al.(36) and Mumma et al.(37) have found that internal residues of the chlorinated hydrocarbons were not quantitatively removed by repeated blending of the field-treated crops with 2:1 mixture of hexane–isopropyl alcohol. A subsequent 12-hr Soxhlet reextraction of the plant tissues with 1:1 chloroform–methanol resulted in essentially quantitative extraction of the remaining labeled pesticide. Apparently the extraction of internal plant residues due to root uptake presents a particularly vexing problem. Since such residues were not completely removed by the mixed hexane–isopropanol blending procedure, an additional exhaustive Soxhlet extraction using a more polar solvent system was advocated by Mumma et al. (37).

Exhaustive extraction procedures using a suitable nonpolar solvent in a Soxhlet apparatus have been reported to be necessary for the effective removal of chlorinated hydrocarbons from dry, finely divided samples. The overall efficiency of this procedure has not been fully evaluated, but there is a possibility that significant losses from fresh crop samples could occur during the heat drying process or by codistillation with water. Removal of the excess sample water before Soxhlet extraction by mixing with anhydrous sodium sulfate is recommended.

In the final analysis, the residue chemist must use considerable forethought and judgment in selecting a practical, yet efficient extraction procedure based on the chemical nature of the pesticide and its surrounding substrate. For example, a simple one-solvent extraction by tumbling or shaking for 1 hr is probably adequate for the removal of surface residues on whole, smooth-surfaced crops but is probably quite inadequate for leafy crops. When dealing with surface and subsurface residues on frozen or well-cut fresh commodities containing significant quantities of water, one recommended extraction procedure is a two-solvent blending technique. The following procedure will usually result in rapid extractions that are free of emulsions and high per cent recoveries of the pesticide: to one part of the sample add two parts of benzene (hexane) and four parts of isopropanol. In general, the immiscible solvent should always be around 2 ml/g of substrate. A relatively large volume of isopropanol is necessary to dehydrate the sample, thereby preventing troublesome emulsions from forming with many high-water content crops.

The mixture is usually blended at high speed for 3 to 5 min. The solid materials can be removed by aspiration through a Buchner funnel, by conventional filtering, or by centrifugation (care should be taken here to avoid excessive loss due to evaporation) and the solvents transferred to a separatory funnel equipped with a TEFLON stopcock. If an emulsion has formed despite the isopropanol, add approximately 100 ml saturated salt solution and shake funnel well. The water layer is discarded and the benzene (hexane) extract washed with more water before drying with a small amount of anhydrous sodium sulfate.

Some pesticide residue workers recommend that the crop be blended with isopropyl alcohol before adding a water-immiscible solvent and reblending. Whether this technique is superior to a single blending of a mixture of the same two solvents has not been definitely established.

The quantitative removal of a pesticide from a fatty material or an oil presents the residue chemist with a particularly vexing problem. Specialized techniques must often be considered in the analysis of residues in dairy products, mammalian tissues of all types, and certain fruits, nuts, etc., containing a high percentage of fats and oils.

Sometimes, when extracting certain fatty materials, it is best to blend with hexane first, centrifuge, and follow with an acetonitrile extraction of the hexane. Direct extraction with acetonitrile often removes too many of the water-soluble contaminants. The pesticide can be taken from the acetonitrile by adding sufficient water so that the acetonitrile concentration is not over 20% and then extracting with hexane. When the acetonitrile extract contains excessive fats, dilute to 65% with water and pass through a polyethylene alumina column.

In general, it can be stated that many non-ionic pesticides of the chlorinated hydrocarbons and organophosphate groups can be extracted from most acid or basic substrates with solvents such as hexane while anionic compounds, such as the phenoxyalkanoic acid herbicides, usually require acidification of the substrate prior to extraction with more polar solvents.

There have been several "universally" applicable extracting systems proposed, one by Schnorbus and Phillips (38) that is reported to be suitable for use on a broad spectrum of commodities and pesticides. Fruits, vegetables, grains, meats, and dairy products were blended with propylene carbonate and filtered through a Buchner funnel. Soils and dehydrated products were shaken with propylene carbonate for 2 hr and then filtered. Fats and oils were extracted by shaking with propylene carbonate in a separatory funnel. The primary advantages of the use of propylene carbonate over other currently used extraction solvents were listed by Schnorbus and Phillips (38) as economy, low mammalian toxicity, non flammability, and lack of the usual requirement of prepurification prior to use with electron-capture detection systems.

In the following sections on extraction and cleanup, primary emphasis is placed on the more successful, widely used procedures. Comprehensive reviews on extraction and cleanup can be found in Refs. *14, 19, 21, 39,* and *40.*

B. General Considerations Involving Cleanup

Following extraction, a cleanup of the sample extract is usually necessary before the final determinative steps can be performed successfully.

Cleanup is a term used in residue methodology for the isolation of the pesticide from interfering extraneous extractives or solvents. The type and extent of cleanup will depend on the nature of the substrate and the final method of pesticide analysis. If an extract is to be analyzed by a method in which nearly any organic material would interfere, the cleanup procedure must be extensive. As a general rule, it is usually best to clean-up an extract only as much as necessary since even the most satisfactory methods tend to remove some pesticide. When evaluating any cleanup procedure, the efficiency of recovery should be at least 75–80% of the pesticide added prior to cleanup. It is mandatory, of course, to test per-cent recoveries of various levels of the pesticide in question thoroughly after they have been carried through the planned cleanup. The levels of pesticide tested should encompass the range of residues expected when the unknown samples are analyzed. A satisfactory cleanup must be achieved before any analysis of the treated samples is initiated. The residue analyst should run one spiked pesticide standard through the cleanup manipulations without the presence of tissue extractives, in order to ascertain the loss of pesticide alone during cleanup.

Most tissue extracts require a cleanup procedure in which the inter-fering compounds are removed by adsorption on an active solvent or adsorption by a liquid impregnated on an inert solid. Cleanup of extracts rich in fats can be achieved by partitioning between polar and nonpolar solvents, thereby separating the pesticides from the large bulk of lipids. Following propylene carbonate extraction of fruits, vegetables, grains, meats, dairy products, fats, and oils, Schnorbus and Phillips(38) re-commended cleanup of the filtered extract by eluting through a FLORISIL column prepared according to Langlois et al.(41) Soils and dehydrated products were also cleaned up by passing through a similar FLORISIL column. Petroleum ether was found by these workers to be a satisfactory partitioning agent for the organochlorine compounds from propylene carbonate, whereas isooctane was found to be more suitable for parti-tioning the organophosphates from propylene carbonate extracts.

Unfortunately, as with the extraction of pesticide samples, there is no precise, universal cleanup procedure applicable to all types of extracts. Most of the modern cleanup methods for pesticide residue extracts in-volve partition, sorption, or chemical methods of some combination of these techniques. However, recently there have been several new clean-up techniques introduced, such as thin-layer chromatography and sweep codistillation or forced volatilization. All of these various cleanup procedures are discussed in more detail in subsequent sections of this chapter. Some previous reviews on general cleanup procedure for pesticide residues in various biological extracts have been made by Gunther and Blinn(14), Gunther(1), and Thornburg(19).

C. Chlorinated Hydrocarbons

1. Extraction of Nonfatty Foods (<2% fat).

a. LEAFY VEGETABLES, FRUITS, AND ROOT CROPS. There are two methods in rather general use for the extraction of chloro-organic pesticides from fresh plant materials. One procedure by Mills et al.(34) calls for the blending of finely chopped crop material with acetonitrile and CELITE 545. A second method utilizes a dual solvent system. The benzene–isopropanol mixture is blended with chopped plant material as described by Thornburg(19) and the aqueous alcohol removed from the benzene by repeated washings in a separatory funnel. Wheeler et al. (36) and Mumma et al.(37) have suggested that the blending procedure using benzene (hexane) and isopropyl alcohol is not capable of quantitatively extracting all the internal residues of certain chlorinated insecticides from fresh forage grown in contaminated soils. This group recommended, in addition to the dual solvent blending, an additional 12-hr Soxhlet extraction with 1 : 1 chloroform–methanol in order to provide quantitative extraction of these pesticides. Goodwin et al.(42) report the use of nonpolar–polar solvent mixtures such as hexane–isopropanol or hexane–acetone as unsatisfactory because in crops containing a high proportion of water, emulsion problems were often encountered. For this reason acetone alone was chosen as the crop-macerating solvent. Direct gas–liquid chromatography of the acetone extracts of leafy vegetables, fruit, and root crops was found to be impractical because the excessive coextracted crop materials adversely affected the detector.

Johnson(31) extracted various chlorinated hydrocarbons in lettuce, spinach, cabbage, and broccoli by blending with acetonitrile and then partitioning the pesticides into petroleum ether in the presence of 30% sodium sulfate and large volumes of water. Baetz(43) and Albert(44) used the same extraction procedure on collards, carrots, okra, and pears, but differed only in their cleanup procedure. Burke and Porter(33) compared nine blending-type extraction procedures for their effectiveness in the removal of field-applied p,p' TDE from kale. These workers found that the acetonitrile blending procedures of Mills et al.(34) extracted 99% of the TDE from the kale as compared to an exhaustive extraction procedure.

b. CEREALS, GRAINS, FEEDSTUFFS, FORAGE (LOW-MOISTURE). A method by Moffitt and Nelson(45) for the extraction of chloro-organic pesticides from cereals is similar to the dual solvent (isopropanol–benzene or hexane) blending procedure described by Thornburg(19). An alternate procedure that can be followed on cereals and grains is almost identical to that used by Moffitt(46) for the extraction of chlorinated hydrocarbons from milk and milk products. In this procedure the cereal in grain is shaken with 5% aqueous potassium oxalate and water. Ethanol or isopropanol, diethyl ether, and pentane (1 : 2 : 1) are added one at a time and shaken vigorously for 2 min each time. The aqueous layer is discarded and the organic phase washed and filtered through anhydrous sodium sulfate. The method of Mills et al. (34) has been found satisfactory for recoveries of pesticides in low-moisture, nonfatty foods, when added just prior to extraction. However, Bertuzzi et al. (32) found that the addition of water to the acetonitrile was a superior solvent combination for the removal of chlorinated hydrocarbons from low-moisture, nonfatty foodstuffs. These workers found that a 35% water–acetonitrile extractant was optimal for the removal of dieldrin from beet pulp.

Wheeler et al. (36) found that the usual 2 : 1 hexane–isopropanol blending procedure did not extract all the radioactive dieldrin present in wheat and corn plants as soil-absorbed residues. In order to remove most of the unextracted dieldrin, it was found necessary to reextract the plant tissue with 1 : 1 chloroform–methanol for 12 hr in a Soxhlet apparatus.

2. Extraction of Fatty Foods and Tissues (>2% fat). Fatty foods such as milk, milk products, cheese, nuts, and avocados require specialized extraction procedures because of their chemical composition and structures. Chlorinated pesticides are usually most effectively extracted from fatty foods by the isolation of the lipid fraction through the combined action of an alcohol and fat solvent or a fat solvent alone. The chloro-organics can then be separated from the lipids by partitioning the extract between a hydrocarbon solvent and a polar solvent. The combined action of the alcohol and fat solvent separates the lipids (with the chloro-organic pesticides) from the proteinaceous and other water-soluble compounds. The second step, liquid–liquid partitioning between the hydrocarbon and polar solvents, results in the isolation of the pesticides from the bulk lipid fractions. The polar solvent extract containing the pesticide is now ready for cleanup prior to the determinative steps.

a. DAIRY PRODUCTS, FATS, OILS, NUTS. In milk, the fat is contained in small globules that are surrounded by membranes composed of phospholipids and proteins. Consequently, in order to extract the fat plus the fat-soluble organochlorines from milk completely, these membranes must be disrupted, thereby causing a coalescence of the globules. An alcohol is the most desirable solvent for shattering the protein phospholipid envelopes since acids, alkalis, or detergents could be destructive to the pesticides. Once these membranes are disrupted, the usual fat solvents such as diethyl ether, benzene, or hexane can be effectively employed.

Chlorinated pesticides in these fatty or fat-containing commodities are sometimes extracted by dissolving in petroleum ether followed by acetonitrile. Repeated extractions in a separatory funnel, according to Mills(47), are required for complete isolation of the pesticides from the lipid fractions. For whole milk, milk products, and cheese, Mills(47) and Johnson(48) prepared the milk sample with initial extractions of potassium oxalate, methanol, and a 1:1 v/v mixture of petroleum ether and ethyl ether prior to acetonitrile partitioning. The procedure proposed by Moffitt(46) for the extraction of milk and milk products may not be as quantitative as that by Mills(47), since only one extraction is recommended rather than three.

To achieve quantitative recovery of most organochlorine pesticide residues from milk, the milk fat has to be completely extracted. Since the phosphate pesticides are usually more polar than the chlorinated hydrocarbons, and since they are therefore more likely to be found in the aqueous phase than in the fat phase, complete fat extraction may not always be necessary or even desirable. Beroza and Bowman(49) found low recoveries when nonpolar pesticides, such as aldrin and HEPTACHLOR, were added to milk, either immediately or 5 hr prior to extraction with (1:1) hexane–ether, a procedure that resulted in about 10% fat removal only. This same low-fat extraction technique resulted in excellent (85–95%) recoveries of milk similarly spiked with the more polar organophosphates such as methyl parathion and IMIDAN. A second milk extraction procedure, involving an initial 1:1 ethanol mixing and followed by a hexane–ether (1:1) extraction, resulted in a more complete fat extraction and high recovery of the more nonpolar chlorinated hydrocarbon pesticides. These studies of Beroza and Bowman(49) illustrate the affinity of the polar organophosphate pesticide molecules for the aqueous phase, from which they may be extracted more readily than from the fat phase. In general, it was noted that the polarity of the pesticide increased as the p-value decreased.

This study emphasized the present lack of knowledge concerning the adsorptive binding forces that interfere with the adequate recovery of pesticides from milk. For butter extraction, merely warm the sample at about 50°C until the fat separates and decant fat through a dry filter prior to acetonitrile partitioning. In the extraction of fats proposed by de Faubert Maunder et al.(50), a finely ground sample is boiled gently in hexane before filtering. The sample is cooled and brought to volume with hexane prior to cleanup and analysis.

In the above citations, it should be noted that the extractive phases involve dissolving the lipid fraction (plus the pesticide) into a nonpolar solvent, prior to the cleanup phase involving liquid–liquid partitioning. Various chlorinated hydrocarbons were extracted by McCully and McKinley(51) from vegetable oils, butter, and animal fats. The oils were dissolved in a 1 + 19 mixture of benzene and acetone, whereas butter was first melted and the oily layer decanted off. After filtering, the butter oil was dissolved in the 1 + 19 mixture of benzene and acetone. The animal fats were blended in the same solvent mixture in the presence of anhydrous sodium sulfate as reported by McCully and McKinley(51) and McCulley et al.(52).

b. MAMMALIAN TISSUES (MUSCLE, LIVER, HEART, KIDNEY, BRAIN, ETC.). Tissue samples obtained from freshly killed animals should be completely frozen. On sampling, care should be taken not to allow the gross sample to thaw to any extent to avoid the possible loss of blood and/or other fluids. When dealing with large samples, the frozen specimen should be chopped with dry ice in a foodcutter. The dry ice content should be approximately twice that of the sample, and sublimation overnight in the storage freezer is recommended.

Extraction can be with warm hexane as described by de Faubert Maunder et al.(50) or by blending with acetonitrile and Skellysolve B as suggested by Olson(53). The acetonitrile is normally the lower layer, but in fat extractions inversions may occasionally occur. Mills(54) has described an extractive procedure for chlorinated organic pesticides in animals tissues by paper chromatography. The tissue sample is extracted by repeated shaking in petroleum ether and the resultant extract evaporated at low temperatures to isolate the fat fraction. A rapid procedure has been developed by Onley and Bertuzzi(55) for the extraction of several organochloride residues from animal tissues, fats, and meats. This method combines the use of a mixture of acetone, methyl cellosolve, and formamide to extract the pesticide, using calcium stearate to coagulate the lipid fractions.

c. POULTRY PRODUCTS. Chlorinated hydrocarbons that are not pesticides are frequently found in eggs in this country. These compounds can lead to high backgrounds when the determinative steps are by electron-capture or microcoulometric gas chromatography. A procedure for removing these interferences is described by Onley and Mills(56).

Extraction of a fluid or dehydrated egg can be accomplished by the de Faubert Maunder et al.(50) procedure which involves two hours of Soxhlet extraction with an acetone–hexane solvent mixture. Prior to extraction, sufficient anhydrous sodium sulfate is mixed with fluid egg to yield a granular mass.

Stemp et al.(57) report the successful modification of the one-step
FLORISIL column by Langlois et al.(41) to analyze egg yolk and poultry
tissue samples for determination of chlorinated hydrocarbon residues by
electron-capture gas chromatography. In this unique procedure, 1–2 g
of poultry tissue or egg yolk are incorporated into 25–30 g FLORISIL and
placed as the top layer in a chromatographic column. The FLORISIL mixture is eluted with 20% methylene chloride in petroleum ether. The eluant
was evaporated and picked up in hexane prior to injection. Cummings
et al.(58) report a procedure where the organochlorines were determined
in eggs by combining certain features of the one-step FLORISIL procedure
and the conventional elution from FLORISIL. The Johnson(31) procedure
was modified by Sawyer(59) to provide a rapid and satisfactory method
for the cleanup of fresh eggs prior to EC–GLC analysis. In this procedure,
acetonitrile extraction was followed by centrifugation and chromatography on a layered FLORISIL–sodium sulfate column. A rapid method
has been developed by Beckman et al.(60) for the extraction of DDT
residues from eggs by blending the sample with acetone.

3. **Extraction of Miscellaneous Samples.**

a. BODY FLUIDS. Pesticide residues in body fluids that are attributable
to the intake of pesticides in foods are usualy found in very low concentrations. The extraction of urine samples for the removal of non-ionic
pesticides can be accomplished normally by the subsequent procedures
described for water. Usually centrifugation will break stubborn emulsions formed by shaking urine samples and water-immersible solvents too
vigorously.

Johnson et al.(61) have developed a procedure for the analysis of drug
metabolites in urine that could be adapted to the extraction and determination of pesticide residues.

The chlorinated hydrocarbons can be effectively extracted from blood
plasma by a water-immiscible solvent in which the pesticides are soluble.
However, plasma extractions are more difficult than those with urine
because of difficult-to-break emulsions caused by large amounts of proteins. Whole blood can also be extracted directly with water-immiscible
solvents provided the sample is first diluted with seven volumes of water
in order to rupture the cells. Better results probably would be achieved
by the use of a mixed solvent technique recommended for the extraction
of milk.

b. WATER. Direct measurement and identification of organic pesticides in water is not always feasible without resorting to extraction
techniques to concentrate the pesticide to detectable levels. With the
development of the halogen-sensitive titration cell microcoulometry, it

is now possible to detect many organochlorines as low as 25 to 50 ng/liter of water. By using an electron-capture detector, this sensitivity can be increased at least ten-fold. The simplest technique available is to extract one liter of water completely with an organic solvent. In a continuous extraction method used by Kahn and Wayman(62), the water in each unit is extracted continuously and independently with petroleum ether or diethyl ether, while a constant flow of water is maintained through each extractor. By employing this procedure it is possible to isolate, purify, and identify very small quantities (nanogram or picogram range) and to collect samples large enough for infrared analysis. There are many other water extraction techniques for the detection of chlorinated hydrocarbons and their metabolites available in the literature. An excellent review of pesticides in water by Faust and Suffet(63) describes the extraction of water, the preliminary separation of pesticides from interferences, and various carbon absorption, liquid–liquid extraction, and cleanup techniques. They also discuss the separation and identification of parent compounds, metabolites, and hydrolysis products by various chromatographic and spectrophotometric means. Another useful review by Burchfield(64) covers various chromatographic and infrared methods for the analyses of pesticides in water.

c. SHELLFISH, MUD, AND SEDIMENT. Shellfish can be extracted by blending them in acetonitrile and then subjecting them to liquid–liquid partitioning with petroleum ether. Robertson and Tyo(65) have adapted the procedures of Mills(47) and Johnson(31) for the extraction and cleanup of oysters prior to electron-capture determination of DDT, DDE, heptachlor, and its epoxide. Mud and sediment can be extracted for pesticide residues by blending with 10% acetone and 90% acetonitrile prior to partitioning the residues into petroleum ether. Schutzmann et al.(66) extracted soil, water, and sediment for the removal of various chlorinated hydrocarbon and organophosphate pesticides by rotating the sediment for 4 hr with a 3:1 mixture of hexane–isopropanol.

d. SOILS. Satisfactory extraction methods have not been fully developed for the extraction of many chlorinated pesticides and their metabolites from soil. This is partly due to the fact that the soil environment is very complex, resulting in many known and unknown interactions with the pesticides. Another reason for the lack of adequate soil extraction methods is that not until recently has adequate emphasis been placed on soil residues that have been accumulating for the past two decades. As with plants, but probably to even a greater extent in soils, it becomes progressively more difficult to extract quantitative amounts of residues from aged soils. Most fertile soils contain diversified flora and fauna which are capable of retaining substantial amounts of chloro-organic pesticides. Micro-organisms in soil attack pesticides and often convert them to metabolities. Lichtenstein(67) reports that as a result of microbiological activity, aldrin is oxidized to dieldrin and heptachlor to heptachlor epoxide. Apparently the percentage of soil moisture is very critical when extracting soils. Recoveries tend to be poor from very dry soils or exceptionally wet soils. Lichtenstein(68) reports that the best recoveries were obtained from a loam soil having 12 to 16% moisture.

In a study on aldrin and dieldrin in soils and their translocation into carrots, Lichtenstein et al.(69) report the use of 1:1 v/v acetone and hexane blending for extraction prior to gas-chromatographic detection. Barthel and Dawsey(70) indicate that a 3:1 hexane–isopropanol combination has proven effective for the extraction of chlorinated organic pesticides from soil, water, and sediment. In this procedure the extraction is maintained for 4 hr on a concentric rotator. Shell Chemical Company(39) advocates vigorous shaking of the slurried soil in 3:1 hexane–isopropanol for 20 min. Then the soil extraction should be repeated two additional times and any excess alcohol washed from the hexane extracts with water. Harris et al.(71) extracted chlorinated hydrocarbons from various types of soils, ranging from sands to mucks using a 1:1 acetone–petroleum ether solvent mixture. The petroleum ether fraction was removed from the acetone by water washing prior to analysis by EC–GLC. Three extraction procedures for DDT- and endrin-contaminated soils were compared for efficiency, reproducibility, and precision by Teasley and Cox(72). The extraction procedures included a Shell Chemical Company procedure(39), a Soxhlet and Immerex† method. These workers concluded that the most efficient and precise results were obtained by extracting the soils in the Immerex extractor with a 9 + 1 hexane–acetone solvent mixture for 16 hr.

4. Cleanup Methods for Chlorinated Hydrocarbons. McKinley et al. (73) have reported six cleanup procedures for the isolation of several chloro-organic pesticides in various crops and biological media by partitioning between solvents, column chromatography, saponification, acetone precipitation, oxidation, and sulfonation. Because of their present common usage, the first three cleanup procedures used by the above workers are discussed in more detail in the following sections. In addition, several new and promising cleanup procedures are also described. For ease of reference, the cleanup procedures for the various classes of pesticides are separated. In some instances, cleanup procedures between

†Immerex Extractor – A. H. Thomas, Cat. No. 68, p. 31.

the chlorinated hydrocarbons and organo phosphate pesticides may be interchangeable.

a. PARTITION. Extracts containing significant quantities of fats can be partitioned between immiscible polar and nonpolar solvents in order to separate the pesticide from the lipid fractions. Acetonitrile saturated with hexane is often used as the polar solvent, while petroleum ether or hexane is used as the nonpolar solvent. Jones and Riddick(74) found that pigments and waxes were preferentially soluble in n-hexane, while the pesticides, because of their high polarities, were extracted into the acetonitrile. The acetonitrile can then be evaporated or diluted with water and reextracted with hexane in order to recover the pesticides. By this liquid–liquid partitioning, separation from lipids is adequate but not complete since some nonsaponifiable lipids and pigments are carried over into the acetonitrile layer. Usually cleanup by adsorption is necessary before or following the partitioning step. If animal tissue samples are extracted with acetonitrile, the partition step is not necessary since lipids are not very soluble in this solvent. In general, acetonitrile partitioning is not very satisfactory as the only cleanup of samples containing more than 5 g of fat, due to the accumulation of lipid residues in the acetonitrile layer.

A refinement of the original Jones and Riddick(74) procedure was made by Erwin et al.(75), who substituted paraffin supported on aluminum oxide for the hexane. Hoskins et al.(76) suggested using polyethylene-coated alumina in place of the paraffin. Using this technique, they could elute the pesticides from the column with a 65 : 35 acetonitrile–water mixture, while the fats and waxes remained impregnated in the polyethylene-coated alumina. Thornburg(19) describes the "wet benzene" cleanup procedure, which is a technique to remove interfering pigments from a plant extract. The solvent for extraction is benzene saturated with water. The extract is then passed through a cleanup column composed of 5 parts ATTACLAY and 1 part activated NUCHAR C-190N. Apparently, the water in the saturated benzene sufficiently modifies the ATTACLAY and charcoal so that the pesticide is not adsorbed on the column. If plant or animal tissue were first extracted with "wet benzene" passed through the ATTACLAY charcoal column and then through the polyethylene alumina column, the extract should be reasonably free of pigments, waxes, and fats.

Burchfield and Storrs(77) partitioned the sample between N,N-dimethylformamide and hexane to separate pesticides from lipids. Eidelman(78, 79) used dimethylsulfoxide for the extraction of pesticides from butterfat, vegetable, and fresh oils and claimed that the procedure was applicable to samples up to 100 g. In this partitioning procedure, a mixed solvent system is used, consisting of dimethylsulfoxide, acetone, and petroleum ether. There have been reports of troublesome emulsions and high analytical backgrounds by this technique when samples of 15 to 20 g of fat were extracted.

De Faubert Maunder et al.(50) report 92% recovery of DDT when they used N,N-dimethylformamide and hexane, as compared with only 64% when dimethylsulfoxide, acetone, and petroleum ether were used. In cases where the extracts are dissolved in solvents other than acetonitrile or dimethylformamide, these solvents are often evaporated to dryness and then taken up in petroleum ether. The petroleum ether is then extracted several times with acetonitrile, thereby transferring the pesticide into the acetonitrile phase and leaving many waxes, fats, pigments, etc., behind in the petroleum ether. In the final step of this particular partitioning, as described by Mills(47), the acetonitrile is shaken with a large excess of water and then partitioned back into petroleum ether. After removal of excess water, the petroleum ether is evaporated prior to analysis.

Webster and McKinley(28) report on a comparison of the following cleanup procedures, all followed by FLORISIL column chromatography: acetonitrile–hexane, N,N-dimethylformamide–hexane, and acetone precipitation. All of these techniques were found to cleanup the extracts satisfactorily with only very small amounts of colorless residue remaining. However, the acetone precipitation at − 70°C resulted in the highest per cent recoveries of DDT from both apple and lettuce extracts. Acetonitrile–hexane partitioning resulted in excellent DDT recoveries from the lettuce extracts but fell down somewhat on the DDT recoveries from the apple extracts.

Beroza and Bowman(80, 81) using the countercurrent distribution technique of King and Craig(82), have illustrated the usefulness of using p-values to partition pesticides and other halogenated compounds in one solvent phase while leaving the unwanted material in the other phase. When the p-value of a pesticide is known, it is a simple matter to calculate the number of extractions that will be necessary to remove a given insecticide into the upper solvent layer. If interference from a food extract appears to have a great affinity for the upper phase, the analyst can select an extraction system which produces a low p-value. Beroza and Bowman(81) report the extraction behavior of 25 pesticides in 19 binary solvent systems and the distribution of extractives from five representative foods in 11 of the solvent systems. Although the separating power of a five-plate countercurrent distribution (CCD) system is not great, it is much better than a single separatory funnel extraction. Cleanup would go much more quickly with a battery of CCD apparatus than with the currently used chromatographic columns. Not only can CCD be used as a cleanup procedure but the compound's characteristic p-value in a given binary solvent system is constant and remains so in the presence of food extractives. Therefore, the p-value of a compound should be useful in identifying or confirming the identity of a questionable pesticide.

Of the six binary solvent partition systems that have been reported by Bowman and Beroza(83), the best cleanup and recovery of several non-oxygenated and oxygenated chlorinated hydrocarbon pesticides on 15 assorted foodstuffs was obtained by Stanley and Post(84) with the DMF cycle. These workers divided the foodstuffs into three groups (low, medium, and high fat) for cleanup purposes. The low fat group was cleaned up by eluting the isooctane extract through a prepacked column composed of layers of sodium sulfate, 1:35 NUCHAR–FLORISIL, deactivated FLORISIL, and sodium sulfate. The group containing the moderate-to-high fat content were partitioned in an isooctane–DMF cycle before the additional column cleanup described.

b. SORPTION. Although liquid–liquid partitioning between immiscible solvents is a convenient technique for separating pesticides from lipids, it is not very useful for removing pigments and unsaponifiable lipids. Unfortunately, these interferences tend to partition into the polar phase along with the pesticides. Therefore, an additional cleanup is often required in the analysis of extracts from plant and animal tissues. In pesticide residue analysis, selective adsorption of the interferences on an active solid is the procedure most commonly followed. The active solids most commonly used are FLORISIL, activated carbon, activated alumina, and magnesium oxide. Polyethylene-coated alumina is probably effective in both adsorptive and absorptive processes for the removal of lipids and other interfering substances. In some cases, a single adsorbent is not capable of removing all of the interfering materials in an extract, and consequently, mixtures of different adsorbants and filter aids are often employed. These can be mixed together as a "batch" or packed serially in a chromatographic column so that there will be a sequential elution of the sample extract and eluting solvent through the adsorbents. In the batch cleanup procedure advocated by Thornburg(19), a benzene extract of the tissue is shaken in the presence of a 1:5 ATTACLAY–NUCHAR mixture. Although considerable pigmentation and some other interferences are removed by this technique, additional cleanup is often required.

Normally, cleanup by adsorption can be more efficiently carried out by utilizing column chromatography. After the column has been properly packed with adsorbents(s), it is rinsed with the eluting solvent and the concentrated tissue extract applied to the top of the column. During the solvent washing, most pesticides move down the column faster than the pigments and other impurities. Therefore, if elution is discontinued after a predetermined solvent volume has been collected, the pesticide residues will be relatively free of impurities. Morley(85) has presented a comprehensive review of the adsorbents and their application to column cleanup of various pesticide residue extracts. This review covers the standardization and activation of adsorbents and their application to alumina, silica gel, FLORISIL, carbon, magnesia, and other columns for cleanup of extracts. The suitability of six adsorbents for the cleanup of extracts containing many of the chlorinated hydrocarbon pesticides is presented. The order of elution for 15 organochlorine pesticides from an alumina adsorption column was illustrated using 10% ethyl ether and n-hexane as the eluting solvent.

Unfortunately, adsorption is not particularly reproducible because of inherent adsorbent variations introduced during processing and activation. The nature and number of active sites of various adsorbents can be varied considerably by the method of activation. It is mandatory that each new batch of adsorbents be pretested with the pesticides in question before any unchecked samples are attempted. Elution of certain adsorbents, such as activated carbon, with a particular solvent can result in certain impurities being released that are often detectable by electron-capture gas chromatography. Therefore, care should be taken to check routinely the purity of all adsorbents to be used in the cleanup of residue extracts.

Chromatographic columns are usually packed dry, the adsorbent being added in small increments and tamped down uniformly with a clean dowel having a diameter slightly less than the inside of the column. If the column packing is not uniform, channeling of the solvent through the voids will result, producing inadequate separations. It is advisable to place a 1-in. layer of anhydrous sodium sulfate in the bottom of the column and over the top of the adsorbent. This will assure the removal of any trace moisture from the tissue extract after it is applied to the column. When the concentrated tissue extract is introduced to the top of the column, the eluting solvent level should be kept above the top of the sodium sulfate layer once chromatography has commenced. It is necessary that the tissue-extract solvent be the same as the solvent used for the first stage of sorption chromatography. If the extract is in a more polar solvent, it should be evaporated off and replaced with the solvent to be used for eluting. An extract dissolved in wet acetonitrile, for example, would probably deactivate a column intended for development by petroleum ether. Burchfield and Johnson(21) indicate that the tissue extract should be concentrated to 5 ml or less before placing on top of the column in order to achieve more compact zoning and separations. Others such as Mills et al. (*34*) found that it was not necessary to concentrate the extract before chromatography through a FLORISIL column. Regardless of whether the tissue extract is introduced to the column while the adsorbent is still dry or has been prerinsed, the solvent layer should not be permitted to recede below the top of the column packing once the chromatographic run has commenced.

Regarding elution of the column adsorbents, a single solvent of measured volume is sometimes adequate. The predetermined solvent eluant should allow the pesticide to pass through more or less quantitatively without removing most of the interfering impurities. When the tissue extract contains multiple pesticides with varying polarities, Mills et al. (*34*) suggest eluting with a series of solvents that increase in polarity to remove different groups of pesticides. To elute aldrin, DDT, etc., these workers eluted a FLORISIL column with 6% diethyl ether in petroleum ether. After collecting this fraction, the FLORISIL column was reeluted with 15% diethyl ether in petroleum ether in order to separate dieldrin, endrin, etc. Unfortunately, the pesticides eluted through FLORISIL with highly polar solvents may also contain considerable interfering substances. Therefore, it is often necessary to rechromatograph the fractions eluted with the more polar solvents on a column containing a different active solid. For example, the fractions that were eluted with the 15% diethyl ether through FLORISIL could be eluted through a second cleanup column of magnesium oxide and CELITE 545 prior to analysis. Moats (*86*) reports a one-step chromatographic cleanup for chlorinated pesticides where a NUCHAR–CELITE mixture was selectively eluted with 20% acetone in ethyl ether. This cleanup technique was found to be particularly effective on samples containing up to 0.8 g of fat and for cleanup of green leafy plant materials.

Cleanup procedures in which 20% methylene chloride in petroleum ether were used to elute chlorinated hydrocarbons from a FLORISIL column were reported by Moats (*87*) and from a silicic acid column also by Moats (*88*). Acid-washed NORIT A charcoal was used by Baetz (*43*) as a rapid cleanup of 13 chlorinated hydrocarbons extracted from collards, carrots, okra, and pears by the Johnson (*31*) procedure.

A rapid cleanup of dairy products for analysis of various chlorinated hydrocarbons by EC–GLC was reported by Langlois et al. (*41*). Dairy products containing not more than 1 g of butterfat were ground with 25 to 30 g FLORISIL. The residue–FLORISIL mixture was then packed in a column and prewashed with 1:1 methylene chloride–petroleum ether and the column eluted with 20% methylene chloride in petroleum ether. Stemp and Liska(89) presented a simplified method for cleanup of milk samples containing chlorinated pesticides prior to analysis by EC–GLC. Ten milliliters of whole milk were pipeted directly onto 25 g of 5% deactivated FLORISIL and slurried in an Erlenmeyer flask containing 200 ml 20% methylene chloride in petroleum ether. The supernatent was decanted through a 5% deactivated FLORISIL column. The deactivated FLORISIL column removed the milk fat and other contaminants that would eventually have reduced the detector sensitivity. Giuffrida et al.(90) reported a rapid cleanup for chlorinated pesticide residues in milk, fats, and oils prior to EC–GLC analysis. The procedure consists of two parts: (a) the isolation of fat from the sample by the addition of acetone, and (b) the elution of the pesticides from a deactivated FLORISIL column with 10% water–acetonitrile followed by partitioning into petroleum ether. For a final cleanup, a FLORISIL column described by Mills et al.(34) was used.

Moats and Kotula(91) have introduced a single-step chromatographic cleanup of chlorinated pesticide residues using high elution rates. It was found that elution rates of up to 250 ml/min could be used with FLORISIL and 100 ml/min with carbon–CELITE without adversely affecting either the cleanup of egg yolk and butterfat or the recovery of DDE, DDT, TDE, endrin, and dieldrin.

Undoubtedly, FLORISIL is currently the most widely used adsorbent for cleanup of commodity extracts containing one or more chlorinated hydrocarbons. In the past, various workers have reported considerable adsorptive variation in various lots of this synthetic magnesium silicate. Burke and Malone(92) reported that there were three primary difficulties encountered with the commercial FLORISIL that had been calcinated at 1200°F for 2 hr: (a) pesticides could not be totally eluted from the column with any eluant and (b) certain pesticides were difficult to remove from the column in any significant amount, and (c) inconsistency in adsorption properties between batches occurred. Because of these difficulties, considerable time had to be spent to verify the performance of a specific lot of FLORISIL before reliable residue analysis could be performed. In an attempt to determine where the variation had occurred in the manufacturing process, Burke and Malone(92) calcined 19 samples of FLORISIL at times ranging from 2 to 16 hr and at temperatures ranging from 1000 to 1400°F. Their results indicate that both the temperature and period of calcination are critical and that close control of these conditions is necessary. Of the 19 test calcinations, it was found that the FLORISIL calcinated for 2 hr at 1250°F was found to be the most desirable for the cleanup of residue extracts tested. It was noted that presently available production line FLORISIL that was calcined at 1200°F is more retentive to pesticides and associated extractive matter than was previously noted.

The per cent water in the activated FLORISIL is apparently quite critical. Moats(87) pretreated FLORISIL by heating overnight at 130°–150°C and then 5 ml water per 100 g FLORISIL were well mixed and allowed to stand in a sealed jar for a day or two. Wood(93) eluted dieldrin, aldrin, DDT, TDE, and DDE from FLORISIL columns under varying conditions in an attempt to improve recoveries. Higher recoveries of these chlorinated hydrocarbons were obtained from the columns when the FLORISIL was partially deactivated after activation at 130°C for 5 hr. The ideal amount of water to be added to the activated FLORISIL was found to be 0.7% by weight.

With the advent of highly selective detection systems, most pesticides stable enough to pass through gas-chromatographic columns can be identified and quantitated satisfactorily if the tissue-extract cleanup has been reasonably effective. However, adsorption techniques for cleanup purposes remain difficult to reproduce, particularly with certain types of pesticides and tissue extracts. Fortunately, there has been a considerable degree of standardization of adsorbents used for column cleanup of residue extracts containing organochloride pesticides, but the picture is not as bright for extracts containing organophosphate and carbamate pesticide residues.

c. SAPONIFICATION. Fats, oils, and other high triglyceride-containing pesticide residue samples can be effectively cleaned up by saponification. By refluxing the sample with alcoholic potassium hydroxide, the triglycerides are converted to soaps and glycerine. The soap solution is then extracted with hexane, petroleum ether, etc., with the pesticides partitioning into the hydrocarbon phase, while the soaps remain in the aqueous layer. The hydrocarbon layer is washed with aqueous alcohol to remove most traces of saponifiable soap, but most fatty samples will contain unsaponifiable lipids which will not be removed by this procedure. Fortunately, the unsaponifiable lipids are usually a small percentage compared with the easily separated triglycerides. The greatest disadvantage of saponification is that only alkali-resistant pesticides such as aldrin, dieldrin, and endrin can be cleaned up by this technique. Schafer et al.(94) have used saponification to hydrolyze milk fats and simultaneously dehydrochlorinate DDT to DDE, which then can be measured readily by gas chromatography. This method does not distinguish between DDE originally present in the milk due to *in vivo* dehydrochlorination and DDE produced from DDT during the alkali digestion. This procedure should also be applicable to other chlorinated hydrocarbons that yield well defined dehydrochlorination products. A method has been developed by Albert(44) in which a KOH-CELITE column is used instead of the longer, saponification procedure for cleanup of acetonitrile extracts on several vegetables prior to GLC analysis for endrin.

Sulfuric acid or fuming sulfuric acid can sometimes be used for cleaning up fat samples. The acid is impregnated on an inert solid support, such as CELITE 545, packed in a chromatographic column through which the residue sample, previously dissolved in an inert solvent, is eluted. The fats are strongly bound to the stationary phase, but some limited number of pesticides can be removed from the column by solvent rinsing. Pesticides are more conveniently separated from fats and oils by dissolving the sample in methylene chloride and repeatedly extracting with concentrated sulfuric acid. By following this procedure, the lipids preferentially remain in the acid layer while some chlorinated pesticides remain in the organic layer. This procedure has been employed by Graupner and Dunn (95) for the separation and cleanup of toxaphene (chlorinated camphene) in high fat samples. Stanley and LeFavoure(96) have employed a perchloric–acetic acid digestion of animal tissues containing lindane, heptachlor, and heptachlor epoxide, and isomers of DDE, TDE, and DDT. Following digestion, the fat is extracted in hexane and destroyed on a sulfuric acid-CELITE column.

d. THIN-LAYER CHROMATOGRAPHY. In the past few years, thin-layer chromatography (TLC) has become a widely used analytical tool among residue analysts, but its unique capabilities have been utilized primarily as a screening, separating, and identifying technique in the pesticide residue field. Very little has been done to utilize TLC as a cleanup procedure prior to GLC separation and quantitation. Walker and Beroza(97) discussed the use of this technique as a possible method for tissue-extract cleanup. Morley and Chiba(98) implemented TLC as a cleanup procedure prior to analysis by GLC as well as a rapid screening method of pesticide residues in plant extracts with prior cleanup. These workers were able to cleanup wheat extracts (0.5 g crop) successfully by TLC prior to GLC detection of o, p-DDT, p,p'-DDT, and p,p'-DDE.

Schutzmann et al.(66) found that cleanup of pesticide extracts by TLC and other methods prior to GLC analysis of surface soils and soil sediment was often more rapid and less tedious than the usual column-cleanup techniques. These workers also noted that for crop and biological samples containing less than 2% fat, a suitable cleanup can be achieved by simple extraction with acetonitrile, thereby eliminating the need for TLC cleanup. For 1 g or more of crop and biological samples containing over 2% fat, a double acetonitrile partitioning followed by TLC cleanup proved satisfactory prior to subsequent GLC determination. Moye(99) is developing the use of unbound thin layers as a cleanup step prior to analysis by GLC. The unbound layer is 15 cm long, 2 cm wide, and 2 mm deep. It is held in place by a trough formed from five microscope slides which were epoxy-glued onto a glass plate. Two slides form each side of the trough and one the bottom. The unbound layer is made by placing a second glass plate over the trough, clamping it down, and holding the assembly upright while pouring in the loose adsorbent. The top glass plate is then removed for development of the trough in a developing tank. Up to 10 g equivalents of leafy vegetables have been cleaned up by this procedure, which eliminates the tedious application step by collecting the extract on the thin-layer substrate during concentration.

e. CHANNEL-LAYER CHROMATOGRAPHY (CLC). Channel-layer chromatography (CLC) cleanup is based on the principles of both thin-layer and column chromatography. It combines the capacity of column chromatography with the simplicity and speed of thin-layer chromatography. Matherne and Bathalter(100) used CLC to clean up extracts of peanuts, onions, carrots, and cabbage containing residues of aldrin, dieldrin, endrin, and endosulfan. The capacity for cleanup by CLC at this time is rather limited (5 g of sample). Burke(101) has reported that a channel depth of 4 mm (instead of 2 mm) may increase the cleanup ability of CLC and that it may have some application as a supplementary cleanup of extracts. For example, the use of CLC has removed electron-capture interferences remaining in a FLORISIL column eluate of carrot extract.

In this procedure, aluminum oxide G is used to separate the pesticides from the plant extractives. A square glass plate was prepared with recessed, discrete parallel channels (10 mm wide and 2 mm deep) filled with aluminum oxide. The plant extractives were applied to the channel-layer strip and then the strip was chromatographed with 1 : 1 acetonitrile and tetrahydrofuran followed by acetonitrile alone. As the solvent extracts moved up the strip, the plant extracts remained in place or moved only part way up the column, while the pesticides moved with the solvent front. The portion of the channel-layer strip containing the solvent front (and any pesticides) were mechanically removed from the plate and then the pesticides eluted from the absorbent. This solution was injected directly into the gas chromatograph for identification and quantitation.

f. SWEEP CO-DISTILLATION OR FORCED VOLATILIZATION. Rosenberg and Storherr(102) have adapted the procedure by Storherr et al.(103) as a sweep codistillation cleanup of 14 chlorinated hydrocarbons in edible oils. A micro-FLORISIL column, prepared from a disposable pipet, was used to eliminate extraneous GLC peaks. This cleanup technique is reported to be simple and more rapid than conventional adsorption-column chromatography. Ott and Gunther(104) and Gunther et al.(105) have reported a similar procedure of forced volatilization for the cleanup of chlorinated hydrocarbons in butterfat prior to microcoulometric gas-chromatographic measurement. The described device physically separates volatile materials from those which are either nonvolatile or considerably less volatile at a given operating temperature. Mestres and Barthes(*106*) have described a similar forced draft or sweep extraction apparatus for the cleanup of organochloride pesticides from various vegetable extracts. The vegetable sample was extracted with acetonitrole and the residue partitioned into petroleum ether. The ether was concentrated 10 : 1 before introduction into the system. The purified extract was reduced to 1 ml and a 2 μl injection made into the electron-capture gas chromatograph.

D. *Organophosphates*

The volume of research on extraction procedures for the removal of organophosphate pesticides from various fatty and nonfatty commodities and from soil and water is considerably less than what is available for the chlorinated hydrocarbons. Many of the extractive procedures that have already been discussed for the organochlorine pesticides can be applied generally to the nonpolar organophosphates. However, many of the procedures are often unsatisfactory for the extraction of water-soluble organophosphates or their oxidative products. Water or an organic solvent is usually adequate to extract most organophosphorus compounds. When these pesticides are almost insoluble in water they do not penetrate deeply into the aqueous internal tissues although there can be some diffusion into the cuticle and oily-associated substances. For safety, it is advisable to blend the plant tissues since surface extraction may not remove all of the residues diffused into the surface area. For those organophosphates that are fairly water soluble and are considered to be somewhat systemic, a vigorous tissue extraction, such as by high-speed blending in the appropriate solvents, is always mandatory in order to remove the compound from within its substrate quantitatively. A general method for the extraction of commodities containing organophosphate pesticides is by blending the material in a single solvent such as chloroform or acetonitrile. All non-ionic organophosphates appear to partition favorably into chloroform from water owing to exceptionally strong molecular forces between P $\rightarrow$ O and $CHCl_3$. Organophosphate oxidative metabolites produced in a plant are more favorably partitioned into water and therefore, more exhaustive chloroform extraction is necessary when these oxidative metabolites are determined according to Chilwell and Hartley(*107*). Partitioning from water into chloroform is not sufficient to remove all interferences and additional cleanup is usually necessary before final analysis. Schnorbuṡ and Phillips(*38*) have proposed propylene carbonate as a potentially effective broad-spectrum extracting agent. These workers report good to excellent recoveries of a number of organophosphate pesticides from a wide variety of commodities such as fruits, vegetables, grains, meats, dairy products, soils, dehydrated products, fats, and oils. In addition to its universal application, this extractant has a low toxicity and is not inflammable as most other common pesticide extractants.

1. *Extraction of Nonfatty Foods (< 2% fat).*

a. VEGETABLES, FRUITS, ROOT CROPS. Organophosphate pesticides have been extracted from fresh plant materials by Coffin and Savary(*108*) by blending with acetonitrile. This procedure is probably more effective in extracting some of the more polar organophosphate oxidative products since partitioning with petroleum ether is not employed. Thornburg(*19*) advocates the use of a mixed solvent (benzene–isopropanol) to remove organophosphates from most fresh plant tissues. Van Middelem et al. (*29*) found this procedure effective in the extraction of parathion from various leafy vegetables. A rapid extraction method by Watts and Storherr(*109*) utilized blending of field-treated kale with ethyl acetate for the removal of dimethoate, diazinon, carbophenothion, azinphosmethyl, mevinphos, and parathion. The advantage of this procedure is that it eliminates the lengthy acetonitrile concentration and reduces the volume of solvent used in the acetonitrile procedure. However, Burke and Porter (*33*) reported the best recoveries of field-applied diazinon and parathion from kale by using the Mills et al.(*34*) procedure compared with that of Watts and Storherr(*109*). The basic acetonitrile extraction of fruits and vegetables developed by Mills et al.(*34*) for the chlorinated hydrocarbons was applied to the thiophosphates by Nelson(*110*). After the petroleum-soluble thiophosphates were partitioned into the petroleum ether layer, the previously discarded aqueous layer was repeatedly extracted with methylene chloride. It was found that dimethoate and azinphosmethyl could be satisfactorily recovered from the water–acetonitrile layer by methylene chloride. Laws and Webley(*111*) describe an extraction scheme for the extraction and separation of organophosphates in vegetables. The first extraction consists of blending with methylene chloride before partitioning between petroleum ether and 15% methanol and water. Separate cleanup procedures followed for the petroleum-soluble and water-soluble organophosphates. The vinyl phosphates were removed from the vegetable material by Lau(*112*) by first blending with chloroform or methylene chloride. Following evaporation of the chloroform or methylene chloride, the residue was picked up in hexane and water. The organophosphates were then partitioned into the aqueous phase prior to additional cleanup and isolative steps.

b. FEED, FORAGE. George, et al.(*113*) extracted dimethoate from range forage by blending with methylene chloride. Lau(*112*) extracted two vinyl phosphates, BIDRIN and AZODRIN, from alfalfa and corn foliage by blending with chloroform or methylene chloride. Bayer 37289 and its oxygen analog were extracted by Katague and Anderson(*114*) from alfalfa by blending, first in acetone and then in benzene and anhydrous sodium sulfate. Ronnel was extracted by Gehrt(*115*) from various cattle feeds, using 2:1 methanol–petroleum ether and hydrolyzed to 2,4,5-trichlorophenolate with sodium hydroxide.

2. Extraction of Fatty Foods and Tissues (< **2% fat).** The extraction procedures for the chlorinated hydrocarbon pesticides in fatty foods are often applicable to the non-ionic organophosphates. Some of these procedures, which utilize fat solvents such as hexane, probably would not be satisfactory for the recovery of water-soluble organophosphates or certain water-soluble oxidative products. Usually however, the water-soluble organophosphates or their oxidative by-products are not lipid soluble and, therefore, are not residue problems by accumulating in fat depots.

a. DAIRY PRODUCTS, FATS, OILS, NUTS. For colorimetric analysis of IMIDAN, milk was repeatedly extracted by Bowman and Beroza(*116*) with 1:1 v/v hexane–diethyl ether. Following evaporation to dryness, the residue was picked up in acetonitrile and hexane, and partitioned into the lower layer with three hexane extractions. For gas-chromatographic analysis, the acetonitrile solution is shaken with water, saturated NaCl, and 10 ml benzene. After a repeated washing with water and saturated HCl, the benzene layer is separated, dried, and injected into the instrument. Everett et al.(*117*) describe a procedure for the spectrophotofluorometric and radiometric measurement of azinphosmethyl residues extracted from milk and animal tissues. The whole milk was first blended with acetone, filtered, and the extract blended with benzene. George et al.(*113*) extracted dimethoate from whole milk by shaking at 10°C with cold methylene chloride.

b. MAMMALIAN TISSUES. Lean and fatty tissues are first extracted by blending with acetonitrile and subsequently with SKELLYSOLVE B as suggested by Olson(*53*). The organophosphate residue in the SKELLYSOLVE B is then partitioned into acetonitrile. Adams and Anderson(*118*) describe a procedure very similar to that described by Everett et al.(*117*) for the extraction of azinphosmethyl from milk and animal tissues. The tissue was blended with acetone followed by blending in benzene. Claborn and Ivey(*119*) extracted ronnel from various mammalian tissues by blending with acetone and hexane. After transfer of the pesticide to the hexane fraction, the aqueous layer was discarded following 5% sodium sulfate washes, and the pesticide analyzed by gas chromatography. Smith et al.(120) extracted ronnel from poultry egg yolks by blending in acetonitrile and partitioning with petroleum ether.

3. Extraction of Miscellaneous Samples.

a. BODY FLUIDS. The most widely used analytical procedure for the determination of organophosphate effects on body fluids has been the measurement of cholinesterase inhibition in blood plasma and whole blood. Because of the widespread application of this technique and considerable available literature, it will not be elaborated upon or cited. The principal advantage of the technique is its simplicity and its disadvantages are lack of specificity and sensitivity for low levels of pesticides. The extraction of urine samples for the removal of nonpolar organophosphates can probably be accomplished by the procedure of Teasley and Cox(121). During the extraction of urine samples with water-immiscible solvents, stubborn emulsions can result if the sample is shaken too vigorously. Moye and Winefordner(122) analyzed for p-nitrophenol, a major metabolite of parathion, by subjecting the urine to acid hydrolysis prior to extraction with ether. Extraction of blood plasma with water-immiscible solvents is more difficult than with urine, due to formation of stable emulsions during the extraction procedure. Mild acid or base hydrolysis can eleviate the emulsion problem but can also result in serious losses of labeled organophosphates. Most nonpolar organophosphates can probably be extracted from whole blood by resorting to a mixed solvent extraction similar to that recommended by Moffit(46) or Mills(47).

b. WATER. No individual organic solvent efficiently extracts all organic pesticides from natural water, and certain organic interferences are often coextracted from grossly polluted waters. The pH of the water sample is an important variable for quantitative extractions, and slightly acid-to-neutral conditions are optimum for the removal of organophosphates. Significant losses of organophosphates can occur during certain extractions and during evaporation of the water sample. All things considered, it appears that liquid–liquid extractions of water samples are the most practical and result in the highest recoveries of organophosphates from water. Faust and Suffet(63) have made an excellent review of pesticides in natural and potable water. Of the rather meager literature available on extraction of organophosphates from water, there appears to be little agreement on any common solvent system. Sumiki and Matsuyama(123) extracted parathion from water for infrared spectrophotometric analysis with a 1 : 1 mixture of benzene and *n*-hexane. A 1 : 1 ether–petroleum ether or chloroform mixture was found satisfactory by Teasley and Cox(*121*) for the quantitative extraction of parathion and diazinon from water. El-Refai and Giuffrida(*124*) proposed a simple, rapid, and sensitive method for the determination of the hydrolysis rate of trichlorfon and dichlorvos in water. The water samples were extracted twice with redistilled ethyl acetate prior to analysis with a sodium thermionic detector in a gas chromatograph. Warnick and Gaufin(*125*) achieved excellent recoveries of methyl parathion and parathion by extracting water samples with highly purified hexane. ABATE was extracted by Blinn and Pasarela(*126*) from natural waters. The water samples were acidified with 6 *N* sulfuric acid prior to filtering and extraction with chloroform.

c. SOILS. Organophosphates are considered to be relatively nonpersistent in soils and, therefore, very little research has been published on soil extraction techniques for this class of pesticides. In general, however, benzene or a mixture of benzene and isopropanol is considered a good solvent system for the removal of nonpolar organophosphates from soils. However, benzene is often not a suitable solvent for the highly polar phosphates or their oxidative metabolites. For these types of compounds methylene chloride has been suggested by Burchfield and Schuldt(*127*) as a more efficient extractive solvent. It is recommended that dry soils be brought up to 10–12% moisture before extraction is attempted in order to liberate bound pesticides. Parathion, methyl parathion, and malathion were extracted from soils by Lichtenstein and Schulz(*128*) with a 3 : 1 mixture of benzene and redistilled acetone. Lichtenstein(*129*) also reports using a 4 : 1 mixture of redistilled benzene and acetone to extract soils containing residues of parathion, phorate, and disulfoton. To remove diazinon and ZINOPHOS from the soil, Getzin and Rosefield(*130*) employed a 1-hr Soxhlet extraction with redistilled acetone, whereas Coahran(*131*) used an overnight Soxhlet extraction with hexane to quantitatively remove ZINOPHOS residues from soil.

4. Cleanup Methods for Organophosphates. As with the organochloride pesticides, the extracts containing organophosphates usually are cleaned up by partition, sorption or chemical methods or some combination of these techniques. The general procedures already outlined for the cleanup of organochloride-containing extracts are usually applicable, with minor modifications, to the nonpolar organophosphates. Therefore, emphasis in this section is on general cleanup techniques and on special modifications, primarily for the organophosphates.

a. PARTITION. The acetonitrile–petroleum ether partitioning method of Mills(47) and Mills et al.(34) is applicable for the separation of a number of nonpolar organophosphates from nonfatty and fatty foods. Further cleanup by passing through a FLORISIL column is usually required following the partitioning procedure. Following blending in methylene chloride, Laws and Webley(111) proposed a partitioning extraction between petroleum ether and 15% methanol–water on vegetable extracts to separate various organophosphate pesticides into water-soluble and petroleum ether-soluble groups. Such compounds as disulfoton, diazinon, parathion, methyl parathion, and phorate were found in the petroleum layer, whereas demeton, phorate oxygen analog sulfone, mevinphos, phosphamidon, and dimethoate were found in the water-soluble fraction.

b. SORPTION. Cleanup by adsorption can usually be carried out efficiently by the use of a chromatographic column to remove interferences such as pigments, waxes, etc., from the commodity extract. The comments previously made under Sorption Methods for chlorinated hydrocarbons are generally applicable also for the organophosphates. Probably the most widely used cleanup column for organophosphates is that of Mills et al. (34) where the FLORISIL is heated at 130°C for at least 5 hr prior to use. Burke and Porter(33) have reported satisfactory cleanup of parathion and diazinon from field-sprayed kale by elution through the Mills et al.(34) FLORISIL column.

Coffin and Savary(108) report the use of polyethylene-coated alumina and magnesol for the cleanup of 25 organophosphates from acetone-blended lettuce. The pesticides were eluted through the poly-alumina column with slightly acidified 40% acetone, partitioned from the aqueous layer into chloroform, and then eluted through magnesol column with successive washings with chloroform, acetone, and methanol. The suitability of six common adsorbents for cleanup of tissue extracts containing organophosphate pesticides have been compared by Morley(85) and the most useful generally would appear to be polyethylene-coated alumina or activated carbon. Caution should be used, however, with the use of poly-alumina when the final determinative step involves highly sensitive ionization detectors such as electron capture, since various workers have found polyethylene to be a source of artifacts for the electron-capture detector.

Carbon column chromatography has been applied primarily to the cleanup of organophosphorus residues. Several brands of activated carbon have been reported but NUCHAR C-190N has been found to be

generally satisfactory. Certain solvents have been found to remove interferences even from pretreated, activated carbon and these interferences
can be particularly noticeable in extracts injected into gas chromatographs
equipped with an electron-capture detector. Thornburg(*19*) reports the
use of a NUCHAR–ATTACLAY column mixture for removing pigments and
other interfering materials from benzene–isopropanol extracts of plant
tissues containing a number of nonpolar organophosphates. However,
the corresponding oxons of most thiophosphates will not pass through
this column. It was indicated by Moddes and Cook(*132*) that acetonitrile
would extract organophosphates efficiently without including too much
waxy plant material. Getz(*133*) attempted to decolorize acetonitrile-
extracted kale containing six phosphate pesticides by testing various char-
coals and combinations of eluting solvents. It was found that NORIT A
decolorizing carbon, when treated with an ethanolic hydrochloric acid
solution, proved to be an efficient decolorizing agent for the kale extract
when eluted with chloroform or benzene plus ethyl acetate. Thornburg
(*134*) has recommended eluting the FLORISIL column with 50% ethyl
ether for anesthesia in hexane for organophosphates such as parathion or
malathion. If the residue sample is high in chlorophyll, the samples are
put through the "wet-benzene" procedure prior to the FLORISIL column
cleanup. If the sample is high in oil, Thornburg(*134*) recommends passing
the extract through polyethylene-coated alumina prior to final cleanup in
a FLORISIL column.

Barry et al.(*40*) have outlined a more or less "universal" cleanup of
tissue extracts containing ten different organophosphates. Following acetonitrile–petroleum ether partitioning, the petroleum ether fraction is eluted through a FLORISIL column followed by elution with 6:94 ethyl ether–
petroleum ether. Organophosphates such as ethion, ronnel, phorate, and
carbophenothion come through in this latter eluate. Then the column is
again eluted with 15:85 ethyl ether–petroleum ether, which releases such
organophosphates as diazinon, methyl parathion, parathion, EPN, tetradifon, and endosulfan from the FLORISIL column. If the column is finally
eluted with 1:1 ethyl ether and petroleum ether, malathion can be quantitatively recovered in the terminal eluate.

In order to separate water-soluble oxidized organophosphates from
methylene chloride extracts, Laws and Webley(*111*) partitioned between
petroleum ether and 15% aqueous methanol. The water-soluble organophosphate extracted was then eluted through an acid-washed active carbon column with chloroform. Polar organophosphates that can be cleaned
up by this procedure include dimethoate, methyl demeton and metabolites, mevinphos, phorate oxygen analog sulfone, phosphamidon, and
dimethoate.

Undoubtedly, there is an urgent requirement for additional research toward greater standardization of adsorbents used for column cleanup of tissue extracts containing organophosphorus and carbamate pesticide residues. Unlike the chlorinated hydrocarbons, the organophosphorus compounds present a far more complicated problem. Because of their inherent complexity, no standardized procedure has established itself as the preferred column-cleanup mixture for tissue extracts containing one or more organophosphates. FLORISIL has been proven to be an excellent adsorbent for column cleanup of the extracts containing organochlorine pesticides but often it is far from suitable for use with the organophosphates. The reverse tends to be true for the carbon adsorbents.

c. THIN-LAYER CHROMATOGRAPHY. Schutzmann et al.(66) report effective cleanup of methyl parathion and parathion extracts by thin-layer chromatography prior to analysis by electron-capture gas chromatography. These two organophosphates were satisfactorily recovered by use of TLC cleanup from extracts of the following samples: pollen, mice, frogs, bees, algae, rabbits, soybean plants, and fish.

d. CODISTILLATION. Probably the most recent promising "universal" cleanup procedure for crop extracts containing various organophosphates was introduced by Storherr and Watts(135). In this study, sweep codistillation was compared with the column adsorption method of Storherr et al.(136), which involved a lengthy acetonitrile extraction and NORIT charcoal column cleanup. Recoveries of diazinon, methyl parathion, parathion, and malathion were higher following cleanup by the sweep codistillation procedure as compared to eluting through the adsorption column. The sweep codistillation method has been reported to be a simple, rapid, and effective means of cleanup of these organophosphates from crops such as kale, carrots, apples, strawberries, and potatoes prior to final determination by GLC equipped with a sodium thermionic detector. Storherr et al.(103) and Watts and Storherr(137) have expanded the sweep codistillation technique to include cleanup of organophosphates in edible oils and milk.

E. Carbamates

The general procedure for the extraction of carbamate insecticides involves the blending of the commodity with a solvent followed by centrifuging and often reblending. Most of the published work on extraction and cleanup of this class of insecticides involves carbaryl, for which a number of tolerances have been established on various commodities.

352   C. H. VAN MIDDELEM

1. Extraction of Nonfatty Foods (< 2% fat). Carbaryl residues were extracted from fruits and vegetables by Johnson(*138*) by blending three times with methylene chloride. The method is based on alkaline hydrolysis of the carbamate followed by colorimetric determination of the resulting 1-naphthol. Gajan et al.(*139*) extracted carbaryl from fruits and vegetables with methylene chloride blending prior to quantitative measurement in the presence of alpha naphthol by oscillographic polarography. A rapid analytical procedure for carbaryl residues by Benson and Finocchiaro(*140*) utilized repeated blending of various fruits and vegetables with methylene chloride. In this procedure the previous time-consuming evaporation of methylene chloride is improved by steam evaporation in the presence of diethylene glycol. Ralls and Cortes(*141*) extracted carbaryl from green string beans by blending in methylene chloride prior to cleanup and bromination for detection by electron-capture gas chromatography. Carbaryl was hydrolyzed, brominated, and esterified by Gutenmann and Lisk(*142*) following the extraction of several fruits and vegetables by blending with acetone. Van Middelem et al.(*143*) extracted carbaryl from snap beans by blending with methylene chloride prior to hydrolysis and bromination to phenol derivatives which were highly sensitive to electron-capture gas-chromatographic measurement. TEMIK was extracted by Johnson and Stansbury(*144*) from raw fruits and vegetables by blending with chloroform, and Johnson and Stansbury (*145*) removed TRANID another methylcarbamate pesticide, from fresh fruits and vegetables by blending in benzene.

2. Extraction of Fatty Foods and Tissues (> 2% fat). Gyrisco et al. (*146*) removed carbaryl residues from cream by extracting three times with 1:1 hexane–ether and acetonitrile prior to three terminal extractions with diethyl ether. Johnson et al.(*147*) extracted carbaryl and its metabolites from poultry tissue and eggs by blending twice with methylene chloride. The residue was transferred to petroleum ether and partitioned with acetonitrile. The acetonitrile was evaporated and the residue picked up again in methylene chloride before being eluted through a FLORISIL column. Whitehurst et al.(*148*) reported a procedure for the extraction of 1-naphthol (hydrolysis product of carbaryl) and conjugated detoxification products of 1-naphthol in milk, urine, and feces of dairy cows. Carbaryl and 1-naphthol were extracted from cream, urine, and feces by three extractions with methylene chloride. The 1-naphthol is then removed by two extractions with 0.5 N sodium hydroxide, acidified, and equilibrated twice with methylene chloride. The 1-naphthol conjugates in skimmed milk and urine were subjected to acid hydrolysis prior to extraction with methylene chloride. The methylene chloride layer was equilibrated with a sodium hydroxide solution to remove the hydrolyzed conjugate residues. Claborn et al.(*149*), used essentially the same extractive procedures outlined by Whitehurst et al.(*148*) to remove carbaryl, 1-naphthol, and conjugates of 1-naphthol from animal tissues such as omental and renal fat, muscle, liver, heart, kidney, and brain. Nir et al.(*150*) extracted carbaryl residues from poultry tissues by extraction with chloroform prior to colorimetric measurement.

3. Cleanup Methods for Carbamates.

a. VEGETABLES, FRUITS. Following blending of fruits and vegetables with methylene chloride, the extract was evaporated, dissolved in acetone (*139, 141, 144*), and coagulated before passing through a Buchner packed with a hyflosupercel layer. Ralls and Cortes(*141*) cleaned up their extracts by passing them through an activated FLORISIL column with methylene chloride rinsing. Gutenmann and Lisk(*142*) partitioned their aqueous acetone extract with distilled chloroform, and if the sample had a low moisture content, water was added to the acetone extract before partitioning. After adding the precipitating solution, these workers filtered through a Buchner funnel before partitioning with redistilled carbon tetrachloride. The carbon tetrachloride extract was evaporated to dryness before hydrolysis, bromination, and esterification. The cleanup procedure for TEMIK(*144*) in chloroform extracts of fruits and vegetables includes coagulation and filtering through a Buchner with hyflosupercel prior to evaporation. Benzene extracts of fruits and vegetables containing TRANID (*145*) residues were coagulated, filtered, extracted with petroleum ether, and the aqueous layer was extracted with chloroform before evaporating.

b. DAIRY PRODUCTS, EGGS, MAMMALIAN TISSUES. Cleanup of the fatty commodities, such as eggs, animal tissues, and milk, usually involve partitioning the methylene chloride extracts with petroleum ether followed by additional cleanup by passing through a FLORISIL column(*146–148*). Sometimes methylene chloride or hexane–acetonitrile partitioning was employed prior to FLORISIL column cleanup(*147, 148*).

F. Herbicides

It is necessary to classify herbicides into their respective groups in order to determine methods of extraction effectively. Following the classification of Brian(*151*), the organic herbicides are divided into two main categories: no-nitrogen and nitrogen-containing compounds plus a few miscellaneous herbicides. The herbicides containing no nitrogen included the following acids: phenoxyacetic, phenoxypropionic, phenoxybutyric, phenylacetic, benzoic, and halogenated aliphatic. The nitrogen-containing herbicides include the amides, ureas, urethanes, thiol and dithiocarbamates, triazines, substituted phenols, bipyridylium salts, and toluidines. The miscellaneous class includes amitrol and dichlone. A comprehensive listing of herbicides, including structural and empirical formulas, chemical, biological, and physical properties as well as methods for residue and formulation analysis, has been edited by Zweig(152).

The most widely used herbicides, with which a great share of the herbicide research has been conducted, are the phenoxyacetic acids, particularly 2,4-D, 2,4,5-T, and MCP. Other acids such as silvex, 2,4-DB, and amiben have also received considerable attention in the literature. These phenoxyalkanoic acid herbicides are applied to crop lands, sidings, and waterways as free acids, salts, and esters. Therefore, it is necessary for the analyst to detect them in the anionic and neutral form when extracted from samples.

There is considerable evidence available that indicates that esters are hydrolyzed rapidly to the free acids by plant tissues. Klingman et al.(153) found that the butyl ester of 2,4-D was almost completely hydrolyzed to the acid form in samples of forage taken within one-half hour after spraying. In clams and oysters, however, Coakley et al.(154) noted that the residue was primarily in the form of the butoxyethyl ester rather than the acid.

Esters of 2,4-D and other related herbicides can be extracted and isolated from interferences by methods similar to those already described for the nonionic chlorinated hydrocarbons and organophosphate pesticides. The extraction and cleanup of the acid forms, however, present the residue analyst with different extraction and cleanup problems. In general, all samples containing anionic pesticides must be acidified prior to extraction. Often it is convenient to separate compounds into polar and nonpolar fractions by adjusting the pH of the residue sample between extractions. In certain media, the extraction of 2,4-D and related herbicides is complicated by their existence in the carboxylate form in alkali extracts, binding to proteins, and formation of conjugates.

1. Extraction of Phenoxyalkanoic Acids.

a. VEGETABLES, FRUITS. Frozen spinach was extracted by Yip(155) to remove 2,4-D acid, esters, and salts. The crop was blended with a mixture of sulfuric acid, ethyl alcohol, petroleum ether, and ethyl ether. Erickson and Hield(156) blended and refluxed citrus fruit with acetone to extract 2,4-D and, following filtration, the pulp was reblended in acetone. After concentrating, acidifying, and adding water, the solution was extracted three times with diethyl ether. At this point the samples were separated into acid and ester fractions by extracting the ether solution with phosphate buffer. The phosphate buffer which contained the acid fraction was acidified, extracted with carbon tetrachloride and evaporated; the 2,4-D acid residue was now ready for esterification.

Erickson and Hield(*156*), after applying the 2,4-D to citrus in the form of the isopropyl ester, recovered both acid and ester fractions following extraction. The acid fraction partitioned to organic solvents at a low pH and to water in a neutral solution. The fraction considered to be the ester comprised only a small percentage of the total 2,4-D residue, but was found to be water-insoluble in a neutral solution; however, alkaline hydrolysis yielded the 2,4-D acid fraction. Unpeeled, chopped potatoes were blended with benzene and HCl by Bevenue et al.(*157*) to extract several 2,4-D esters. After filtering and several washings with acidified benzene, the bottom water layer was discarded. The benzene layer was concentrated to near dryness and any water remaining was removed by adding ethanol and reconcentrating.

Meagher(*158*) reports analyzing for 2,4-D and 2,4,5-T in citrus by electron-capture gas chromatography. The citrus sample was first blended with hot acetone, filtered, and the residue resuspended in acetone, boiled briefly, and again filtered. Following the removal of the acetone, the mixture was acidified and repeatedly extracted with hexane. The aqueous phase was saved for determination of the conjugated forms. The hexane extracts were extracted with 0.2 M K_2HPO_4, partitioning the free acid into the aqueous phase. The hexane phase was then saved for determination of the ester forms. The aqueous K_2HPO_4 phase containing the free acid was extracted with chloroform.

Bache et al.(*159*) blended peas with distilled acetone (acidified with O-phosphoric acid) to extract MCP prior to nitration and measurement by electron-capture gas chromatography. CHLORAMBEN (amiben) was extracted from tomatoes prior to thin-layer chromatography by saponification with 2 N NaOH for 30 min by Bache(*160*). After adjusting the pH to 7.5, the solution was extracted four times with diethyl ether and once with benzene. The aqueous solution was adjusted to pH of 2.0, extracted four times with benzene and, after evaporation, the residue was picked up in acetone. Bache et al.(*161*) determined amiben by electron-affinity gas chromatography following extraction of tomatoes. The method involves extraction at pH 2.0 into benzene from an alkali-hydrolyzed tomato extract. The acid was then methylated with boron trifluoride–methanol and the ester detected by EC–GLC. Nonfatty commodities such as asparagus and strawberries were extracted by Smith et al.(*162*) to remove dicamba. The plant materials were blended with ether and 10% sulfuric acid.

b. CEREALS, GRAINS, FEED, FORAGE. Extraction of MCP in timothy was described by Bache et al.(*159*). Wheat was extracted by Yip(*163*) to remove various chlorophenoxyacetic acids and esters by blending with a mixture of petroleum ether, ethyl ether, and sulfuric acid. Stanley(*164*) developed conditions for the gas-chromatographic separation of the methyl and ethyl derivatives of 2,4-D, 2,4,5-T, etc., from each other on one column and the methyl, ethyl, and butyl phosphoric acids from each other on another column. Bevenue et al.(*165*) extracted 2,4-D from various grains, dried forage, seeds, beans, and nuts. Very low moisture content samples were first mixed with water and then mechanically shaken for 1 hr with a high ratio (5 or 6:1) of the following solvent A mixture: 10 ml 10% H_2SO_4, 15 ml 95% ethanol, 25 ml redistilled petroleum ether, and 75 ml ethyl ether. Materials with high fat content such as walnut meats were blended first with water and the solvent A mixture prior to the 1-hr shaking period. Hagin and Linscott(*166*) removed 2,4-D and 2,4-DB from frozen forage plants by first bringing just to a boil and then blending the plant slurry three times with 2-propanol. Smith et al.(*162*) extracted dicamba from corn forage, oat foliage, and barley by blending these nonfatty plant samples with ether and 10% H_2SO_4. Residues of trichloro- and polychloro-benzoic acids were extracted from sorghum, wheat, and barley grain by Kirkland and Pease(*167*). The grains were ground in a wet mill with methylethyl ketone and 85% phosphoric acid. Gutenmann and Lisk(*168*) removed 2,4-D and 2,4-DB from timothy grass forage by blending with 70:1 acetone and 85% orthophosphoric acid.

c. DAIRY PRODUCTS, FLUIDS AND FECES. Gutenmann et al.(*169, 170*) and Lisk et al.(*171*) extracted 2,4-DB and 2,4-D from milk, feces, and urine by blending with 70:1 distilled acetone and 85% phosphoric acid. Extraction of 2,4-D from feces, urine, and blood from sheep was reported by Clark et al.(*172*). Feces was extracted by stirring with a 50:50 v/v mixture of 95% ethanol and diethyl ether plus 1% by volume of concentrated HCl. The slurry was filtered and the residue washed twice with the above solvent mixture and once with chloroform. Urine samples were diluted 1:100 with 95% ethanol before assaying the ^{14}C-labeled 2,4-D by paper chromatography and paper electrophoresis. The blood samples were extracted with hot 70% ethanol (adjusted to pH 1.0 with HCl) followed by chilling and filtration. The herbicides, MCP and MCPB, were extracted from cow urine by blending as per previously cited procedure (*169*). Milk and urine were extracted by St. John et al.(*173*) with the acetone–phosphoric acid mixture as per procedure(*169*) to isolate silvex (or fenoprop) and 2,4,5-T prior to final determination by electron-affinity gas chromatography. The procedure for the extraction of 2,4-D by Yip

(*163*) has been modified for milk and forage by Yip and Ney(*174*) for application to the acid, ester or bound form of 2,4-D. Milk was extracted with an acidified mixture of 80:20 ethyl ether–petroleum ether.

d. WATER. Silvex and 2,4-D are often used as aquatic herbicides; therefore, their extraction from water is an important consideration. As with soil extractions, the water must be acidified prior to extraction with water-immiscible solvents. Phosphoric acid is recommended over HCl unless specified otherwise. Moderately polar solvents such as diethyl ether, methylene chloride, or chloroform should be used for extraction, whereas petroleum ether and hexane are not considered to be satisfactory solvents for 2,4-D or related herbicides.

A rapid screening method for the detection of phenoxyalkanoic acid herbicides in water has been outlined by Gutenmann and Lisk(*175*). The water is acidified and then extracted with ethyl ether for 2,4-D and related herbicides. Gutenmann and Lisk(*175*) have proposed a micromethod for the extraction of silvex from water and quantitation by electron-capture gas chromatography. The filtered water sample is saturated with sodium sulfate, then acidified prior to extraction with diethyl ether. This is a rapid screening procedure which should be applicable to all phenoxyalkanoic acid herbicides in water. Pope et al.(*176*) have determined silvex in water and muds by converting the free acid into the methyl ester using methanolic boron trifluoride. The esters were determined by microcoulometric and electron-capture gas chromatography.

Woodham et al.(*177*) extracted 2,4-D from water by first acidifying the sample below pH 3.0 and then extracting the water with redistilled ethyl ether by rotating for 20 min.

e. SHELLFISH, FISH. The butoxyethanol ester of 2,4-D was extracted from shellfish and fish by Coakley et al.(*178*). This method involves hydrolysis of the ester by shaking with dilute base, acidification, and extraction of the 2,4-D with benzene. The benzene is then extracted with a phosphate buffer (pH 6 to 7) followed by a carbon tetrachloride extraction of the buffer.

f. SOILS. The following extraction procedures for the removal of various phenoxyalkanoic acid herbicides were primarily developed for final analysis by gas chromatography using either microcoulometry or electron-capture detectors. For less sensitive determinative steps, the sample size would have to be increased as much as tenfold and the entire extract evaporated for analysis. However, this would undoubtedly increase the amount of impurities in the extract, particularly in soils high in organic matter, and necessitate some cleanup prior to detection. The level of impurities could be reduced somewhat by extracting the soil with a water-immiscible solvent such as diethyl ether. If benzene or a mixture of benzene–isopropanol is to be used to extract the soil, it is recommended that the soil be pretreated with 1 ml concentrated phosphoric acid per gram of soil. This soil acidification step would suppress ionization of the herbicide during the extraction with the water-immiscible solvent system. The pH of the soil slurry should be about 1 to 2 for adequate extraction of the phenoxyalkanoic acids.

Woodham et al. (177) have proposed a procedure for the extraction of 2,4-D residues from soil, sediment, and water. Soil or sediment were rotated for 4 hr with redistilled ethyl ether and 1:1 solution of sulfuric acid. The pH of the slurry should be below 3.0 in order that the herbicide can be removed from the soil. MCP was extracted from soil for electron-affinity detection of the 2-chloroethyl ester by Gutenmann and Lisk (179) by blending the soil in acetone containing 1% concentrated phosphoric acid. The extracted MCP was then esterified with 2-chlorethanol to increase its response to the electron-capture detector. Kirkland and Pease (167) extracted polychlorinated benzoic acids 2,3,6-TBA [TRYSBEN] from soil, shaking the sample in an Erlenmeyer flask with methylethyl ketone and 85% phosphoric acid. Dicamba residues were extracted from soil by the procedure of Smith et al. (162).

2. Cleanup Methods for Phenoxyalkanoic Acids. Cleanup of extracts for this class of herbicides is performed only when necessary. Free acids are often cleaned up by partitioning into an aqueous alkaline solution followed by washing with an organic solvent which removes most of the interfering substances. For additional cleanup, a FLORISIL column has been utilized on seafood by Coakley et al. (178). The residue extract in carbon tetrachloride solution was placed on the column and the impurities eluted through by washing with carbon tetrachloride followed by ethyl ether. When cleanup is required for this class of herbicides usually partitioning and/or column cleanup are employed.

a. PARTITION. Yip (155, 163) extracted ether solutions of 2,4-D and 2,4-D esters with aqueous sodium bicarbonate, the 2,4-D and related anionic metabolites partitioning into the water solution and the esters remaining in the organic layer The aqueous phase was extracted with chloroform to remove impurities while the 2,4-D remained in the aqueous phase. Depending on the crop extract, the aqueous alkaline phase must sometimes be extracted repeatedly with chloroform to reduce the interferences to tolerable levels. Following this cleanup step, the bicarbonate solution is acidified to liberate CO_2 and suppress ionization of the polar compounds. Often the acid fraction is cleaned up sufficiently by partitioning so that additional cleanup by adsorption is unnecessary. Schlenk and Gellerman(*180*) and Rogozinski(*181*) esterified the free acids prior to the determinative step.

The analysis for total 2,4-D content can be accomplished by saponifying the sample to convert all of the esters to free acids (salts). These can be measured directly by paper or gas-chromatographic means or they can be converted to methyl esters prior to the determinative step. Burchfield and Storrs(*182*) eliminated the saponification step by esterifying a mixture of free acids and esters with methanol and hydrochloric acid in order to obtain the methyl esters. Marquardt et al.(*183*) obtained free phenoxyalkanoic esters and acids in the same fraction by partitioning the extract between hexane and acetonitrile. The acetonitrile phase containing the herbicides is then back-extracted several times with hexane to reduce the amount of lipids to a minimum. This procedure, however, will usually result in low recoveries, particularly of the esters, but can be compensated for by the addition of an appropriate internal standard to the sample before extraction and gas-chromatographic analysis. Recovery of the internal standard need not be quantitative as long as it has chemical properties similar to 2,4-D and behaves similarly during extraction, cleanup, and esterification.

b. SORPTION. It is sometimes necessary to clean up extracts of phenoxyacetic acids, and it is usually necessary to use column adsorption cleanup on the esters before quantitation. Cleanup by adsorption usually follows liquid–liquid partition. The free acids can be chromatographed directly on FLORISIL, as by Coakley et al.(*178*), since they are held more tenaciously to the FLORISIL than most of the impurities. The interferences are eluted from the column with diethyl ether, and then the 2,4-D is displaced by elution with methanol. Bevenue et al.(*157*) converted the 2,4-D to its methyl ester prior to chromatography on FLORISIL by eluting the sample on the column first with benzene and then eluting the 2,4-D with diethyl ether in benzene. Yip(*163*) noted that higher esters of 2,4-D, when separated from the free acids by liquid–liquid partitioning, must be cleaned up thoroughly by eluting through a FLORISIL column prior to quantitation. Yip(*163*) accomplished this by chromatographing the extract on FLORISIL, eluting the esters, and extracting them from petroleum ether with aqueous acetonitrile. The acetonitrile was then diluted with saturated salt solution and the esters back-extracted into petroleum ether. The phenoxyalkanoic acid herbicides are subsequently eluted with methyl alcohol. Yip(*155, 163*) has reported the use of a combination of alumina and/or FLORISIL adsorbents for the cleanup of the chlorophenoxy acid esters. Other investigators such as Bevenue et al.(*157*) have also reported the use of FLORISIL for extract-cleanup of phenoxyalkanoic acids. The free acids and esters are often separated by partitioning between an aqueous alkali solution and a nonpolar solvent as recommended by Yip(*163*). The free acids are then esterified prior to analysis as by Schlenk and Gellerman(*180*) and Rogozinski(*181*).

3. Extraction of Substituted Ureas. Diuron, monuron, fenuron, and neburon residues in fruits and vegetables were analyzed colorimetrically by Dalton and Pease(*184*). The herbicides were extracted by refluxing in a strong alkali medium, which thereby hydrolyzed the urea compound to its corresponding amine.

Three substituted urea herbicides in fruits and vegetables were analyzed by Gutenmann and Lisk(*185*) using electron-capture GLC. The plant materials were extracted with acetone and partitioned into hexane prior to hydrolysis and bromination. Katz(*186*) extracted five substituted ureas from surface water with redistilled chloroform and then hydrolyzed the extract under reflux conditions with 6 N HCl prior to colorimetric measurement. HERBAN; a dimethylurea herbicide, was determined colorimetrically in cottonseed, potatoes, sorghum, soybeans, and sugar cane by Ford et al.(*187*). The herbicide was removed from oil-producing crops by a 2-hr Soxhlet extraction with hexane. Noruron was extracted from vegetable and grain crops by 1-hr tumbling with isopropanol.

4. Cleanup Methods for Substituted Ureas. Dalton and Pease(*184*) partitioned the corresponding amine into an aromatic solvent which in turn was washed with dilute acid before colorimetric measurement. Acetone extracts of several fruit and vegetables were partitioned into hexane by Gutenmann and Lisk(*185*) before direct hydrolysis and bromination of substituted ureas. Ford et al.(*187*) cleaned up hexane extracts of oil producing crops by partitioning into acetonitrile. Vegetables and grain crops extracted with isopropanol were cleaned up by shaking with a large excess of water and extracting the aqueous alcohol with chloroform.

5. Extraction of Triazines. St. John et al.(*173,188*) extracted atrazine and simazine from whole milk and urine by blending with acetone and shaking with chloroform prior to electron-affinity detection. Various triazine herbicides were extracted from soil by Henkel and Ebing(*189*) by shaking with acetone prior to analysis with a flame ionization detector. Mattson et al.(*190*) extracted various triazines from vegetables, forage, and grains by mechanically shaking the plant material with chloroform prior to microcoulometric detection. Atrazine and several of its metabolites were extracted from pea plants by Shimabukuro et al.(*191*) by homogenizing and boiling with 95% methanol. Sikka and Davis(*192*) removed atrazine from soil by refluxing with absolute methanol in a Soxhlet for 3 hr.

6. Cleanup Methods for Triazines. Urine samples containing atrazine and simazine residues were cleaned up by St. John et al.(*173, 188*) by dissolving the residue in chloroform and shaking with NORIT A carbon. The milk samples dissolved in hexane were partitioned with redistilled acetonitrile. Additional cleanup was accomplished by eluting the residue through a FLORISIL column with chloroform. Mattson et al.(*190*) cleaned up various vegetable, forage, and grain extracts containing triazine residues by eluting through an aluminum oxide (WOELM) column with *n*-hexane. In the radioactive atrazine experiments by Shimabukuro et al. (*191*), chlorophyll and other residues were removed from aqueous extracts of pea plants by high-speed centrifugation. In the nonradioactive atrazine experiments, the aqueous layer containing the chlorophyll, etc., was filtered through glass wool and filter paper. After repeated chloroform washings to remove the atrazine and chloroform-soluble metabolites, Shimabukuro et al.(*191*) subjected the residue to additional cleanup by use of thin-layer chromatography prior to separation of the atrazine and its major metabolites by gas chromatography. Sikka and Davis(*192*), following 3 hr of Soxhlet extraction of soil with absolute methanol, cleaned up the extracts containing atrazine in carbon tetrachloride by passing through a column packed with basic alumina oxide WOELM, prior to ultraviolet spectroscopy.

7. Extraction and Cleanup of Substituted Phenols. Gutenmann and Lisk(*193*) blended several polar phenolic herbicides (DNOC, DNOSBP, and ioxynil) with acetone to remove their residues from varied agricultural samples. The extracts were filtered through HYFLO SUPER-CEL on a Buchner funnel covered with filter paper.

8. Extraction and Cleanup of Carbamates. Potatoes were blended in their own liquid by Gutenmann and Lisk(*185*) and then with acetone to extract residues of CIPC. The extract was cleaned up by shaking with a solution of aqueous sodium sulfate, followed by partitioning the herbicide into SKELLYSOLVE B.

9. Extraction and Cleanup of Amides. Potatoes, tomatoes, peppers, and peanuts were blended with benzene by Boyack et al.(*194*) to extract diphenamid. Cleanup included hexane–acetonitrile partitioning followed by elution through an alumina column.

10. Extraction and Cleanup of Bipyridyliums. Since diquat and paraquat are held quite strongly to plant material, it has been found necessary to boil the macerated plant material with dilute sulfuric acid for several hours in order to release the herbicides into solution.

Calderbank(*195*) has applied the use of cation-exchange resins for the separation of bipyridylium herbicides from plant interferences. They are highly polar compounds and cannot be extracted into water-immiscible organic solvents. Due to their cationic properties, however, these herbicides are readily retained on a cation-exchange resin. Therefore, the use of ion-exchange resins is perhaps the only satisfactory technique that could be employed for separating these herbicides from their impurities prior to analysis.

The use of ion-exchange resins to cleanup of extracts is practical if the pesticide is ionic or can be converted into an ionic form. A large volume of crude extract containing traces of pesticide can be passed through a column containing the appropriate resin and the pesticide will be adsorbed under ideal conditions, no matter how dilute the applied extract. The pesticide can then be eluted from the resin with a small volume of acid or salt solution, thereby eliminating considerable impurities from the original extract.

11. Extraction and Cleanup of Miscellaneous Herbicides. Meulemans and Upton(*196*) extracted dichlobenil and its metabolite, 2,6-dichlorobenzoic acid, from fruit, berries, fish, soil, and water by blending the acidified sample macerate in a mixture of benzene and isopropanol. Cleanup of the dichlobenil was achieved by shaking with FLOREX. The acidified aqueous alkaline layer containing the metabolite was cleaned up by extracting with benzene and passing through a column containing FLOREX and anhydrous sodium sulfate. The metabolite was removed from the FLOREX column by eluting with 1% benzoic acid. Racusen(*197*) first used ion-exchange resins to cleanup plant extracts containing aminotriazole and its metabolites. The plant material was extracted with water and the homogenate heated to precipitate the protein. The clear-filtered extract was passed through a column packed with AMBERLITE cation exchanger. The basic herbicide salt was retained by the resin which was water washed prior to elution with a basic solution.

This ion-exchange technique can be used also for concentrating substances from dilute solutions of plant extracts that are more weakly basic and have limited solubility in water. It might also be used to concentrate residues of the weakly basic triazine herbicides prior to analysis. Anion-exchange resins might be utilized to concentrate acidic herbicides such as MCP or 2,4-D. Thornburg(*19*) states that anion-exchange resins can occasionally be used to adsorb acidic herbicides from water extracts of plant tissues, but he has found it difficult to obtain quantitative recoveries of 2,4-D from anion resins.

G. Fungicides

In general, because of their lack of mammalian toxicity, the fungicides are not considered as important a problem to the residue chemist as are the insecticides and herbicides. An excellent review of methods of analysis for fungicide residues, edited by Zweig(*198*), includes chemical and physical properties plus residue and formulation methods of analysis. Probably one of the most important classes of fungicides on the market at present are the dithiocarbamates (ferbam, ziram, thiram, maneb, zineb, and nabam). A carbon disulfide evolution technique for determining dithiocarbamate residues on food crops previously published by Lowen (*199*) was modified by Pease(*200*). The modification insured complete transfer of the carbon disulfide into a Viles reagent trap. Other refinements in the original technique resulted in greatly improving the percentage recovery of the pesticide. This procedure was further revised by Cullen(*201*) by reacting the evolved carbon disulfide with the yellow cupric salt of N,N-bis(2-hydroxyethyl) dithiocarbamic acid which is measured colorimetrically. The speed of decomposition of the dithiocarbamate and the copper–carbon disulfide ratio was found to be critical. Recoveries of 85–100% were generally obtained on all dithiocarbamates for every crop tested, and the method is sensitive to 20 μg of carbon disulfide. Thin-layer chromatography was used by Hylin(*202*) to distinguish between the residues of dithiocarbamate fungicides and their transformation products. Samples were extracted with chloroform on a mechanical shaker, and highly pigmented extracts were decolorized with a minimum of Darco G-60 activated charcoal. Analysis of organo-metallic fungicides, such as ferbam, ziram, maneb, and zineb, by atomic-absorption spectroscopy were made by Gudzinowicz and Luciano(*203*). Chloroform and pyridine were each selected as the organic solvent stripping agents in order to solubilize preferentially specific active organo-metallic fungicide complexes with subsequent solvent evaporation. Wax bean and cucumber leaves were heated slightly below the boiling point of each of the two solvents, and then rinsed with one of the solvents. The fungicide was then determined by assaying the metal content of the complex following acid hydrolysis, which converted the cation to a highly soluble aqueous salt suitable for atomic-absorption analysis.

Captan was extracted from green vegetables by Klayder(*204*) by blending with benzene. Following centrifugation, the decanted extract was cleaned up by shaking with the AOAC cleanup mix (10 parts NUCHAR, 5 parts HYFLO SUPER-CEL, and 5 parts anhydrous sodium sulfate). The above procedures remove the water, pigmentation, and some waxes prior to colorimetric determination. Kilgore et al.(*205*) extracted dichloran from various canned fruit by rolling the chopped fruit in benzene. The syrups from the canned fruit were extracted by first diluting tenfold with distilled water and then shaking with ethyl ether. Both extracts were cleaned up by passing through nonactived FLORISIL. Glyodin residues were extracted from pears and peaches by Kleinman(206) by tumbling or shaking the whole fruit in isopropanol. The extract was filtered, evaporated to dryness, and picked up in $CHCl_3$ prior to spectrophotometric measurement. Klein and Gajan(207) extracted PCNB from vegetables by blending first with ethyl alcohol and then with petroleum ether. Essentially no cleanup was required for the polarographic method, but an activated FLORISIL column was used on the extracts destined for colorimetric or gas-chromatographic measurement. These workers compared a colorimetric, a polarographic and a microcoulometric gas-chromatographic method for the determination of PCNB on lettuce, cabbage, and string beans. They found that the colorimetric method by Ackermann et al.(208), although the slowest of the three methods, was the most accurate in the 0 to 5 ppm range. The polarographic method of Bache and Lisk(209) and Klein and Gajan(207), although the least specific of the three methods, proved to be the most rapid because of the minimum cleanup required. The microcoulometric gas-chromatographic method was found to be more rapid but less accurate than the colorimetric method and, therefore, it was recommended that both procedures be used on a sample in order to verify results.

Dodine residues were extracted from apple and cherries by Steller et al.(210) by rolling for 1 hr in methanol. Following acid hydrolysis, the aqueous–alcohol phase is extracted with carbon tetrachloride. The plant material interferences are left in the discarded carbon tetrachloride phase and, after adjusting to pH 5.5, the dodecylquanidine–bromcresol purple complex is extracted into chloroform. The bromcresol purple which is proportional to the dodecylquanidine cation is extracted from the chloroform into aqueous alkali and measured colorimetrically. Frear et al.(211) used the Steller et al.(210) procedure to extract and cleanup apple extracts. Miller(212) extracted dichlone residues from fruits and vegetables with benzene, and the extracts were cleaned up by eluting with 1% acetone in benzene through an activated FLORISIL column. An electron-capture gas-chromatographic procedure was presented by Thornton and Anderson(213) for the determination of trichloronitrobenzene (a fungicide with an isomeric mixture of dinitrochlorobenzenes) on and in various vegetable crops, cottonseed, cottonseed oil, and soils. The crop samples were initially extracted with acetone, filtered, and the fungicide partitioned into hexane. For the cottonseed samples, an acetonitrile partitioning was necessary to eliminate the oil following an initial Soxhlet extraction with SKELLYSOLVE B. For the crop and soil extracts, no additional cleanup was required other than partitioning into hexane. A rapid, automated determination of the fungistat biphenyl in citrus fruit rind has been proposed by Gunther and Ott(214). The system uses redistilled cyclohexane for extraction of the distillate and reagent-grade sulfuric acid for the cyclohexane wash series. The previously chopped rind is homogenized in water and then steam distilled in the presence of sulfuric acid to liberate the biphenyl plus citrus oils and waxes. These steam volatiles are trapped in cyclohexane, the oils and waxes quantitatively extracted into concentrated sulfuric acid, and the biphenyl left in the cyclohexane to be measured with a UV spectrophotometer. KARATHANE residues were extracted from several fruit by Kilgore and Cheng(215) by rolling for 1 hr in mixed hexanes (bp 60°–80°C). The crop extracts were purified by eluting through a FLORISIL column or washing with sulfuric acid. Anderson and Adams(216) extracted DEXON residues from corn, pineapple, sorghum seed, cottonseed, sugar cane, and juice. Because of the wide variation in the physical nature and moisture content of the pesticide samples analyzed, several different blending procedures had to be employed. Most of the crops, other than cottonseed were blended with a 1% sodium sulfite solution. Because of the high oil content of cottonseed, it was found that an extraction mixture of aqueous sodium sulfite and benzene proved to be satisfactory. The slurries obtained by blending were dialyzed for 20 hr in the dark at room temperatures. Since DEXON is water-soluble, the dialysis furnish a convenient technique for the separation and cleanup of the pesticide from some of its crop interferences. Final cleanup was made by repeated alkali washing of the benzene extracts. Gutenmann and Lisk (217) extracted diphenylamine from apples by blending the chopped fruit with acetone and hexane followed by direct bromination to yield derivatives that could be successfully determined by electron-capture gas chromatography.

5-4   SEPARATION, IDENTIFICATION, AND QUANTITATION

Since there is a volume of information on analytical procedures for the detection of pesticide residues, this section is detailed only in the areas of detection and quantitation that have shown considerable promise in the past five years. Wherever a detailed review is already available it is cited without further elaboration, despite the fact that the procedure is still in current use on pesticide residue analysis. A review by Gunther(1) outlines the final determination of the pesticide into three basic areas of measurement: biological, chemical, and physical. In this review, although the primary biological, chemical, and physical means are all covered, emphasis was placed on the more widely used detection and quantitation procedures presently in the pesticide residue field. Undoubtedly the physical means of measurement such as chromatography and spectrometry dominate the present pesticide residue analytical field. There are, however, other physical means of measurement that are also rather widely employed at the present time. Therefore, the arrangement of the subsequent sections is not necessarily in order of their current usage.

Generally, however, it can be stated that the physical means of pesticide measurement are becoming increasingly dominant, with the chemical methods more or less holding their own. The biological methods appear to be fading out of the pesticide residue picture, except perhaps as parallel screening techniques.

A. Chromatography (Gas, Thin-layer, Paper)

Chromatography is the technique universally available and most used by the residue analyst for the qualitative separation of complex pesticidal mixtures and their quantitation. It is also a powerful aid in the identification of compounds.

In the past five years thin-layer and gas chromatography have become accepted widely in the residue field for segregation, identification, and measurement of micro amounts of pesticides in various biological substrates.

In gas–liquid chromatography (GLC) the stationary phase is usually a liquid impregnated on an inert solid, while the mobile phase is an inert gas such as nitrogen or argon. Gas chromatography is rapid and is one of the most sensitive tools now available to the residue analyst. Several disadvantages to this procedure are "fouling" of the detector due to ineffective cleanup of the injected extracts and the breakdown of certain pesticides at the high temperatures often required for adequate separation.

Thin-layer chromatography (TLC) is a relatively new technique which is rapidly replacing paper chromatography in pesticide residue research. The stationary phase is an active solid, bound to a glass plate or strip, whereas the mobile phase is an organic solvent or a mixture of an organic solvent and water. Thin-layer chromatography is more rapid, provides better resolution, and is more sensitive than paper chromatography. This is because its detection technique is more sensitive and TLC spots are more confined. Moreover, the TLC-adsorbent layer can often accept larger amounts of sample extract than can the paper; and this capability makes TLC a useful cleanup tool for samples destined for gas-chromatographic measurement or for infrared analysis following isolation by gas chromatography.

In paper chromatography (PC), the stationary phase is a modified filter paper, whereas the mobile phase is an organic solvent or a mixture of organic solvent and water. In the past PC has been a very useful screening technique for pesticides at the microgram level. However, its primary disadvantages are that it is less sensitive, slower, and exhibits less resolution than either GLC or TLC. Moreover, sample cleanup requirements are usually more stringent for paper than for gas chromatography. Table 5-1 is useful for comparing these three basic chromatographic procedures from Burchfield and Johnson(21)

TABLE 5-1

	PC	TLC	GLC
Initial cost	Inexpensive	Moderate	Expensive
Sensitivity range	$1-10\,\mu g$	$0.1-1.0\,\mu g$	$10\,pg-10\,\mu g$
Resolution	Fair	Good	Excellent
Usefulness for labile compounds	Good	Very good	Poor
Analysis time	8–10 hr	1–2 hr	15–30 min
Suitability for mass production	Excellent	Good	Fair

1. Gas Chromatography. The application of the gas chromatograph to the quantitative determination of pesticide residues, and also as an aid in their qualitative identification, has flourished in the past decade. Unfortunately, this highly sensitive tool has been abused on occasion by the untrained and inexperienced operator who mistakes mere quantitation of a peak for an actual determination of the pesticide under study. This problem is frequently aggravated by the lack of appreciation of the importance of adequate cleanup of sample extract, which can be reflected in retention time, peak shapes, minimum detectability as well as reproducibility. At least a tenfold decrease in sensitivity can be anticipated where a technical pesticide dispersed in a biological extract is compared to a purified pesticide in a distilled solvent. For adequate evaluation and interpretation of the data, the pesticide peak height or area should be at least twice the variation in the background "noise."

Most of the gas chromatographs presently in use in pesticide residue analyses have the following types of detectors: (i) microcoulometric, (ii) electron-capture, or (iii) thermionic.

New nonionization detectors currently employed successfully, but to a more limited extent, include the flame photometric, the microwave emission, and electrolytic conductivity detectors. An excellent review of current gas-chromatographic detectors has been published by Westlake and Gunther(218).

a. MICROCOULOMETRIC DETECTORS. The Dohrmann microcoulometric gas chromatograph (MCGC), introduced in 1959, was the first instrument designed specifically for pesticide residue analysis. This system was widely used for the specific detection of organic pesticides containing halogen or sulfur. The operation of the MCGC has been described by Coulson and Cavanagh(8), Coulson et al.(219), Coulson(220), Challacombe and McNulty(221), Stamm(222), and Burchfield and Johnson(21). In this system, the sample components are passed through a combustion tube where the organic compounds are oxidized to water, carbon dioxide, hydrogen chloride, and sulfur dioxide.

In order to detect halogen-containing pesticides selectively, the effluent from the combustion tube is passed through a titration cell composed of four electrodes that function as anode–cathode generator pair and a sensor-reference. For the detection of halide ion, the sensor and generator anodes are silver, the generator cathode is platinum, and the reference electrode is silver in saturated silver acetate. When HCl enters the cell, silver chloride is precipitated. The electrical imbalance created is sensed by the microcoulometer, and silver equivalent to the amount precipitated is regenerated from the generator anode. The current required to regenerate the silver ion is measured on the recorder.

Sulfur can be measured by replacing the titration cell with one containing platinum electrodes and an electrolyte containing acetic acid, potassium iodide, and water. The reference electrode is platinum and saturated tri-iodide. Sulfur dioxide entering the cell is titrated by iodine to sulfate and the electrical imbalance is sensed and iodine regenerated again. The current required to restore equilibrium is recorded. Therefore, through the use of this titration system, almost absolute specificity for organic pesticides containing chloride or sulfur atoms can be obtained. Microcoulometric detection eliminates interfering peaks, reduces enhanced baselines, and minimizes tailing due to the solvent peak. MCGC also allows programming without column bleed and permits the use of more volatile and less stable liquid phases than is possible with ionization detectors.

Burchfield et al.(223,224) have adapted the microcoulometer to the selective or simultaneous detection of phosphorus, sulfur, and chlorine. The column effluent was passed through a quartz 950°C combustion tube where the pesticides were reduced to water, hydrocarbons, phosphine, hydrogen sulfide, and hydrogen chloride. The reduction products were then passed into a silver–silver microcoulometric titration cell where silver phosphide, silver sulfide, and silver chloride were precipitated. When a short tube containing aluminum oxide was placed between the reduction tube and the cell, HCl and H_2S were quantitatively removed from the gas stream, while the phosphine passed into the cell unchanged. Therefore, when the MCGC was operated in this mode, the detector possessed absolute specificity for phosphorus. When a short gas–solid chromatography column containing silica gel was substituted for the aluminum oxide tube between the quartz tube and the titration cell, HCl gas was removed completely and peaks for phosgene and hydrogen sulfide were resolved from one another. Following this mode of detection, parathion yielded two peaks (one for its sulfur atom and one for its phosphorus atom) while paroxon produced only a single phosphorus atom peak. Therefore, by employing the three MCGC modes described, Burchfield et al.(*223, 224*) were able to measure either the sum of chlorine, sulfur and phosphorus, phosphorus alone, or to resolve sulfur and phosphorus when they are present in the same compound. This method has been compared directly with the MCGC oxidation system using pesticides containing both chlorine and phosphorus as reference compounds under identical column conditions. Quantitative agreement between the two methods was obtained.

The limit of detectability of the more recent MCGC titration systems is approximately 1×10^{-9} g of chlorine or sulfur. Precision at the minimum detectable limit is $\pm 10\%$ and $\pm 1\%$ at amounts 100 times greater than the minimum detectable limit. The current cells are sensitive to combustion products of all halogen-containing pesticides (except fluorine), sulfides, mercaptans, and organophosphates.

There have been numerous applications of the MCGC system in the pesticide field during the past several years. Only a few of the most recent applications are cited in addition to the basic publications to which reference has already been made.

Mattson et al.(*190*) used MCGC to determine several chloro- and thio-methyl-triazine residues in crops. Of the various liquid phases tested, CARBOWAX 20 M appeared to resolve the triazine herbicides satisfactorily. Residues of a dessicant and defoliant (a phosphorotrithioate in cotton-seed) were determined by Thomas and Harris(*225*) by using sulfur titration cell with MCGC. The method was reported to have a sensitivity of approximately 0.05 ppm or better and recoveries at this level were reported to be 80–90%. Terriere et al.(*226*) used a MCGC for the detection of toxaphene in lake water and in plants and fish taken from the lake. Limits of detection by this procedure were reported to be below 1 ppb in the water and below 1 ppb in the aquatic plants or fish tissues. Recoveries were reported to exceed 80%. A bromine-containing herbicide (bromacil) was analyzed by Pease(*227*) using MCGC. The sensitivity reported was about 0.05 ppm based on a 25 g sample with an average recovery of better than 85%. Chiba and Morley(*228*) determined the herbicide (TCA) in wheat grain using microcoulometry.

Satisfactory sensitivity was achieved in the 0.1 to 3.0 ppm range based on 10 g of wheat. Using microcoulometric gas chromatography, Burchfield and Wheeler(229) detected various chlorine, sulfur-, and phosphorus-containing compounds at a reported practical limit of detection of 1 to 100 ng of the pesticides. Naled and dichlorvos residues in fruit and vegetables were reported by Boone(230) to have been analyzed by MCGC.

In summary, although the microcoulometric detector is not as sensitive as the electron-capture detector, it provides an absolute and stoichiometric method for determining the amount of chlorine, sulfur, or phosphorus passing into the titration cell from the combusted pesticide. With microcoulometry, only compounds that contain these elements and that are volatile enough to pass through a gas chromatograph can interfere. The chances of making qualitative and quantitative errors may be greater when using the electron-capture detector than when using the microcoulometric detector. For many pesticide residue problems, however, microcoulometry is simply not sufficiently sensitive, and many workers find it difficult to achieve reliable results using this means of detection.

b. IONIZATION DETECTORS. It has been said that the modern era of pesticide residue analysis was born in 1960 with adaptation to gas–liquid chromatography of the microcoulometric detectors by Coulson and Cavanagh(8) and the ionization detectors by Lovelock and Lipsky(9) and Lovelock(231). Although a considerable number of ionization detectors are commercially available, this review is restricted to the two most widely used in pesticide residue analysis; that is, electron capture and hydrogen flame.

(1) *Electron Capture.* In evaluating the various ionization detectors, the most sensitive and useful in the pesticide residue field has been the electron-capture detector. This detector exhibits very high sensitivity to most chlorinated hydrocarbon pesticides, rendering it possible to detect nanogram and picogram quantities of these compounds. This detector is not sensitive to many other volatile or organic materials, which are invariably present in the crop extracts, although some cleanup of certain types of extracts is often necessary.

The electron-capture detector measures the loss of signal rather than a positively generated current. As the nitrogen carrier gas flows through the detector, an ionization source such as tritium, radium, or nickel ionizes the nitrogen molecules, and slow electrons are formed. These slow electrons migrate to the anode under a fixed potential, and when collected or "captured," the slow electrons produce a steady or standing current. Therefore, when a pesticide (containing electron-capturing moities) is introduced into the cell, the standing current will be reduced. The loss of current is a measure of the degree of electron affinity of the pesticide molecule and when amplified by an electrometer can be presented as a peak on a strip chart recorder. Compounds which capture electrons strongly usually include one or more of the following electrophores: $CO \cdot CO$, $-CO \cdot CH : CH \cdot CO-$, $-CO \cdot O \cdot CO-$, $-CH : CH \cdot CO-$, $-NO_2$, and halogens. According to Lovelock(231), the order of affinity in halogen-substituted hydrocarbons is $I > Br > Cl > F$. The most generally useful gases exhibiting the widest range of discrimination between classes of compounds are hydrogen, nitrogen, and helium. Since these gases have a negative affinity for free electrons, they are passed through a chamber containing a source of ionizing radiation and the current measured at different applied potentials. The pesticides containing one or more of the above functional groups will modify the ionization properties of the carrier gas. When the pesticide vapors are mixed with the inert ionized gas, some of the free electrons will be captured to form negative molecular ions. The effect of ion recombination is the basis of the design of the electron-capture detector.

There have been several recent reviews on the use of the electron-capture detector coupled to gas chromatography for the separation and determination of pesticide residues. Dimick and Hartmann(232) have described the electron-capture cell design, operating techniques, practical applications, and typical analysis in the pesticide residue field. Clark(233) has reported on the critical parameters of the detector and other components of the system closely associated with the detector. Such important variables as applied potential, dc versus pulsed operation, carrier flow rate, temperature of the detector, sensitivity, linearity of response, and effect of column bleed are discussed in detail. Gaston(234) describes two electron-affinity cells and goes into some general considerations for the satisfactory employment of this detector.

Because of the widespread application of the electron-capture detector in the pesticide residue field in the past decade, no attempt is made to cite the numerous publications in the literature. To illustrate the versatility of the electron-capture detector, a review of these publications in the past several years shows that all classes of pesticides, with varying degrees of success, have been separated and determined by this type of gas chromatography. Most of the reported data have been on the chlorinated organic insecticides, with the organophosphates second. However, several investigators have used the electron-capture detector for the detection of certain fungicides, herbicides, and even carbamate insecticides. The carbamates, for example, have to be chemically altered prior to injection, in order to exhibit sufficient sensitivity to electron-capture detection.

There are currently several types of electron-capture detectors which vary as to mode of operation, cell geometry, and radioactive source.

(a) *Mode of operation.* Clark(233) has discussed the advantages and disadvantages of direct current and pulsed modes of operation. Apparently the sensitivity of the detector is approximately the same regardless of the mode of operation. What little difference in sensitivity that may exist is attributable to the nature of the carrier gas used, since dc-polarized detectors usually employ nitrogen carrier gas whereas argon–methane is preferred for the pulsed operation. However, if the highest possible sensitivity is required from a dc-operated detector, argon–methane can be used although it is more expensive than the nitrogen gas. One real advantage of the pulsed operation is its insensitivity to changes in carrier-gas flow rates. However, if a dc detector is operated with an adequate flow controller and at a fairly high flow rate, errors from this source should be minimal. Adjustment for optimum linearity is more critical in the dc system and variable results are obtained if the applied voltage is too high. Another definite advantage to the pulsed system is that during an analysis the sensitivity of the detector can be changed by a factor of as much as 10. It appears, at least for the chlorinated hydrocarbons, that the pulsed system is somewhat less temperature dependent. Clark(233) stresses, however, that if reasonable care is exercised, the dc mode can be used satisfactorily over an extended period of time. Hartmann and Oaks(235) have noted that the pulsed mode would have more advantages in the parallel-plate electron capture detector but very little benefit in the detector having the concentric-tube design. Negative-dipping the concentric tube can be greatly minimized by increasing the dc voltage and pulsing will provide only minimal improvement beyond this.

(b) *Cell geometry.* Lovelock's(236) original electron-capture detector was of parallel-plate design. Although basically of sound design, this detector exhibited poor purge characteristics and was somewhat awkward and bulky in construction. The concentric tube was introduced to overcome these disadvantages of the parallel-plate configuration. Bonelli et al. (237) presented some pesticide residue data obtained with a concentric-tube detector including retention times, sensitivity, and standing current relationships. It was noted that the relationship between the detector potential, standing current, and sensitivity was materially affected by the detector temperature. Hartmann and Oaks(235) have investigated the two designs as to their sensitivity and requirement for pulsing. The concentric-tube design was reported to be more sensitive under all operating conditions. Moreover, the concentric tube is apparently easier to disassemble and returns to equilibrium in the chromatograph much sooner after cleaning. This recovery advantage of the concentric tube was also noted when septums were changed. Therefore, it would appear that the concentric tube is definitely superior to the parallel-plate design for the electron-capture detector.

(c) *Radioactive source.* During the past decade there were commercially available electron-capture detectors having the following sources of irradiation: ^{3}H, ^{63}Ni, ^{226}Ra, and ^{90}Sr. The most common source currently is tritium, but the nickel-63 source has achieved some popularity with residue analysts. Since the ^{63}Ni detector appears to be the latest innovation and has several advantages in the pesticide residue analytical field, some specific references are made in this review.

One of the primary disadvantages of the present, widely used tritium source is that the Atomic Energy Commission has imposed an operational temperature limitation of 220°C. Although tritium leakage does not present a severe physiological hazard to personnel in the laboratory, it can cause considerable interferences in areas where trace counting of labeled compounds is underway. However, at temperatures above 225°C, loss of tritium from the foil is serious.

In the process of repeated injections of inadequately cleaned up biological extracts, a thin film of material will gradually accumulate on the surface of the titanium tritide. This film can also be caused by excessive "bleed off" from the column liquid phase. Since the beta particles are of such low energy, the thin film on the foil surface will inhibit their penetration into the carrier-gas stream. This causes the detector's standing current to decrease, the baseline to become unstable, the sensitivity to decrease, and the linearity range to decrease. Because of the tritium temperature limitation, this "fouling" of the detector cannot be alleviated by raising the detector temperature above 220°C.

The primary advantage of the ^{63}Ni detector would be the Atomic Energy Commission clearance for heating as high as 400°C. The detector temperature can be elevated significantly above the column temperature, thereby preventing condensation of the vaporized column effluent on the surface of the ionization source. In pesticide residue analysis, the column temperature might be maintained at 200°C, whereas the detector can now be operated at 300°C, thereby eliminating the undesirable contamination effect. If the residue chemists want to utilize a high-temperature ^{63}Ni electron-capture detector, Hartmann et al.(*238,239*) have indicated a few disadvantages as compared to the tritium source of ionization energy. Since ^{63}Ni is a more expensive isotope and the cell geometry for its use is more complex, the ^{63}Ni detector will probably be considerably more expensive than the tritium detector. Several other minor disadvantages are that the ^{63}Ni detector is not quite as sensitive and exhibits slightly less linear dynamic range when compared with the tritium detector.

Eight isotopes were originally considered acceptable as radiation sources for an electron-capture detector by Shoemake et al.(240). They concluded that nickel-63 would be undesirable from an economic point of view and concluded that tritium, americium-241, or radium-226 would work equally well at detector temperatures of below 225°C. Ahren and Phillips(241) have reported on the continuous use of two high-temperature ^{63}Ni detectors for over a year. Cleaned-up extracts of a wide variety of samples, such as soils, fruits, vegetables, dairy, and poultry products have been analyzed successfully with high reliability and minimum maintenance. "Down time" has been significantly reduced since the detectors have never had to be disassembled and cleaned or the foil replaced because of "fouling." These investigators operated their columns at 200°C, detector at 310°C, and the injection port at 255°C.

(2) *Hydrogen Flame.* When organic compounds are burned in a hydrogen flame, the resultant increase in electrical conductivity of the flame is the basis of operation of the hydrogen flame ionization detector. This response varies with the number of carbon atoms present but is unaffected by the presence of halogen or phosphorus atoms and, therefore, there is no selective response to compounds containing these particular elements.

(a) *Thermionic.* It was noted originally by Giuffrida(242) and also by Karmen and Giuffrida(243) that when sodium ions are present in a hydrogen flame, the response to phosphorus was greatly enhanced. It was also noted that the response to halogen atoms was increased to a lesser extent and the carbon response remained unchanged. The original modification consisted of a conventional hydrogen flame-detector with a coat of sodium salt fused onto the electrode and was called a "sodium thermionic" detector. The sodium salt-coated electrode superimposes an additional source of ions on the background flame ionization, and the standing current, when the sodium electrode is used, is so much higher than the standing current for the conventional flame that the latter may be negligible in comparison. For typical organophosphates, the ionizing efficiency of the sodium thermionic detector is two to three times higher than that of the conventional flame detector. Giuffrida(242) has reported enhancement of about 20 times for halogen pesticides containing six chlorine atoms by applying the sodium salt coating.

Schmit et al.(244) reports the investigation of various salt coatings for use with the thermionic detector in an effort to develop a low-noise, long-life coating that preserves the sensitivity and specificity of the original detector. The most satisfactory coating found by these investigators was a 1 : 1 mixture of copper nitrate–boric acid, which was applied as a bead to a coil of 27 gauge copper wire. The sensitivity, noise level, and background current are dependent on the hydrogen flow rate. Therefore, considerable care must be exercised in the hydrogen flow rate adjustment, since these three factors can be altered at least tenfold by small changes in the flame temperature. The response of this detector to chlorinated pesticides is less than to organophosphates by a factor of about 100. For example, an equal response for 1×10^{-7} g of heptachlor and 1×10^{-9} g of parathion were observed. Although a potentially powerful tool, the sodium thermionic detector has three inherent disadvantages: (a) the average life of the wire coating is at best only a few days; (b) equilibration time for a newly coated wire is several hours, and (c) the sensitivity constantly decays as the salt vaporizes in the flame.

Oaks et al.(245) and Hartmann(246) have introduced a modification to the single thermionic detector which reportedly overcomes the primary disadvantages of previous sodium thermionic detectors. Their modification has been called a "phosphorus" detector and is simply a conventional flame detector with a "salt tip" composed of a cesium bromide pellet. This pellet consists of about 1 g of cesium bromide with a suitable filler; it is pressed under high pressure to form a rugged ceramic-like material. This phosphorus detector uses 170 ml/min ± 0.1 ml/min of air and 16 ml/min ± 0.01 ml/min of hydrogen. Consequently, a high-quality flow controller is necessary to meet these very precise flow rates. The phosphorus detector is reported to have a sensitivity to parathion down to 12 pg, with a linearity at slightly better than 1000-fold, to be simple in design and construction, and easily interchanged.

(b) *Stacked flame.* Karmen(247,248) has developed a hydrogen flame detector specific for phosphorus- and halogen-containing compounds. A wire mesh, previously treated with sodium hydroxide or salts, was heated in a hydrogen flame. The presence of pesticides containing phosphorus or halogen atoms in the flame gases increased the rate of volatilization of sodium metal vapor from the screen, which in turn was detected by flame ionization in a second hydrogen flame placed above the original. This stacking of two hydrogen flames one above the other is often referred to as the "stacked flame" detector. Organic interferences from the sample are essentially eliminated in this system as they are consumed in the lower flame. The "stacked flame" is reported to be responsive only to phosphorus and halides and sensitive to these compounds at the 1 to 1000 ng level. The single-thermionic flame detector response enhancement is 200 to 700 times that of a conventional flame ionization detector but is responsive also to all organics eluted from the gas-chromatographic column.

The earlier stacked flame detectors had many of the disadvantages already enumerated for the original sodium thermionic detectors. However, modifications have been made to increase the lifetime of the salt reservoir. Phillips(249) has reported the successful employment of a 1:1 w/w mixture of KCl–boric acid as a bead or melt in the top igniter coil of a stacked flame thermionic detector. The salt and acid are ground and mixed with mortar and pestle. The coil upper flame wafer is warmed over a Bunsen flame while the mixed powder is melted onto the top turn of the coil until a solid, hardened bead is formed. The bottom flame is operated and wired in the same manner as a standard flame with a negative polarizing voltage and igniter coil. The top, thermionic flame, is the same except there is no negative polarizing voltage applied and the coil igniter lead on the terminal strip at the back of the flame power supply is shunted to ground.

The response phenomenon has not been elucidated for either the single sodium thermionic or the stacked flame detectors.

c. NON-IONIZATION DETECTORS

(1) *Flame Photometric.* A compact and sensitive detector has been developed by Brody and Chaney(250) that is specific to phosphorus- and sulfur-containing pesticides and their metabolites. It is reported to be sensitive to phosphorus at the parts per billion level, and at the parts per million level for sulfur. The principle upon which the detector is based is the photometric detection of the flame emission of phosphorus and sulfur compounds in a hydrogen air flame and probably functions through a chemiluminescence mechanism. Interferences from the emission produced by other elements present are eliminated by an optical arrangement whereby only the portion above the tip of a normal flame envelope is viewed by the photomultiplier tube. The detector is sensitive to 0.25 ng of malathion and parathion and the response linear between 6 ppb and 60 ppm. This detector is easily adaptable to gas chromatography, operates well at 200 ml/min air flow, and does not utilize water cooling or an afterburner. It does not require a drain for liquid condensation or optical realignment for measuring phosphorus and sulfur. Filters can be used instead of a monochromator to increase the sensitivity of the detectors. Specificity is achieved by utilizing optical filters and a geometric arrangement of the burner that shields the primary flame from view by the detector. Phosphorus- and sulfur-containing pesticides are reportedly analyzed by this detector with minimum cleanup and with no interference from chlorinated hydrocarbon pesticides.

The first reported use of the flame photometric detector on pesticide residues has been made by Bowman and Beroza(251). IMIDAN and its oxygen analog IMIDOXON were quantitatively determined from field-treated sweet corn extracts at detection levels as low as 0.002 ppm. The flame photometric detector used in these analyses was found to be very sensitive and, unlike the electron detector, was not contaminated by uncleaned-up benzene extracts of corn plants.

The flame photometric detector has been found to be quite simple both to operate and to maintain. It is not contaminated, apparently, by either column bleed or unclean extracts of any type, and response can be attenuated readily. The switch from sulfur to phosphorus sensing can be accomplished rapidly — in less than 60 sec — and the relative response of phosphorus to nonphosphorus compounds appears to be very high. It is possible to assess derivatives of carbamate residues containing phosphorus or sulfur quantitatively. This detector, therefore, has a broad applicability to the analysis of multicomponent pesticides and their metabolites, both as a quantitative and as a screen tool.

(2) *Microwave Emission.* McCormack et al.(*252*) have used a microwave-powered argon discharge for excitation and emission detection of organic compounds emerging from a gas-chromatographic column. McCormack(*253*) first suggested its possible application as a sensitive, specific phosphorus detector for the determination of organophosphorus insecticide residues. Using this detector system, Bache and Lisk(*254–258*) have reported the determination of various organophosphate pesticides in various biological extracts. As the pesticide leaves the gas-chromatographic column it becomes fragmental. The fragmentation products are comprised mostly of atoms and diatomic molecules. Excitation takes place due to the extremely high electronic temperature of the discharge. Photon emission occurs when the atoms and molecules return to the ground state. Sharp atomic lines are seen from atomic emissions with broader lines and bands coming from molecular emission.

The limits of detection of emission spectroscopic methods, although quite sensitive, have, in the past, been inadequate for low-level pesticide residue analyses. This lack of sensitivity was largely overcome by the use of a microwave-powered argon discharge as an emission source coupled to a high-resolution spectrometer as part of a gas-chromatography detection system. McCormack et al.(*252*) reported sensitivities for hydrocarbon and organic compounds containing halogens, nitrogen, sulfur, and phosphorus to levels ranging from 10^{-9} to 10^{-16} g/sec. It was reported by Bache and Lisk(*255*) that the response of the emission spectroscopic detector is in the same range as that of the Giuffrida sodium thermionic detector. This detection system appears to be an accurate, sensitive, and selective means of determining many organophosphorus insecticides in food samples. Apparently, sensitivities of 100 ppb for many organophosphorus residues in crops is feasible and depends largely on the effectiveness of the accompanying isolation procedure. By using the microwave-powered argon emission detector, selectivity was determined by retention time, wavelength, and microwave power setting.

Bache and Lisk(255) report recoveries of six organophosphorus pesticides (diazinon, dimethoate, disulfoton, ethion, parathion, and ronnel) in the low-nanogram range from such diverse extracts as alfalfa, lettuce, milk, soil, chickens, halibut, bees, and urine. Bache and Lisk(256) determined the oxidative metabolites of dimethoate and phorate in soil by emission spectroscopic gas chromatography. Preparative thin-layer chromatography was used for prior separation of phorate oxygen analogs, sulfoxide, and sulfone. Bache and Lisk(257) found that the system could be made more sensitive and selective to organophosphates if the discharge were operated at reduced pressure. Using a microwave-powered helium plasma, Bache and Lisk(258) report the analysis of phorate in sugar beets, thiodan in potatoes, and disyston in soil. A mixture of carrier gases, approximately 50% argon in helium, has been reported by Moye(259) to give a tenfold increase in sensitivity for six organophosphate pesticides over the sensitivity observed for argon. This gaseous mixture also was observed to have decreased the continuum radiation of the discharge.

(3) *Electrolytic Conductivity.* Coulson(260,261) described a gas-chromatographic system that is reported to be highly selective for organic nitrogen compounds. The system includes an electrolytic conductivity detector that is equally responsive to all types of organic nitrogen. The detector is sensitive to as little as 1 ng of organic nitrogen and is insensitive to aliphatic and aromatic hydrocarbons, alcohol, ketones, and sulfur, and phosphorus- and halogen-containing compounds such as pesticides. The system operates by converting organic nitrogen to ammonia by reduction with hydrogen over a nickel catalyst in a pyrolyzer. The effluent stream from the GC column is delivered into a high-temperature microtube furnace, fitted with a quartz tube where hydrogenation of the effluent gases takes place. The hydrogenated products are passed through an acid absorption packing which removes any hydrogen sulfide and hydrogen halides and the resulting gas stream passed through the electrolytic conductivity detector cell. Coulson(261) reports sensitivities of from 200 to 400 ng for simazine and carbaryl but no response to various chlorinated pesticides. Although not as sensitive as electron capture or certain flame ionization detectors, the electrolytic conductivity detector may fill an important instrumental gap for the residue analyst. To date, there is no other sensitive detector available that will detect nitrogen in the carbamates, an important class of pesticides.

(4) *Other Non-ionization Detectors.* Thermal conductivity detectors have been used to a limited extent in the pesticide residue field. Zweig and Archer(262) and Zweig et al.(263) employed this detector for the determination of endosulfan residues. Berck(264) reports the use of a TC detector for the determination of ethylene bromide, etc., in cereal products. The primary disadvantage of this detector in the pesticide residue field is its insensitivity compared to electron-capture detection. Winefordner and Glenn(265) have reviewed other nonionization detectors that are currently available. However, excepting those already mentioned, most of the other nonionization detectors have had little application to the pesticide residue field.

d. PRECAUTIONS IN THE USE OF GLC. In the past few years, pesticide residue analysis laboratories have been initiated almost as rapidly as the number of gas chromatographs in use by residue analysts. Therefore, in some instances there are relatively inexperienced personnel using highly sensitive, but fallible, instruments. Rarely is it possible to remove interfering extractives from plant or animal tissues to the extent that the results of gas chromatography are completely reliable without rather elaborate analytical precautions. Undoubtedly, some of the more inexperienced residue analysts have adapted gas chromatography somewhat short sightedly and have not always been fully cognizant of the difficulties to be encountered. As Gunther(266) has pointed out, more residue misinformation has resulted from inadequately performed gas chromatography than from any other single technique used in the past two decades. Although it is basically an extremely useful tool, perhaps too much emphasis has been placed on the obvious advantages of GLC without sufficient emphasis on the definite limitations of this technique.

Van Valin et al.(267) have reported that contaminants arising from solvents stored in polyethylene containers have induced spurious DDE spots appearing in chromatographic separations. Burke and Giuffrida (268) have stressed that considerable precautions must be taken in order to obtain zero reagent blanks by electron-capture GLC detection. First and most important, all solvents must be carefully redistilled unless specially redistilled commercial solvents are purchased. No solvent should come in contact with polyethylene prior to injection in the electron-capture detector. All glassware, particularly volumetrics, pipets, separatory funnels, Danish–Kuderna concentrating tubes, etc., must be scrupulously cleaned and finally rinsed with a redistilled solvent such as acetone. Blendor blades, shafts, and associated glassware must be carefully cleaned. There have been reports of spurious peaks due to the detergents used in glassware washing.

During the past few years, a special class of nonpesticide compounds called polychlorinated biphenyls (PCBs) has received considerable attention, not only by pesticide residue analysts but also many ecologists. No attempt will be made to review all the published information on this potentially serious interference problem involving the gas chromatographic analysis of environmental samples for chlorinated pesticides. Several of the most recent review articles on the present status of the PCB problem have been by Risebrough and Brodine(366), Peakall and Lincer(367), and Gustafson(368). When PCBs are present in a sample, they exhibit a pattern of GC peaks with retention times very similar to those exhibited by DDT, DDE, aldrin, dieldrin and heptachlor. Because of the possibility of PCB contamination in certain environmental samples, some previous chlorinated hydrocarbon pesticide analyses may be questionable. The PCB interference problem can be largely alleviated by prior treatment of the sample in order to separate the chlorinated pesticides being measured from any contaminating PCBs that may be present. One recommended procedure is to extract the biosample initially with hexane, followed by a acetonitrile partitioning to remove interfering fats. Then the cleaned-up extracts can be successively eluted through FLORISIL and silicic acid chromatographic columns before analyzing by gas chromatography. This technique results in the chlorinated pesticides being retained on the silicic acid column whereas the PCBs pass through the columns and are quantitated separately.

Burke and Holswade(269,270) have shown that the separation of compounds with relative retention times of 1.01, usually cannot be achieved with currently available columns. Littlewood(271) has pointed out that for compounds with relative retention times of 1.05 the number of theoretical plates required is 2800 in order to yield a separation with an impurity fraction of 10%.

A still more difficult problem that may face the residue chemist using GLC, is the presence of artifacts or interferences that arise from the technique itself, as opposed to natural interfering components. If the analyst is fortunate enough to have control samples that are known to have had no previous contact with the particular pesticide under investigation, the natural interfering components can usually be tolerated and removed from the final calculation of the pesticide entity. If the pesticide can be isolated in a pure state and in sufficient quantity, it is advisable to verify the compound by determining its infrared spectrum. A mass spectrum of the compound would also yield considerable information on its identity. On the basis of current investigations, Robinson et al.(272) have postulated that the following parameters be considered to be independent criteria for the identity of pesticides: (a) retention times (GLC) and either

R_f values (TLC) or Beroza–Bowman p-values, (b) retention times and chemical reactions, and (c) retention times and insecticidal activity. The independence of the following parameters has either been disapproved or are still under investigation: (d) retention times on different stationary phases, (e) R_f values (TLC) and p-values, and (f) responses given by electron-capture detectors and microcoulometric detectors for halogenated pesticides. Robinson et al.(272) have suggested that, in view of the dependent characteristics of the parameters in (d), (e), and (f) above, the use of these in combination with the independent parameters (a), (b), and (c) requires considerable care.

2. Thin-Layer Chromatography. Between the level of sensitivity encompassed by paper and gas chromatography is an area of application to the residue chemist which has been termed thin-layer chromatography (TLC). Like paper chromatography, TLC is essentially a microchromatographic technique usually involving a glass plate, layered with a thin film of a suitable adsorbent. The glass acts as a physical support for the inert medium holding the stationary phase. The adsorbents usually used are finely graded silica gel or aluminum oxide containing binding agents, such as starch or gypsum, which add greater mechanical stability to the adsorbent layer coating the glass plate. Powerful reagents can be used for locating spots on the plate without destroying the stationary phase, as would be the case when using paper. Prewashing of the plates is not as critical as with paper and TLC development; solvent evaporation and ultraviolet treatment are more rapid. The overall savings in time required to spot the chromatogram, to complete chromatography, and to develop and to read the R_f values is about 1½ hr by TLC compared to about 12 hr with paper chromatography. As with GLC, thin-layer chromatography has been widely applied to the pesticide residue field of research and, therefore, no attempt is made in this review to cite even a fraction of the publications pertaining to this subject. Conkin(273) has compiled an excellent review of current literature on the subject as well as descriptions of essential TLC techniques employed. Burchfield and Johnson(21) have described the preparation of the TLC adsorbents and their application for the separation and detection of the organochlorine- and phosphate-containing pesticides. A review of the application of TLC techniques to the analysis of pesticide residues has been made by Abbott and Thomson (274) which reviews the current art of preparation of chromatoplates, preparation, and application of sample extracts, development techniques, and quantitative evaluations. Special applications are covered such as preparative-layer, wedge-layer, multi-band, and gradient-layer chromatography. A sample of several recent applications of TLC to the various classes of pesticides are as follows: chlorinated hydrocarbons by Kovacs (275–278), Faucheux(279), and Moats(280); organophosphates by Bazzi et al.(281), and carbamates by Finocchiaro and Benson(282).

Channel-layer chromatography (CLC) has been introduced by Matherne and Bathalter(100) as a cleanup procedure of crop extracts prior to the analysis of several chlorinated hydrocarbons by GLC. Channel-layer chromatography cleanup is based on the principles of both column and thin-layer chromatography. It has been reported to be a suitable, generalized procedure for separating residues from coextracted plant material, with the final detection of the pesticide being made by gas chromatography.

3. Paper Chromatography. Paper chromatography in the past has been an excellent method for determining pesticide residues, for regulatory work, and for screening purposes when high sensitivity and accuracy were not important. This procedure has some of the advantages already outlined for TLC but its major drawback is the limited quantity of purified extract that can be spotted on the paper. The requirement for cleanup of the extract prior to spotting of the paper is often as stringent as that required prior to injection into a gas chromatograph, or more so. McKinley (283) has thoroughly reviewed the applications of paper chromatography for pesticide residue applications. Coffin(284) has reviewed its advantages and disadvantages and has suggested PC applications and modifications that might extend its usefulness in the separation and detection of minute pesticide residues.

B. Spectrometry

As in the past, the residue chemist will continue to rely on spectrophotometric procedures to aid in meeting the requirements of adequate qualitative and quantitative residue methodology. When a pure pesticide sample is available, this type of instrumentation produces specific identification and also greater absolute accuracy than many types of chromatography.

1. Visible. A chemical colorimetric method of analysis includes the treatment of compound in solution with a reagent that causes the development of a color, the absorbance of which is directly related to the determined substance concentration. Visual colorimetry can be distinguished from visible spectrophotometry since it generally refers to the quantitative measurement of light-absorbing material by comparing it with a colored standard. Visible spectrophotometry, however, implies that a photoelectric cell is used in the instrument, and the response from the cell is transformed to a measurable signal.

A decade ago, colorimetric methods employing spectrophotometric equipment were the mainstay of the residue analyst. Recently, this type of measurement has been augmented, or even replaced, by more versatile and sensitive tools such as thin-layer chromatography and gas chromatography equipped with the various types of discussed detectors. Colorimetric procedures, however, are still occasionally applied for residue analyses and, in some instances, colorimetry may actually be more practical than the more sensitive detection techniques introduced more recently.

The primary difficulty with colorimetric methods, after the basic color reaction has been established, is to develop adequate cleanup of the biological extracts. The most common difficulties with colorimetric procedures attributable to faulty cleanup are incomplete or false color development, turbidity, color instability, or excessive loss of the pesticide during the cleanup phase.

Gunther(1) and Beckman et al.(285) have outlined the theory, principles, and reasons for using colorimetric methods of analysis. Disadvantages in the use of colorimetry are the resultant destruction of the sought parent compound and considerable loss of specificity during the color-producing reaction. Final color reactions are often strongly pH dependent in intensity and character and are usually affected by heavy metal ions, oxygen, and light. A review of the literature pertaining to residue analyses for the past decade reveals numerous, relatively specific, and sensitive colorimetric procedures available for pesticide residue analysis. Because of their general application in the past and continued application at present, no attempt can be made in this review to cite any specific procedures. In summary, it might be reemphasized that simple, relatively specific, and sensitive colorimetric procedures will probably continue to be employed as a useful technique by residue chemists, in conjunction with, or even occasionally in place of, more sophisticated instrumentation.

2. Ultraviolet. The ultraviolet region (185–400 mμ) should be broken down into two subregions: the far-ultraviolet from 170 to 220 mμ and the near-ultraviolet from 220 to 380 mμ. The region from 10 to 170 mμ has not been exploited to any extent because of instrument limitations. As with colorimetry, ultraviolet determinations are usually designed to be specific for the pesticide being determined, but are generally somewhat more sensitive than colorimetric measurements. Compounds whose absorbance maxima are near the lower wavelength scale may be detected with greater sensitivity than those whose absorbance maxima are near the visible spectrum. However, this increased sensitivity is accompanied by greater background absorbance problems. Blinn and Gunther(286) have summarized the important principles and applications of ultraviolet spectrophotometric procedures for the assay of pesticide residues. The principles of ultraviolet spectrophotometry and recent techniques for achieving increased sensitivities are also discussed by Blinn(287).

3. Infrared. Despite the use of more recently introduced analytical techniques such as nuclear magnetic resonance, mass spectrometry, gas and thin-layer chromatography, the unequivocal identification of trace pesticide residues is still best handled by infrared spectroscopy. In the past, most of the applications of infrared involving pesticides has been in the analyses of formulations and technical materials. Infrared spectro-photometry has not been applied to a great extent in the pesticide micro-residue field due to the apparent lack of sensitivity of presently available instruments. However, with the eventual development of equipment capable of handling the very small samples necessary for most residue determinations, IR spectrophotometry will be employed increasingly as a routine analytical tool for pesticide residue determinations. Blinn and Gunther(288,289) have noted that definitive new applications in infrared instrumentation and methodology have brought this powerful and data-rich technique within the reach of every residue laboratory. Beckman et al. (285) have summarized the theory of available instruments for qualitative and quantitative analyses using infrared spectrophotometry and have reviewed the application of IR to pesticide residue analyses. The primary difficulty involved is the relatively high concentrations necessary in order to obtain a suitable spectrum. This problem has been overcome to some extent by the use of microcells and beam-condensing systems. A second major problem in the pesticide residue field is the necessity for using microliter-sized volumes, which are difficult to handle and cannot be measured easily.

Therefore, at present, even with the latest techniques using unusually small, micro-liquid cells, nanogram amounts of pesticide residues cannot be attained. However, it is possible to obtain satisfactory spectra of as little as 1 μg of many organopesticides.

As pointed out by Blinn(287), increased infrared sensitivity for residue analysis can be achieved either by increasing the energy path length of the sample cell or by increasing the volume of the absorbing solution through which the energy passes. With increased sensitivity, there must also be increased awareness of interferences due to inadequate cleanup of the substrate, to solvents, to reagents, and to handling. Miniaturization of sample size can be improved by the use of beam-condensing optics to concentrate sufficient energy through the small apertures of ultramicro cells designed to hold microliter solutions or ultramicro KBr pellets. In contrast to solvents, KBr pellets have the advantage of being transparent through the conventional infrared region (2.5–25 μ). There is a commercial fraction collector using KBr pellets for collection and IR identification of gas-chromatographic fractions in 50 μg quantities. However, there is currently no IR instrumentation sensitive enough for the identification of GLC pesticide residue gaseous fractions in the nanogram range.

Frehse(290) has written an extensive review on the use of infrared in pesticide residue analysis. Although this review covers all classes of pesticides, the greatest emphasis is on organophosphates. The need for thorough cleanup is discussed, as well as extraction procedures, cells, solvents, analysis of solid substances, special equipment, and the infrared characteristics of the organophosphates. Crosby and Laws(291) used gas chromatography as an additional cleanup step in preparing organophosphate residue extracts for infrared analysis. The entire gas-chromatographic effluent was trapped by passing it through methylene chloride. After evaporation of the methylene chloride, the residue was dissolved in carbon disulfide and the infrared spectrum determined over the 5–15 μ range. Satisfactory spectra were obtained with as little as 1 μg of organophosphate residue. Recoveries of 50 to 80% were obtained at 0.2 to 4.0 ppm levels and 40 to 50% at the 0.1 ppm level. Boyle et al.(292) describe a procedure that consisted of a hexane extraction of fish tissue, partition between hexane and acetonitrile, column chromatography, thin-layer chromatography cleanup, and final identification by micro-infrared in a KBr disk. The practical limit of sensitivity for a number of chlorinated insecticides was found to be about 1 ppm in fish tissue. An infrared method for determining residues of demeton and its insecticidally active metabolites from fruits and vegetable extracts has been reported by Giang and Schechter(293). Methylcarbamate pesticide residues in grapes were determined by Broderick et al.(294) using infrared absorption of the N—H stretching bond at 2880 mμ. The method of cleanup described removed 93% of the interfering methyl anthranilate, thereby permitting detection of as little as 50 μg of carbamate. Payne and Cox(295) report micro-infrared analysis of dieldrin, endrin, and other chlorinated pesticide residues in complex substrates such as sludge, soils, fish tissue, aquatic fauna, and aqueous industrial effluent samples. These workers combined paper, column, and thin-layer chromatography to isolate the insecticide residues from the substrates. Micro-quantities of these residues were then identified by IR spectroscopy using KBr micro-disks and the appropriate micro-sampling equipment. When scale expansion is used with a

0.5 mm disk, 0.5 μg, or slightly less, of most persistent pesticides should be capable of identification by the procedure outlined.

Therefore, for most residue investigations, it is now possible to identify positively microgram amounts of pesticides that have been identified in the past by GLC or TLC. In recent years several workers such as Chen(296) and Giuffrida(297) have reported infrared spectra of 1–3 μg samples of pure chlorinated hydrocarbons by use of the micro KBr disk technique. Apparently the minimum sample size is largely determined by unremovable impurities to be found in certain foods containing considerable fats. Blinn(298) has described techniques which have been successfully employed for the quantitative measurement and qualitative characterization of micro-quantities of pesticide residues. It is concluded that, at present, quantitative IR measurements are most reliable when made on solutions, using long energy path length cells, whereas qualitative characterization can be best accomplished with micro-pellets.

4. Fluorescence and Phosphorescence. Compounds which are capable of absorbing electromagnetic radiation and then at a later stage, emitting at least a portion of the absorbed radiant energy as visible light, are said to be fluorescent or phosphorescent. Fluorescent and phosphorescent compounds are differentiated based upon the time delay between absorption and re-emission of light. Compounds that emit light only microseconds after absorption are said to fluoresce, whereas phosphorescent chemicals require longer periods of time before light energy is emitted.

As a technique for the determination of pesticide residues, fluorescence has not been very widely applied. One primary reason, of course, for the lack of utilization of this technique is that very few pesticides fluoresce naturally or can be chemically converted or coupled to form a compound which exhibits fluorescence. This technique does have certain advantages over the classical methods of residue analysis, and certainly its applicability has not been thoroughly explored. The advantages and disadvantages of fluorescent methods have been noted by MacDougall(299, 300). Its principal advantage over colorimetry, for example, is its sensitivity, which could approach 1000-fold with certain compounds. One disadvantage of photofluorimetry is that it is limited in its applicability. A second major disadvantage is that many naturally occurring materials fluoresce and therefore considerable cleanup must be undertaken to obtain background values sufficiently low to be practical. The actual cleanup conditions for any compound or its metabolites must be carefully worked out for each type of biological extract involved. Another factor to consider when using this procedure is that fluorescence intensity is dependent upon a variety of environmental conditions such as the solvent used and the pH of the solution during measurement. The presence of a small amount of water or polar solvent in a nonpolar solvent may greatly affect the intensity of fluorescence. Concentration quenching is also a serious problem and therefore the relationship between fluorescence and concentration is linear only over a limited concentration range. Care must be exercised to ensure that measurements are made only in the range where linearity exists. Often an internal standard technique must be employed on all samples in order to control these external and environmental factors. Anderson et al.(*301*) have reported the use of a photofluorometric method for the determination of coumaphos residues in animal tissues.

The application of phosphorimetry to the pesticide residue field has been as limited as that of fluorimetry. Keirs et al.(*302*) emphasized the fact that the existence of an aromatic system, or a highly conjugated system within the compound molecule is necessary for phosphorescence to be exhibited. The utilization of phosphorimetry as a quantitative analytical tool for pesticide residues is a recent innovation. Moye and Winefordner(*303*) surveyed 52 pesticides by phosphorimetric measurement including their limits of detection and spectral characteristics. In this study it was found feasible to detect the following aromatic carbamates at concentrations of 5–10 ng/ml by phosphorimetric means: carbaryl, ZECTRAN, MATACIL, MESUROL, carbofuran, and UC 10854 (3-isopropyl phenyl N-methylcarbamate). The hydrolysis of carbaryl was found to produce a 16-fold enhancement in phosphorescent intensity. Para-nitrophenol, a metabolite of parathion, was detected in urine by Moye and Winefordner(*122*) using thin-layer chromatography and phosphorimetry. The same two techniques were used by Winefordner and Moye(*304*) for the analysis of nicotine, nornicotine, and anabasine in tobacco. The first reported use of phosphorimetry to detect a crop residue was by McCarthy and Winefordner(*305*) for biphenyl in oranges. Moye (*306*) reports research involving the use of phosphorimetry as a final detection tool of intact pesticides or their hydrolysis products and metabolites in the presence of extracts from soils and plant tissues. Thin-layer chromatography is being employed for the cleanup of various extracts.

The advantages and disadvantages in the employment of phosphorimetry are similar to those already mentioned for fluorimetry. Phosphorimetry under ideal conditions is a highly sensitive analytical tool. As with photofluorometry, the chief difficulty in phosphorimetry is to secure a test compound that exhibits an adequate phosphorescent intensity derived from a substrate extract that can be cleaned up sufficiently so that incidental background phosphorescence does not obscure the reading. Unfortunately, many plant extractives, if not removed by an effective prior cleanup, tend to exhibit considerable phosphorescence.

5. Mass Spectrometry. Spectroscopic analysis coupled with separational techniques have recently been under investigation for the unequivocal identification of pesticide residues. Mass spectrometry potentially offers several advantages over many other forms of spectroscopy, such as high sensitivity, ability to determine molecular weights, as well as the number and type of heteroatoms. Analysis of the mass spectra can also be used for positive identification and structural information. Kantner and Mumma(*307*) have reviewed the applications of mass spectroscopy to pesticide residue analysis. This review includes the basic principles and practical consideration as well as performance characteristics. They discuss several ways in which the combination of the separating abilities of the gas chromatograph and the identification capabilities of the mass spectrometer results in a powerful analytical tool for pesticide residue assay. Mass spectra of a large group of organochlorine and carbamate pesticides are also presented.

A general technique for collecting gas-chromatographic fractions for introduction into the mass spectrometer has been introduced by Amy et al.(*308*). The combination of gas chromatography and mass spectrometry has been recognized as a potentially useful tool for the separation and unequivocal identification of the components of complex molecular mixtures such as the organopesticides. Direct continuous introduction of the GC effluent requires rapid recording of the spectrum and therefore lowers sensitivity because of the large carrier-gas dilution of the separated components. Recently developed techniques can rapidly concentrate this effluent, although care must be taken to avoid decomposition and ion source pressure fluctuations. Mumma and Kantner(*309*) report the practical application of mass spectroscopy in the residue analysis of a number of chlorinated hydrocarbon pesticides. These pesticides all yielded easily recognizable molecular ion peaks and characteristic ion fragments. The number of chlorine atoms in each ion, including the mass ion peak, was easily determined by the specific isotopic distribution pattern.

Despite the apparent advantages of mass spectroscopy, there has been limited application of this method to the analysis of pesticide residues. The major problem appears to lie in the coupling of mass spectrometry to the currently available separation techniques, such as gas or thin-layer chromatography, at required levels of detection. Damico and Benson(*310*) reported recently on the use of mass spectra to characterize 14 pure carbamate pesticides, which were divided into three classes (N-methyl, N-phenyl, and N,N-dimethyl). Another obvious disadvantage of this technique is the excessively high initial cost of the equipment for many residue laboratories.

6. Nuclear Magnetic Resonance (NMR). NMR has been primarily employed for the comparative identification of compounds and previously lacked sufficient sensitivity for most quantitative pesticide residue analyses. However, with recently improved instrumentation, NMR is approaching usefulness as a practical tool in the pesticide residue field. Fukuto et al.(*311*) have reported the use of NMR for ^{31}P spectra for the detection and identification of fenthion (BAYTEX) in plants and its breakdown on plant surfaces upon exposure to sunlight and air. The limit of sensitivity of the NMR to various organophosphates was at concentrations of 3 to 10%.

7. Atomic Absorption. A pesticide residue method has been proposed by Gudzinowicz and Luciano(*312*) for the analysis of organo-metallic fungicides (derivatives of thiocarbamic acids) and related compounds by atomic absorption spectroscopy. Calibration curves were established for the determination of the metal content of maneb, ferbam, ziram, and zineb. The most favorable concentration ranges for the analysis by atomic absorption of zinc, iron, and manganese ranged from 0.07 to 5.0 ppm depending on the particular metal. Wax beans and cucumber leaves were extracted with chloroform and pyridine with recoveries of zinc ranging from 95 to 100%. The extracts were subjected to an initial acid hydrolysis to convert the cation to a highly soluble aqueous salt suitable for atomic absorption analysis.

C. Radioisotope Tracers

1. Pesticide Metabolism. By far the greatest use of radioactive tracers in this field has been made to elucidate the fate of pesticides after their application in the environment. These fate studies are conducted to: (a) relate the duration of toxic action to the rate and manner in which the pesticide is metabolized; (b) investigate the rate of pesticide elimination from the mammalian body; (c) evaluate the rate of metabolism, sites of accumulation, and the character of the metabolic products in plants, animals, and soils, and (d) study the increase or decrease in original toxicity upon conversion to various metabolites. All of these interrelated aspects emphasize the importance of knowing the pathways of metabolism as well as the degree of accumulation of metabolic products in plant and animal tissue. Menzie(*313*) has classified the metabolic pathways into seven groups that cover the majority of biotransformations that pesticides undergo. This worker has reviewed, by pesticide, many of the metabolism studies involving tagged compounds.

By far the most common radiotracer method in pesticide metabolism research is to label the toxicant itself. This procedure results in a rapid and sensitive method of pesticide detection and provides a total pesticide accounting of the radiolabeled residue dissipated by physical loss and chemical changes. Radiolabeled compounds are also advantageous for evaluating residues when other suitable methods are unavailable or when a number of metabolites must be determined simultaneously. Only a few of the other workers who have reviewed the use of radioisotopes in pesticide research are cited here.

Casida(314) has reviewed the use of labeled organophosphate for metabolism studies in plants. Dahm(315) has discussed the use of all classes of radiolabeled pesticides in metabolism experiments conducted with insects, plants, animals, and soils. Arthur(316) has made a comprehensive review of the use of tracer pesticides involving the metabolism of systemic insecticides in animals. He states that radioisotopes are indispensable tools for increasing the wealth of knowledge concerning the animal metabolism of insecticides, particularly organophosphates. O'Brien and Wolfe (317) have reviewed the use of radioactive tracers to study the absorption of pesticides and their effect on insects. Smith(318) has outlined how to establish a radioisotopic laboratory, selection of the proper isotope, instrumentation, preparation, and counting of samples. Of the various instrumental counting devices available to detect radioactive emissions, liquid scintillation counting has become the method of choice for the measurement of low-energy beta emitters. This method of counting is based on the fact that certain organic compounds emit photons when excited by irradiation. The photons (light) produced are then detected by a photomultiplier tube, which utilizes this photoelectric effect to produce a pulse that can be amplified and recorded. A recent review of the "state-of-the-art" of scintillation spectrometry has been made by Kaiser(319). This report describes the present uses of, and instrumentation involved with, scintillation spectrometry. Special attention is given to new scintillator materials, advances in photomultiplier tubes, and improvements in resolution.

Most of the past and present studies involving the use of tagged pesticides have been conducted to study the effect of the biological system on the pesticide. Conversely, Winteringham(320) has presented a "labeled pool" technique, which is suitable for isotope studies involving the effect of the pesticide upon a suitably labeled biological system. The term "labeled pool" has been used to refer to all biochemically related metabolites in a biological system that have become labeled in varying degrees due to the original introduction of a suitably labeled parent compound into the system. This technique has not been widely used to date, but shows considerable promise in detecting and identifying "water-soluble" metabolites.

Several of these workers have employed isotopic dilution techniques for the determination of pesticide residues. The term "isotopic dilution analysis" may be applied to any method of analysis that involves diluting the sought pesticide initially present in the sample with a precisely known quantity of a particular isotopic form of the same substance. The primary advantage of the isotopic dilution procedure is that it involves only a small amount of isotope since it is added only to the analytical sample, rather than to the entire experimental plot. Therefore, for pesticide residue work, isotopic dilution procedures not only are distinctly more economical than sample tracer techniques, but are safer to handle because of the small quantities of tracers used. Bogner(321) has stressed the unique features of isotopic dilution as well as some associated problems when using this technique for the measurement of residues.

2. Residue Methodology. Some limited use of tracer techniques has been employed to test the efficiency of extraction of field-treated samples. As pointed out repeatedly by Gunther(1, 17) Gunther and Blinn(14), and others, the true efficiency of any extraction procedure is very difficult to evaluate and verify. When evaluating their extraction efficiencies, practically all residue chemists follow one of two procedures. The first and most common technique is to fortify or "spike" the pesticide-free material to be processed with known quantities of pesticide(s), followed by immediate extraction. If necessary the extract is cleaned up, analyzed, and per cent recoveries determined. The second procedure that may be used is merely to fortify the solution after extraction and cleanup. The former procedure, although reasonably realistic, is nevertheless equivocal; the latter method totally invalid. Regardless at which stage the test pesticide is added to the sample in the laboratory, it is quite unrealistic to extrapolate laboratory-spiked sample extractions directly to extraction of field-weathered residues.

The only way to determine, in absolute terms, a true evaluation of the extraction efficiency of field-weathered "unknowns" is to employ the following scheme: (a) expose the biological materials to labeled pesticides under exactly the same environmental conditions that the pesticide would normally be exposed to, (b) process the residue samples containing the labeled pesticides in the usual manner for this pesticide and commodity, (c) determine the quantity of labeled pesticide removed from the sample by the extraction procedure in question, (d) determine the quantity of labeled pesticide remaining in the tissue or unextracted by the procedure followed and, (e) identify both the extracted and the unextracted radioactivity.

Using the proper isolative techniques and counting equipment, an investigator is able to determine specifically whether the labeled pesticide is removed from the biological tissues in question. If all of the detectable, labeled pesticide is not extracted, a percentage efficiency may be determined, and other, more rigorous measures can be employed in order to remove the labeled material more quantitatively.

Although literally hundreds of investigators have reported extraction procedures for pesticide residues from biological materials and subsequent per cent recoveries, there have been only very few quantitative studies such as reported by Klein et al. (26) in which extraction efficiencies were verified using a labeled pesticide. Three different extraction procedures were studied on spinach that had been experimentally sprayed in pots with radioactive (^{14}C) methoxychlor. A few other workers in the residue field such as Webster and McKinley (28), Redemann and Meikle (322), and Wheeler et al. (36) have recognized the potential value of employing labeled pesticides in evaluating extraction procedures. Wheeler and Frear (35) have stated that if they had not resorted to the use of labeled compounds, they would not have discovered that multiple solvent systems were required to extract all of the dieldrin present in the interior parts of the plants analyzed.

Undoubtedly, the entire subject of extraction efficiency studies has been sadly neglected by residue chemists. The almost total lack of the use of such an indispensable tracing mechanism as a labeled pesticide in this vital area of pesticide residue analysis is glaringly apparent. Reluctance to use radioisotopes in solvent extraction and cleanup studies is probably attributable to the dearth of necessary equipment and lack of training of the residue analysts in their correct use. It is suggested that considerably more research along these lines is urgently needed.

D. Total Halides and Phosphorus

Organochlorine and organophosphate pesticides may be quantitated by measuring the total organic chlorine or phosphorus in the pesticide molecule. These techniques generally do not distinguish between two or more pesticides of the same classes and are usually considered to be nonspecific. These procedures have been and continue to be very useful to regulatory laboratories for screening purposes. An excellent review of total halide analysis has been made by Dunn et al. (323).

Both classes of pesticides may be determined by combustion employing various means. One of the most common and useful techniques for chlorinated hydrocarbons is through the application of the microcoulometric instrument previously described (8, 218, 219) that employs a combustion furnace to decompose the effluent from a gas-chromatographic separation. The hydrogen halide is absorbed and measured continuously by a coulometric silver ion generator having a high degree of sensitivity and accuracy. Sensitivity of this instrument is considerably better than that of other electrometric methods of halid measurement. Gunther and Barkley (324) have reported the conversion of a microcoulometric gas chromatograph to a convenient and rapid total chlorine unit. A conventional MCGC was "short-circuited" by bypassing the use of a column entirely and using the injection block to release volatile components. The combustion unit was used to convert covalently bonded chlorine ion and the total chlorine measured microcoulometrically, regardless of the time required.

A rapid combustion technique for measuring total chlorine residues has been reported by Lisk (325) and St. John and Lisk (326) by using a modified and improved Schöniger method. Following extraction and cleanup of the chlorinated pesticide residues, the solvent is evaporated and the residue burned completely in an atmosphere of oxygen. A total phosphorus technique, using Schöniger flask combustion, for organophosphorus pesticides on treated crops, has been introduced by Blinn (327). A simple cleanup procedure, prior to combustion, involves shaking the chloroform extract with activated carbon, ATTACLAY, HYFLO SUPER CEL, and anhydrous sodium sulfate.

Total phosphorus of several organophosphate pesticide residues has been determined colorimetrically by Anderson (328) and Loeffler (329) using a modification of the method described by Martin and Doty (330). Naturally occurring phosphorus compounds are removed by chromatography on an activated carbon column. Anliker and Menzer (331) report the use of a total phosphorus method (phosphomolybdenum blue) for the determination of phosphamidon residues in various vegetable extracts. A digestion reagent by Saliman (332) has been used to determine microgram quantities of phosphorus in pesticides such as parathion and mevinphos. The molybdenum blue procedure is sensitive to 0.2 μg and is applicable to organic phosphorus in a wide variety of solvents.

Total bromide procedures have been used to determine residues of bromo-organic nematocides. Castro and Schmitt (333) have used neutron activation analysis to determine quantitatively and simultaneously the Na, K, Mn, Cl, and Br content of raw oranges from the same set of gamma-ray spectra. Guinn and Potter (334) have also used neutron activation for the determination of total bromine residues in a variety of crop samples and plant extracts. Bromine content in carrot, asparagus, and lemon peel extracts were determined down to 0.1 ppm by activation analysis, which tended to be more sensitive and accurate than results obtained by chemical analyses.

E. Enzymatic

Enzymatic methods for pesticide residue analysis have been primarily centered around modifications of the basic cholinesterase inhibition procedure. There are four major modifications for assaying cholinesterase activity and the effect of inhibitors: potentiometric, titrimetric, colorimetric, and manometric. The most common procedures used in the past on organophosphate residues have been the potentiometric and the colorimetric determination of acetylcholine. The primary advantage of the cholinesterase inhibition procedures is their simplicity, but their nonspecificity is a major disadvantage. Although these procedures in the past were in general usage for certain organophosphate residues, their usefulness is now much more limited with the advent of TLC, GLC and other more specific and sensitive techniques. Since Archer(335) has covered enzymatic methods very comprehensively, no attempt is made here to review this area of residue methodology again.

The most recent innovation involving acetylcholine esterase in the pesticide residue field is the adaptation of an automated system by Winter and Ferrari(336). The obvious advantages of such a technique are that many more samples can be processed in a given time and residue data can be permanently recorded for later reference. Further adaptations of this basic automated analyses of anticholinesterase organophosphates are discussed in section G.

F. Screening Methods

Screening methods for multiple pesticide residues are usually conducted by state and federal enforcement agencies and the pesticide and food manufacturer, all of whom have similar analytical problems facing them. The selection of suitable screening methods depends, to a large extent, upon the pesticides and commodities to be analyzed as well as the equipment and personnel required. The methods ideally should be simple, rapid, accurate, reproducible, and relatively inexpensive to perform. Since the chlorinated hydrocarbons and organophosphate pesticides are two of the most important classes of pesticides, a parallel screening procedure has been presented by Phillips et al.(337), which consists of determining organic chlorine, organic phosphate, and bioassay on aliquots of the same crop extract. Phillips(338) has outlined pesticide residue screening procedures into three primary techniques: (a) bioassay, (b) organic chlorine, and (c) acetylcholinesterase inhibition. Bioassay screening techniques have usually utilized insects as the test organism. The insect bioassay technique has been adequately covered by Sun(339) and is not elaborated further in this review. Screening procedures for organic chlorine can utilize the modification of the Schöniger combustion method of St. John and Lisk(326) or a rapid total chloride analyzer such as the microcoulometric gas chromatograph described by Gunther and Barkley (324). Parallel screening of pesticides that are adequate cholinesterase inhibitors can be accomplished by the delta pH procedure described by Patchett and Batchelder(340), which employs peracetic acid oxidation of extracts.

Until the relatively recent introduction of gas chromatography and thin-layer chromatography, paper chromatography was rather widely used as a screening procedure by regulatory agencies. Coffin(284) has reviewed the advantages and disadvantages of this technique. A chromogenic spray reagent has been described by Watts(341) for the multiple detection of organophosphates by paper and thin-layer chromatography. As stated previously, thin-layer chromatography is rapidly replacing paper chromatography as a simple, rapid, sensitive, and versatile screening procedure for organic chlorinated and thiophosphate pesticide residues. Kovacs (275–278) has reported on the use of thin-layer chromatography as a rapid, semiquantitative screening tool and a confirmatory method when used in conjunction with gas chromatography of organochlorine and organothiophosphates. Moats(342) has reported a TLC procedure that could be adapted to a useful screening procedure for chlorinated hydrocarbons in dairy products. A screening procedure for organothiophosphate pesticide residues by microcoulometric gas chromatography has been successfully employed by Nelson(343, 344). Three cleanup steps have been proposed by Samuel(345) to screen chlorinated and thiophosphate organic insecticides from fruits, vegetables, green feeds, dairy products, egg yolk, butter, and animal products.

It appears that future pesticide residue screening techniques will tend toward more automation because of the critical shortage of trained analysts and the fact that screening techniques usually favor automatic schemes of analysis.

Out of the 400 major agricultural pesticides available today, approximately half of them contain halogens (mostly chlorine and a few with bromine) and about one-third are cholinesterase-inhibiting organophosphates or carbamates. Therefore, techniques that automate organohalogens or cholinesterase inhibitors would be extremely useful for screening purposes. As pointed out by Gunther and Ott(346), it is clear that automated screening procedures for the two halogens, for phosphorus, sulfur, and nitrogen, and for cholinesterase activity before and after oxidation, for example, would be immensely helpful in providing valuable and rather complete residue information. Automated screening will be required for the rapid detection of chlorine, phosphorus, sulfur, nitrogen, and unoxidized-oxidized ChE activity at levels selected to approximate the tolerance value of the most toxic compound being sought. This type of automated screening could separate the samples into those below tolerance and those possibly above legal tolerance categories. The below tolerance values would then be disregarded, while the latter would be examined in greater detail and by more specific procedures for the verification of the above-tolerance residue.

G. Automation

As pointed out in the two previous sections, automation in pesticide residue analyses, and particularly for screening purposes, has developed rapidly in the past several years. The basic reasons for automation in this field have been stressed by Gunther and Ott(346) and have been briefly covered in the previous section on screening methods.

The first automatic wet chemical analysis, as applied to pesticide residue analyses, was reported by Winter and Ferrari(336) and was derived from a clinical application of an acetylcholinesterase activity determination. Ott and Gunther(347) have introduced a modification of an automated system for anticholinesterase organophosphates to include an automated elution–filtration technique for the analysis of these compounds after thin-layer chromatography. This procedure offers a degree of specificity previously unobtainable by automated anticholinesterase methods and is rapid when combined with a multiple spotter apparatus for TLC. The adaptation of an AUTO ANALYZER for the determination of orthophosphates by Ott and Gunther(348) will apply to the combusted product from the Schöniger oxygen flask, combustion tube, Parr bomb, or wet oxidation techniques. A totally automated analytical procedure for determining the fungistat biphenyl in citrus fruit rind has been developed by Gunther and Ott(214). Small pieces of rind were automatically homogenized in water and steam distilled to liberate the biphenyl. The biphenyl was trapped in cyclohexane solution which was exhaustively extracted to remove interfering steam volatiles. The remaining biphenyl was read in a continuous-flow recording spectrophotometer. Time required from the introduction of the first sample to "read out" of biphenyl present was 9 min between samples, after allowing about 15 min for the first sample to pass through the system. Ott and Gunther(349) have also introduced an automated analysis of organophosphorus insecticides by wet-digestion oxidation and colorimetric determination of the derived orthophosphate.

H. Oscillographic Polarography

Polarographic theory and instrumentation has been thoroughly discussed in recent reviews and books. This technique represents a method of pesticide residue analysis based on the electrolysis of a small amount of solution, and the voltage necessary to achieve electrolysis indicates the nature of the reacting substance. The magnitude of the current required is a function of the concentration of the pesticide undergoing electrolysis. The degree of cleanup required varies, however, with the pesticide and/or crop to be polarographed. The residue extracts to be polarographed often must be carefully freed from contaminants, or the responding contaminants must be accurately evaluated and properly compensated. When polarography can be applied to pesticide residue determinations, it is highly reproducible, fairly sensitive, and can provide qualitative and quantitative information simultaneously. Sensitivities on standard solutions of less than 0.1 μg/ml have been reported. When cleanup is unnecessary, two analyses can be conducted in less than $2\frac{1}{2}$ hr; in cases where cleanup is required, two analyses can be completed in less than 4 hr, according to Gajan(350). An application of oscillographic polarography to pesticide residues was reported by Gajan(351). This research involved the detection of demeton, disulfoton, and phorate residues from field-sprayed kale by both polarography and paper chromatography. An excellent review of early applications of polarography to insecticides and fungicides was presented by Martens and Nangniot(352). Further work with polarography by Gajan(353) resulted in a rapid screening procedure for parathion on a number of fruit and vegetable extracts. Gajan(354) has indicated the chemical structure of the pesticide must be examined when developing a polarographic method for pesticide residues. One must look for the presence of oxidizable or reducible groups such as a nitro group, halogens, carbonyl groups, etc. If a pesticide does not contain one of these groups, perhaps a derivative can be made that will. Each pesticide that is to be polarographed requires individual attention with regard to the selection of solvents, electrolytes, sample preparation, etc.

Gajan(355) has reviewed the application of oscillographic polarography to the following classes of pesticides and their residues: chlorinated hydrocarbons, dithiocarbamates, nitro compounds, and organophosphates. Oscillopolarography was used for determination of DDT and its analogs in cleaned up crop extracts by Gajan and Link(356). Five carbamate insecticides were separated by Eberle and Gunther(357) using TLC and GLC, and were then quantitatively characterized using a combination of these techniques with fluorescence, infrared and ultraviolet spectrometry, and oscillographic polarography. In this operation, oscillopolarography was used only on the pure carbamate standard, but it was concluded that this technique, combined with thin-layer chromatography, has considerable promise as a rapid, sensitive, and specific residue method. Gajan et al. (358) have determined carbaryl residues on certain fruits and vegetables by polarography. Using a modified cleanup method, residues as small as 0.2 ppm were detected on extracts of broccoli, lettuce, potatoes, and apples by this technique. A polarographic method has been proposed by Engelhardt and McKinley(359) for the determination of bipyridylium herbicides in several vegetables. Ion-exchange-type elimination of interfering materials in the vegetable extracts was achieved according to the procedure by Calderbank(195) and the limit of residue detection was from 0.01 to 0.1 ppm. A method combining oscillopolarography and thin-layer chromatography was described by Hearth et al. (360) for the micro-determination of a miticide, oxythioquinox (MORESTAN), in citrus rind.

In the described method, TLC was used for isolation and cleanup of the miticide residues from the rind extracts. There was general agreement in residue values obtained by the proposed TLC oscillopolarography and a colorimetric method. Residues of 0.5 ppm were readily detectable but accurate quantitation was found to be difficult below 1.0 ppm residue values using the TLC oscillopolarographic method.

I. Neutron Activation

Neutron activation analysis is an analytical technique involving the bombardment of samples and standards with neutrons in order to convert stable elements to unstable radioactive isotopes. The radioactive elements can then be identified and assayed quantitatively. Quantitative measurements are accomplished by counting and sorting the characteristic radiations emitted by the radioactive isotopes formed during bombardment. These data can then be combined electronically to yield a characteristic spectrum. When two or more radioactive nuclides emit radiations of similar energy levels, the half-life of each of the nuclides is used to differentiate between elements. Interferences can be reduced or eliminated entirely by chemically isolating and purifying the required radioactive element after the radiation. An excellent review of neutron activation analyses was made by Schmitt(361), in which theory, procedures, applications to pesticide residues, instrumentation, and neutron sources are discussed. Guinn and Schmitt(362) have reviewed the determination of pesticide residues by neutron activation analyses and discussed applications of this technique to determine bromine residues in crops, chlorinated hydrocarbons in various foods, and mercury content in wheat. Schmitt and Zweig(363) achieved chloride sensitivity of 10 ppb using neutron activation analysis for the determination of total organic chloride content in milk products. Total bromine residues in several vegetables were analyzed by neutron activation as reported by Guinn and Potter(364). Bromine levels as low as 10 ppb were detectable in the crop extracts and 1 ppm in the crop samples. Szkolnik et al.(365) compared a colorimetric method and neutron activation for the analyses of mercury residues on apple fruit and the latter technique was found to be sensitive to 0.1 ppb of mercury.

J. Miscellaneous

There are other electrometric methods that have in the past been applied to pesticide residue analysis such as: amperometric, potentiometric, conductimetric, and high frequency (oscillometry). The introduction of more specific, sensitive, and versatile instrumentation has resulted in a general lack of interest in application of the abovementioned electrometric methods during the past decade. The use of ion exchange, electrophoresis, and counter-current distribution have been mentioned as having occasional useful applications in residue methodology.

5-5 CONCLUSIONS

In summary, there is at present an amazing diversity of instrumentation and techniques available to the analyst for various pesticide residue evaluation programs. The residue chemist, of course, cannot be a specialist in use of each of these wide array of tools and techniques, but should be cognizant of their relative capabilities and limitations. Gunther (266) has listed most of the major residue analytical instruments or techniques, and their residue applications. The arrangement in approximate order of frequency of use might be subject to question since this would depend on one's personal experience and general knowledge of what instruments and techniques are being used in all other pesticide residue laboratories. It should be emphasized to any prospective residue chemist, however, that there is unfortunately no one instrument or technique that can, by itself, be used to prove the identity of the measured chemical. Two or more instruments and/or techniques will be required in parallel, and which ones to use will depend on the nature of the particular residue problem and the specific equipment available. In pesticide residue investigations, identifications and determinations must be based solely upon detailed qualitative and quantitative comparisons of the analytical behavior of the unknown sample with that of known standards.

REFERENCES

1. F. A. Gunther, in *Advances in Pest Control Research*, (R. L. Metcalf, ed.), Vol. V, Wiley (Interscience), New York, 1962, pp. 191–319.
2. F. A. Gunther, *J. Assoc. Offic. Agr. Chemists*, **44**, 620 (1961).
3. H. Fischbach, Division of Agricultural and Food Chemistry, 141st Meeting, American Chemical Society, Washington, D.C., March 1962.
4. A. S. Crafts and C. L. Foy, *Residue Rev.*, **1**, 112 (1962).
5. W. Ebeling, *Residue Rev.*, **3**, 35 (1963).
6. F. A. Gunther, in *Chemical and Biological Hazards in Foods* (J. C. Ayres, A. A. Kraft, H. E. Snyder, and H. W. Walker, eds.), Chap. 7, Iowa State Univ., Ames, 1963, pp. 77–88.
7. D. J. Lisk, *N.Y. State J. Med.*, **65**, 1026 (1965).
8. D. M. Coulson and L. A. Cavanagh, *Anal. Chem.*, **32**, 1245 (1960).
9. J. E. Lovelock and S. R. Lipsky, *J. Am. Chem. Soc.*, **82**, 431 (1960).
10. S. Williams and J. W. Cook, *Anal. Chem.*, **39**, 142R (1967).
11. M. J. Garber, in *Analytical Methods for Pesticides, Plant Growth Regulators and Food Additives* (G. Zweig, ed.), Vol. 1, Academic, New York, 1963, pp. 491–530.
12. F. W. Poos, T. N. Dobbins, and R. H. Carter, *Bull. No. E-793*, U.S. Department of Agriculture, Bureau of Entomology and Plant Quarantine, 1950.
13. J. E. Fahey, *Digest of Literature, 1925 to 1952*, U.S. Department of Agriculture, Agricultural Research Service, 1954.
14. F. A. Gunther and R. C. Blinn, in *Analysis of Insecticides and Acaricides*, Wiley (Interscience), New York, 1955.
15. C. H. Van Middelem, J. W. Wilson, and W. D. Hanson, *J. Econ. Entom.*, **49**, 612 (1956).
16. E. W. Huddleston, K. H. Thompson, G. G. Gyrisco, D. J. Lisk, T. W. Kerr, and C. E. Olney, *J. Econ. Entom.*, **53**, 1078 (1960).
17. F. A. Gunther, in *Instrumental Methods for the Analysis of Food Additives* (W. H. Butz and H. J. Noebels, eds.), Chap. IV, Wiley (Interscience), New York, 1961, pp. 47–74.
18. L. Lykken, *Residue Rev.*, **3**, 19 (1963).
19. W. W. Thornburg, in *Analytical Methods for Pesticides, Plant Growth Regulators and Food Additives* (G. Zweig, ed.), Vol. 1, Academic, New York, 1963, pp. 87–108.
20. J. A. Burke, P. A. Mills, and D. C. Bostwick, *J. Assoc. Offic. Anal. Chemists*, **49**, 999 (1966).
21. H. P. Burchfield and D. E. Johnson, *Guide to the Analysis of Pesticide Residues*, Vol. 1, U.S. Government Printing Office, Washington, D.C., 1965.
22. W. W. Thornburg, *Residue Rev.*, **14**, 1 (1966).
23. W. W. Thornburg, *J. Assoc. Offic. Agr. Chemists*, **48**, 1023 (1965).
24. J. M. Bann, Division of Agricultural and Food Chemistry, 131st Meeting, American Chemical Society, Miami, April 1957.
25. A. K. Klein, *J. Assoc. Offic. Agr. Chemists*, **41**, 551 (1958).
26. A. K. Klein, E. P. Laug, and J. D. Sheehan, *J. Assoc. Offic. Agr. Chemists*, **42**, 539 (1959).
27. R. E. J. Moddes and J. W. Cook, *J. Assoc. Offic. Agr. Chemists*, **42**, 208 (1959).
28. C. Webster and W. P. McKinley, in *Instrumental Methods for the Analysis of Food Additives* (W. H. Butz and H. J. Noebels, eds.), Chap. 5, Wiley (Interscience), New York, 1961, pp. 71–83.

29. C. H. Van Middelem, R. E. Waites, and J. W. Wilson, *J. Agr. Food Chem.*, **11**, 56 (1963).
30. L. J. Hardin and C. T. Sarten, *J. Assoc. Offic. Agr. Chemists*, **45**, 988 (1962).
31. L. Y. Johnson, *J. Assoc. Offic. Agr. Chemists*, **45**, 363 (1962).
32. P. F. Bertuzzi, L. Kamps, C. Miles, and J. A. Burke, *J. Assoc. Offic. Anal. Chemists*, **50**, 623 (1967).
33. J. A. Burke and M. Porter, *J. Assoc. Offic. Anal. Chemists*, **49**, 1157 (1966).
34. P. A. Mills, J. H. Onley, and R. A. Gaither, *J. Assoc. Offic. Agr. Chemists*, **46**, 186 (1963).
35. W. B. Wheeler and D. E. H. Frear, *Residue Rev.*, **16**, 86 (1966).
36. W. B. Wheeler, D. E. H. Frear, R. O. Mumma, R. H. Hamilton, and R. C. Cotner, *J. Agr. Food Chem.*, **15**, 227 (1967).
37. R. O. Mumma, W. B. Wheeler, D. E. H. Frear, and R. H. Hamilton, *Science*, **152**, 530 (1966).
38. R. R. Schnorbus and W. F. Phillips, *J. Agr. Food Chem.*, **15**, 661 (1967).
39. Shell Chemical Company, *Manual of Methods*, Agricultural Chemical Division, New York, 1964.
40. H. C. Barry, J. G. Hundley, and L. Y. Johnson, *Pesticide Analytical Manual*, Vol. 1, Food and Drug Administration, U.S. Department of Health, Education, and Welfare, Washington, D.C., 1963.
41. B. E. Langlois, A. R. Stemp, and B. J. Liska, *J. Agr. Food Chem.* **12**, 243 (1964).
42. E. S. Goodwin, R. Goulden, and J. G. Reynolds, *Analyst*, **86**, 697 (1961).
43. R. A. Baetz, *J. Assoc. Offic. Agr. Chemists*, **47**, 322 (1964).
44. R. A. Albert, *J. Assoc. Offic. Agr. Chemists*, **47**, 659 (1964).
45. R. A. Moffitt and J. H. Nelson, *Cereal Science Today*, **8**, 72 (1963).
46. R. A. Moffitt, in *Analytical Methods for Pesticides, Plant Growth Regulators and Food Additives* (G. Zweig, ed.), Vol. 1, Academic, New York, 1963, pp. 545–570.
47. P. A. Mills, *J. Assoc. Offic. Agr. Chemists*, **44**, 171 (1961).
48. L. Y. Johnson, *J. Assoc. Office. Agr. Chemists*, **48**, 668 (1965).
49. M. Beroza and M. C. Bowman, *J. Assoc. Offic. Anal. Chemists*, **49**, 1007 (1966).
50. M. J. de Faubert Maunder, H. Egan, E. W. Godly, E. W. Hammond, J. Roburn, and J. Thompson, *Analyst*, **89**, 168 (1964).
51. K. A. McCully and W. P. McKinley, *J. Assoc. Offic. Agr. Chemists*, **47**, 652 (1964).
52. K. A. McCully, D. C. Villeneuve, and W. P. McKinley, *J. Assoc. Offic. Anal. Chemists*, **49**, 966 (1966).
53. T. Olson, *Chemagro Corp., Report No. 10,081*, 1962.
54. P. A. Mills, *J. Assoc. Offic. Agr. Chemists*, **42**, 734 (1959).
55. J. H. Onley and P. F. Bertuzzi, *J. Assoc. Offic. Anal. Chemists*, **49**, 370 (1966).
56. J. H. Onley and P. A. Mills, *J. Assoc. Offic. Agr. Chemists*, **45**, 983 (1962).
57. A. R. Stemp, B. J. Liska, B. E. Langlois, and W. J. Stadelman, *Poultry Sci.*, **43**, 273 (1964).
58. J. G. Cummings, K. T. Zee, V. Turner, and F. Quinn, *J. Assoc. Offic. Anal. Chemists*, **49**, 354 (1966).
59. A. D. Sawyer, *J. Assoc. Offic. Anal. Chemists*, **49**, 643 (1966).
60. H. Beckman, A. Bevenue, K. Carroll, and F. Erro, *J. Assoc. Offic. Anal. Chemists*, **49**, 996 (1966).
61. D. E. Johnson, J. D. Millar, and H. P. Burchfield, *Life Sci.*, **4**, 959 (1963).
62. L. Kahn and C. H. Wayman, *Anal. Chem.*, **36**, 1340 (1964).
63. S. D. Faust and I. H. Suffet, *Residue Rev.*, **15**, 44 (1966).

64. H. P. Burchfield, in *Lectures on Gas Chromatography 1964* (L. R. Mattick and H. A. Szymanski, eds.), Plenum, New York, 1965, pp. 91–107.
65. E. A. Robertson and R. M. Tyo, *J. Assoc. Offic. Anal. Chemists*, **49**, 683 (1966).
66. R. L. Schutzmann, W. F. Barthel, and J. A. Warrington, U.S. Department of Agriculture, Agricultural Research Service 81–12, May 1966.
67. E. P. Lichtenstein, in *Research in Pesticides* (C. O. Chichester, ed.), Academic, New York, 1965, pp. 199–203.
68. E. P. Lichtenstein, private communication, 1966.
69. E. P. Lichtenstein, G. R. Myrdal, and K. R. Schulz, *J. Econ. Entom.*, **57**, 133 (1964).
70. W. F. Barthel and L. H. Dawsey, U.S. Department of Agriculture, *Agricultural Research Service Memo*, June 1965.
71. C. R. Harris, W. W. Sans, and J. R. W. Miles, *J. Agr. Food Chem.*, **14**, 398 (1966).
72. J. I. Teasley and W. S. Cox, *J. Agr. Food Chem.*, **14**, 519 (1966).
73. W. P. McKinley, D. E. Coffin, and K. A. McCully, *J. Assoc. Offic. Agr. Chemists*, **47**, 863 (1964).
74. L. R. Jones and J. A. Riddick, *Anal. Chem.*, **24**, 569 (1952).
75. W. R. Erwin, D. Schiller, and W. M. Hoskins, *J. Agr. Food Chem.*, **3**, 676 (1955).
76. W. M. Hoskins, W. R. Erwin, R. Miskus, W. W. Thornburg, and L. N. Werum, *J. Agr. Food Chem.*, **6**, 914 (1958).
77. H. P. Burchfield and E. E. Storrs, *Contrib. Boyce Thompson Inst.*, **17**, 333 (1953).
78. M. Eidelman, *J. Assoc. Offic. Agr. Chemists*, **45**, 672 (1962).
79. M. Eidelman, *J. Assoc. Offic. Agr. Chemists*, **46**, 182 (1963).
80. M. Beroza and M. C. Bowman, *Anal. Chem.*, **37**, 291 (1965).
81. M. Beroza and M. C. Bowman, *J. Assoc. Offic. Agr. Chemists*, **48**, 358 (1965).
82. T. P. King and L. C. Craig, in *Methods of Biochemical Analysis* (D. Glick, ed.), Vol. 10, Wiley (Interscience), New York, 1962, pp. 201–228.
83. M. C. Bowman and M. Beroza, *J. Assoc. Offic. Agr. Chemists*, **48**, 943 (1965).
84. C. W. Stanley and A. P. Post, Division of Agricultural and Food Chemistry, 152nd Meeting, American Chemical Society, New York, September 1966.
85. H. V. Morley, *Residue Rev.*, **16**, 1 (1966).
86. W. A. Moats, *J. Assoc. Offic. Agr. Chemists*, **47**, 587 (1964).
87. W. A. Moats, *J. Assoc. Offic. Agr. Chemists*, **46**, 172 (1963).
88. W. A. Moats, *J. Assoc. Offic. Agr. Chemists*, **45**, 355 (1962).
89. A. R. Stemp and B. J. Liska, *J. Dairy Sci.*, **48**, 985 (1965).
90. L. Giuffrida, D. C. Bostwick, and N. F. Ives, *J. Assoc. Offic. Anal. Chem.*, **49**, 634 (1966).
91. W. A. Moats and A. W. Kotula, *J. Assoc. Offic. Anal. Chemists*, **49**, 973 (1966).
92. J. A. Burke and B. Malone, *J. Assoc. Offic. Anal. Chemists*, **49**, 1003 (1966).
93. B. J. Wood, *J. Assoc. Offic. Anal. Chemists*, **49**, 472 (1966).
94. M. L. Schafer, K. A. Busch, and J. E. Campbell, *J. Dairy Sci.*, **46**, 1025 (1963).
95. A. J. Graupner and C. L. Dunn, *J. Agr. Food Chem.*, **8**, 286 (1960).
96. R. L. Stanley and H. E. LeFavoure, *J. Assoc. Offic. Agr. Chem.*, **48**, 666 (1965).
97. K. C. Walker and M. Beroza, *J. Assoc. Offic. Agr. Chemists*, **46**, 250 (1963).
98. H. V. Morley and M. Chiba, *J. Assoc. Offic. Agr. Chemists*, **47**, 306 (1964).
99. H. A. Moye, unpublished work, 1966.
100. M. J. Matherne and W. H. Bathalter, *J. Assoc. Offic. Anal. Chemists*, **49**, 1012 (1966).
101. J. A. Burke, private communication, 1966.
102. L. A. Rosenberg and R. W. Storherr, 80th Meeting Association of Official Analytical Chemists, Washington, D.C., October 1966.

103. R. W. Storherr, E. J. Murray, I. W. Klein, and L. A. Rosenberg, *J. Assoc. Offic. Anal. Chemists*, **50**, 605 (1967).

104. D. E. Ott and F. A. Gunther, *J. Agr. Food Chem.*, **12**, 239 (1964).

105. F. A. Gunther, R. C. Blinn, and D. E. Ott, *Bull. Environ. Contam. and Toxicol.*, **1**, 237 (1966).

106. R. Mestres and F. Barthes, *Bull. Environ. Contam. and Toxicol.*, **1**, 245 (1966).

107. E. D. Chilwell and G. S. Hartley, *Analyst*, **86**, 148 (1961).

108. D. E. Coffin and G. Savary, *J. Assoc. Offic. Agr. Chemists*, **47**, 875 (1964).

109. R. R. Watts and R. W. Storherr, *J. Assoc. Offic. Agr. Chemists*, **48**, 1158 (1965).

110. R. C. Nelson, *J. Assoc. Offic. Anal. Chemists*, **49**, 763 (1966).

111. E. W. Laws and D. J. Webley, *Analyst*, **86**, 249 (1961).

112. S. C. Lau, *J. Agr. Food Chem.*, **14**, 145 (1966).

113. D. A. George, K. C. Walker, R. T. Murphy, and P. A. Giang, *J. Agr. Food Chem.*, **14**, 371 (1966).

114. D. B. Katague and C. A. Anderson, *J. Agr. Food Chem.*, **14**, 505 (1966).

115. A. J. Gehrt, *J. Assoc. Offic. Agr. Chemists*, **48**, 296 (1965).

116. M. C. Bowman and M. Beroza, *J. Assoc. Offic. Agr. Chemists*, **48**, 922 (1965).

117. L. J. Everett, C. A. Anderson, and D. MacDougall, *J. Agr. Food Chem.*, **14**, 47 (1966).

118. J. M. Adams and C. A. Anderson, *J. Agr. Food Chem.*, **14**, 53 (1966).

119. H. V. Claborn and M. C. Ivey, *J. Agr. Food Chem.*, **13**, 353 (1965).

120. C. T. Smith, F. R. Shaw, D. L. Anderson, R. A. Callahan, and W. H. Ziener, *J. Econ. Entom.*, **58**, 1160 (1965).

121. J. I. Teasley and W. S. Cox, *J. Am. Water Works Assoc.*, **55**, 1093 (1963).

122. H. A. Moye and J. D. Winefordner, *J. Agr. Food Chem.*, **13**, 533 (1965).

123. Y. Sumiki and A. Matsuyama, *Bull. Agr. Chem. Soc. Japan*, **21**, 329 (1957).

124. A. R. El-Refai and L. Giuffrida, *J. Assoc. Offic. Agr. Chemists*, **48**, 374 (1965).

125. S. L. Warnick and A. R. Gaufin, *J. Am. Water Works Assoc.*, **57**, 1023 (1965).

126. R. C. Blinn and N. R. Pasarela, *J. Agr. Food Chem.*, **14**, 152 (1966).

127. H. P. Burchfield and P. H. Schuldt, *Contrib. Boyce Thompson Inst.*, **19**, 77 (1957).

128. E. P. Lichtenstein and K. R. Schulz, *J. Econ. Entom.*, **57**, 618 (1964).

129. E. P. Lichtenstein, *J. Econ. Entom.*, **59**, 985 (1966).

130. L. W. Getzin and I. Rosefield, *J. Econ. Entom.*, **59**, 512 (1966).

131. D. R. Coahran, *Bull. Environ. Contam. and Toxicol.*, **1**, 208 (1966).

132. R. E. J. Moddes and J. W. Cook, *J. Assoc. Offic. Agr. Chemists*, **42**, 208 (1959).

133. M. E. Getz, *J. Assoc. Offic. Agr. Chemists*, **45**, 393 (1962).

134. W. Thornburg, private communication, 1966.

135. R. W. Storherr and R. R. Watts, *J. Assoc. Offic. Agr. Chemists*, **48**, 1154 (1965).

136. R. W. Storherr, M. E. Getz, R. R. Watts, S. J. Friedman, F. Erwin, L. Giuffrida, and F. Ives, *J. Assoc. Offic. Agr. Chemists*, **47**, 1087 (1964).

137. R. R. Watts and R. W. Storherr, *J. Assoc. Offic. Anal. Chemists*, **50**, 581 (1967).

138. D. P. Johnson, *J. Assoc. Offic. Agr. Chemists*, **47**, 283 (1964).

139. R. J. Gajan, W. R. Benson, and J. M. Finocchiaro, *J. Assoc. Offic. Agr. Chemists*, **48**, 958 (1965).

140. W. R. Benson and J. M. Finocchiaro, *J. Assoc. Offic. Agr. Chemists*, **48**, 676 (1965).

141. J. W. Ralls and A. Cortes, *J. Gas Chromatog.*, **2**, 132 (1964).

142. W. H. Gutenmann and D. J. Lisk, *J. Agr. Food Chem.*, **13**, 48 (1965).

143. C. H. Van Middelem, T. L. Norwood, and R. E. Waites, *J. Gas Chromatog.*, **3**, 310 (1965).

144. D. P. Johnson and H. A. Stansbury, Jr., *J. Assoc. Offic. Anal. Chemists*, **49**, 399 (1966).

145. D. P. Johnson and H. A. Stansbury, Jr., *J. Assoc. Offic. Anal. Chemists*, **49**, 403 (1966).

146. G. G. Gyrisco, D. J. Lisk, S. N. Fertig, E. W. Huddleston, F. H. Fox, R. F. Holland, and G. W. Trimberger, *J. Agr. Food Chem.*, **8**, 409 (1960).

147. D. P. Johnson, F. E. Critchfield, and B. W. Arthur, *J. Agr. Food Chem.*, **11**, 77 (1963).

148. W. E. Whitehurst, E. T. Bishop, F. E. Critchfield, G. G. Gyrisco, E. W. Huddleston, H. Arnold, and D. J. Lisk, *J. Agr. Food Chem.*, **11**, 167 (1963).

149. H. V. Claborn, R. H. Roberts, H. D. Mann, M. C. Bowman, M. C. Ivey, C. P. Weidenbach, and R. D. Radeleff, *J. Agr. Food Chem.*, **11**, 74 (1963).

150. I. Nir, E. Weisenberg, A. Hadani, and M. Egyed, *Poultry Sci.*, **45**, 720 (1966).

151. R. C. Brian, in *The Physiology and Biochemistry of Herbicides* (L. J. Audus, ed.), Chap. 1, Academic, London, 1964, pp. 1–37.

152. G. Zweig, in *Analytical Methods for Pesticides, Plant Growth Regulators and Food Additives*, Vol. IV, Academic, New York, 1964.

153. D. L. Klingman, C. H. Gordon, G. Yip, and H. P. Burchfield, *Weeds*, **14**, 164 (1966).

154. J. E. Coakley, J. E. Campbell, and E. F. McFarren, *J. Agr. Food Chem.*, **12**, 262 (1964).

155. G. Yip, *J. Assoc. Offic. Agr. Chemists*, **45**, 367 (1962).

156. L. C. Erickson and H. Z. Hield, *J. Agr. Food Chem.*, **10**, 204 (1962).

157. A. Bevenue, G. Zweig, and N. L. Nash, *J. Assoc. Offic. Agr. Chemists*, **46**, 881 (1963).

158. W. R. Meagher, *J. Agr. Food Chem.*, **14**, 375 (1966).

159. C. A. Bache, D. J. Lisk, and M. A. Loos, *J. Assoc. Offic. Agr. Chemists*, **47**, 348 (1964).

160. C. A. Bache, *J. Assoc. Offic. Agr. Chemists*, **47**, 355 (1964).

161. C. A. Bache, W. H. Gutenmann, and D. J. Lisk, *J. Agr. Food Chem.*, **12**, 185 (1964).

162. M. Smith, H. Suzuki, and M. Malina, *J. Assoc. Offic. Agr. Chemists*, **48**, 1164 (1965).

163. G. Yip, *J. Assoc. Offic. Agr. Chemists*, **47**, 343 (1964).

164. C. W. Stanley, *J. Agr. Food Chem.*, **14**, 321 (1966).

165. A. Bevenue, G. Zweig, and N. L. Nash, *J. Assoc. Offic. Agr. Chemists*, **45**, 990 (1962).

166. R. D. Hagin and D. L. Linscott, *J. Agr. Food Chem.*, **13**, 123 (1965).

167. J. J. Kirkland and H. L. Pease, *J. Agr. Food Chem.*, **12**, 468 (1964).

168. W. H. Gutenmann and D. J. Lisk, *J. Agr. Food Chem.*, **11**, 304 (1963).

169. W. H. Gutenmann, D. D. Hardee, R. F. Holland, and D. J. Lisk, *J. Dairy Sci.*, **46**, 991 (1963).

170. W. H. Gutenmann, D. D. Hardee, R. F. Holland, and D. J. Lisk, *J. Dairy Sci.*, **46**, 1287 (1963).

171. D. J. Lisk, W. H. Gutenmann, C. A. Bache, R. G. Warner, and D. G. Wagner, *J. Dairy Sci.*, **46**, 1435 (1963).

172. D. E. Clark, J. E. Young, R. L. Younger, L. M. Hunt, and J. K. McLaran, *J. Agr. Food Chem.*, **12**, 43 (1964).

173. L. E. St. John, D. G. Wagner, and D. J. Lisk, *J. Dairy Sci.*, **47**, 1267 (1964).

174. G. Yip and R. E. Ney, Jr., *Weeds*, **14**, 167 (1966).

175. W. H. Gutenmann and D. J. Lisk, *J. Am. Water Works Assoc.*, **56**, 189 (1964).

176. J. D. Pope, W. S. Cox, and A. Grzenda, *Advan. in Chem. Ser.*, **60**, 200 (1966).

177. D. W. Woodham, G. F. Gardner, and W. F. Barthel, United States Department of Agriculture, Agricultural Research Service, 81–17, November, 1967.

178. J. E. Coakley, J. E. Campbell, and E. F. McFarren, *J. Agr. Food Chem.*, **12**, 262 (1964).

179. W. H. Gutenmann and D. J. Lisk, *J. Assoc. Offic. Agr. Chemists*, **47**, 353 (1964).
180. H. Schlenk and J. L. Gellerman, *Anal. Chem.*, **32**, 1412 (1960).
181. M. Rogozinski, *J. Gas Chromatog.*, **2**, 136 (1964).
182. H. P. Burchfield and E. E. Storrs, *Biochemical Applications of Gas Chromatography*, Academic, New York, 1962.
183. R. P. Marquardt, H. P. Burchfield, E. E. Storrs, and A. Bevenue, in *Analytical Methods for Pesticides, Plant Growth Regulators and Food Additives* (G. Zweig, ed.), Vol. IV, Academic, New York, 1964, pp. 95–116.
184. R. L. Dalton and H. L. Pease, *J. Assoc. Offic. Agr. Chemists*, **45**, 377 (1962).
185. W. H. Gutenmann and D. J. Lisk, *J. Agr. Food Chem.*, **12**, 46 (1964).
186. S. E. Katz, *J. Assoc. Offic. Anal. Chemists*, **49**, 452 (1966).
187. J. J. Ford, J. F. G. Clark, Jr., and R. T. Hall, *J. Agr. Food Chem.*, **14**, 307 (1966).
188. L. E. St. John, Jr., J. W. Ammering, D. G. Wagner, R. G. Warner, and D. J. Lisk, *J. Dairy Sci.*, **48**, 502 (1965).
189. H. G. Henkel and W. Ebing, *J. Gas Chromatog.*, **2**, 215 (1964).
190. A. M. Mattson, R. A. Kahrs, and J. Schneller, *J. Agr. Food Chem.*, **13**, 120 (1965).
191. R. H. Shimabukuro, R. E. Kadunce, and D. S. Frear, *J. Agr. Food Chem.*, **14**, 392 (1966).
192. H. C. Sikka and D. E. Davis, *Weeds*, **14**, 289 (1966).
193. W. H. Gutenmann and D. J. Lisk, *J. Assoc. Offic. Agr. Chemists*, **48**, 1173 (1965).
194. G. A. Boyack, A. J. Lemin, F. W. Staten, and A. Steinhards. *J. Agr. Food Chem.*, **14**, 312 (1966).
195. A. Calderbank, *Residue Rev.*, **12**, 14 (1966).
196. K. J. Meulemans and E. T. Upton, *J. Assoc. Offic. Anal. Chemists*, **49**, 976 (1966).
197. D. Racusen, *Arch. Biochem. Biophys.*, **74**, 106 (1958).
198. G. Zweig, *Analytical Methods for Pesticides, Plant Growth Regulators, and Food Additives*, Part I, Vol. III, Academic, New York, 1964, pp. 1–150.
199. W. K. Lowen, *J. Assoc. Offic. Agr. Chemists*, **36**, 484 (1953).
200. H. L. Pease, *J. Assoc. Offic. Agr. Chemists*, **40**, 1113 (1957).
201. T. E. Cullen, *Anal. Chem.*, **36**, 221 (1964).
202. J. W. Hylin, *Bull. Environ. Contam. and Toxicol.*, **1**, 76 (1966).
203. B. J. Gudzinowicz and V. J. Luciano, *J. Assoc. Offic. Anal. Chemists*, **49**, 1 (1966).
204. T. J. Klayder, *J. Assoc. Offic. Agr. Chemists*, **46**, 241 (1963).
205. W. W. Kilgore, K. W. Cheng, and J. M. Ogawa, *J. Agr. Food Chem.*, **10**, 399 (1962).
206. A. Kleinman, *J. Assoc. Offic. Agr. Chemists*, **46**, 238 (1963).
207. A. K. Klein and R. J. Gajan, *J. Assoc. Offic. Agr. Chemists*, **44**, 712 (1961).
208. H. J. Ackermann, H. A. Baltrush, H. H. Berges, D. O. Brookover, and B. B. Brown, *J. Agr. Food Chem.*, **6**, 747 (1958).
209. C. A. Bache and D. J. Lisk, *J. Agr. Food Chem.*, **8**, 459 (1960).
210. W. A. Steller, K. Klotsas, E. J. Kuchar, and M. V. Norris, *J. Agr. Food Chem.*, **8**, 460 (1960).
211. D. E. H. Frear, E. C. Smith, and T. G. Bowery, *J. Agr. Food Chem.*, **8**, 465 (1960).
212. G. A. Miller, *J. Assoc. Offic. Agr. Chemists*, **48**, 759 (1965).
213. J. S. Thornton and C. A. Anderson, *J. Agr. Food Chem.*, **13**, 509 (1965).
214. F. A. Gunther and D. E. Ott, *Analyst*, **91**, 475 (1966).
215. W. W. Kilgore and K. W. Cheng, *J. Agr. Food Chem.*, **11**, 477 (1963).
216. C. A. Anderson and J. M. Adams, *J. Agr. Food Chem.*, **11**, 474 (1963).
217. W. H. Gutenmann and D. J. Lisk, *J. Agr. Food Chem.*, **11**. 468 (1963).
218. W. E. Westlake and F. A. Gunther, *Residue Rev.*, **18**, 175 (1967).

219. D. M. Coulson, L. A. Cavanagh, J. E. Devries, and B. Walther, in *Instrumental Methods of Food Additives* (W. H. Butz and H. J. Noebels, eds), Chap. 14, Wiley (Interscience), New York, 1961, pp. 183–193.
220. D. M. Coulson, in *Advances in Pest Control Research* (R. L. Metcalf, ed.), Vol. 5, Wiley(Interscience), New York, 1962, pp. 153–190.
221. J. A. Challacombe and J. A. McNulty, *Residue Rev.*, **5**, 57 (1964).
222. J. A. Stamm, in *Lectures on Gas Chromatography 1964* (L. R. Mattick and H. A. Szymanski, eds.), Plenum, New York, 1965, pp. 53–58.
223. H. P. Burchfield, J. W. Rhoades, and R. J. Wheeler, in *Lectures on Gas Chromatography 1964* (L. R. Mattick and H. A. Szymanski, eds.), Plenum, New York, 1965, pp. 59–77.
224. H. P. Burchfield, J. W. Rhoades, and R. J. Wheeler, *J. Agr. Food Chem.*, **13**, 511 (1965).
225. R. F. Thomas and T. H. Harris, *J. Agr. Food Chem.*, **13**, 505 (1965).
226. L. C. Terriere, U. Kiigemagi, A. R. Gerlach, and R. L. Borovicka, *J. Agr. Food Chem.*, **14**, 66 (1966).
227. H. L. Pease, *J. Agr. Food Chem.*, **14**, 94 (1966).
228. M. Chiba and H. V. Morley, *J. Assoc. Offic. Anal. Chemists*, **49**, 341 (1966).
229. H. P. Burchfield and R. J. Wheeler, *J. Assoc. Offic. Anal. Chemists*, **49**, 651 (1966).
230. G. H. Boone, *J. Assoc. Offic. Agr. Chemists*, **48**, 748 (1965).
231. J. E. Lovelock, *Anal. Chem.*, **33**, 162 (1961).
232. K. P. Dimick and H. Hartmann, *Residue Rev.*, **4**, 150 (1963).
233. S. J. Clark, *Residue Rev.*, **5**, 32 (1964).
234. L. K. Gaston, *Residue Rev.*, **5**, 21 (1964).
235. C. H. Hartmann and D. M. Oaks, *Aerograph Research Notes*, Wilkens Instrument and Research, Inc., Walnut Creek, Calif., Winter, 1965.
236. J. E. Lovelock, *Anal. Chem.*, **35**, 474 (1963).
237. E. J. Bonelli, C. H. Hartmann, and K. P. Dimick, *J. Agr. Food Chem.*, **12**, 333 (1964).
238. C. H. Hartmann, D. M. Oaks, and T. Burroughs, *Research Bulletin, No. 2*, Varian Aerograph, Walnut Creek, Calif.
239. C. H. Hartmann, D. M. Oaks, and T. Burroughs, Division of Agricultural and Food Chemistry, 152nd Meeting, American Chemical Society, New York, September 1966.
240. G. R. Shoemake, D. C. Fenimore, and A. Zlatkis, *J. Gas. Chromatog.*, **3**, 285 (1965).
241. A. W. Ahren and W. F. Phillips, *J. Agr. Food Chem.*, **15**, 657 (1967).
242. L. Giuffrida, *J. Assoc. Offic. Agr. Chemists*, **47**, 293 (1964).
243. A. Karmen and L. Giuffrida, *Nature*, **201**, 1204 (1964).
244. J. A. Schmit, R. B. Wynne, and U. J. Peters, *Biomedical GC Notes*, F & M Scientific Corporation, Avondale, Pa., June 1965.
245. D. M. Oaks, K. P. Dimick, and C. H. Hartmann, *Publication W-122*, Varian Aerograph, Walnut Creek, Calif., 1966.
246. C. H. Hartmann, *Bull. Environ. Contam. and Toxicol.*, **1**, 159 (1966).
247. A. Karmen, *Anal. Chem.*, **36**, 1416 (1964).
248. A. Karmen, *J. Gas Chromatog.*, **3**, 336 (1965).
249. W. F. Phillips, private communication, 1966.
250. S. S. Brody and J. E. Chaney, *J. Gas Chromatog.*, **4**, 42 (1966).
251. M. C. Bowman and M. Beroza, *J. Assoc. Offic. Anal. Chemists*, **49**, 1154 (1966).
252. A. J. McCormack, S. C. Tong, and W. D. Cooke, *Anal. Chem.*, **37**, 1470 (1965).
253. A. J. McCormack, M.S. Thesis, Cornell University, 1963.
254. C. A. Bache and D. J. Lisk, *Anal. Chem.*, **37**, 1477 (1965).
255. C. A. Bache and D. J. Lisk, *Residue Rev.*, **12**, 35 (1966).
256. C. A. Bache and D. J. Lisk, *J. Assoc. Offic. Anal. Chemists*, **49**, 647 (1966).

257. C. A. Bache and D. J. Lisk, *Anal. Chem.*, **38**, 1757 (1966).
258. C. A. Bache and D. J. Lisk, *Anal. Chem.*, **39**, 786 (1967).
259. H. A. Moye, *Anal. Chem.*, **39**, 1441 (1967).
260. D. M. Coulson, *J. Gas Chromatog.*, **3**, 134 (1965).
261. D. M. Coulson, *J. Gas Chromatog.*, **4**, 285 (1966).
262. G. Zweig and T. E. Archer, *J. Agr. Food Chem.*, **8**, 190 (1960).
263. G. Zweig, T. E. Archer, and D. Rubenstein, *J. Agr. Food Chem.*, **8**, 403 (1960).
264. B. Berck, *J. Agr. Food Chem.*, **13**, 373 (1965).
265. J. D. Winefordner and T. H. Glenn, Jr., in *Advances in Chromatography* (J. C. Giddings and R. A. Keller, eds.), Vol. 5, Dekker, New York, 1967.
266. F. A. Gunther, in *Scientific Aspects of Pest Control*, Publication No. 1402, National Academy of Sciences, National Research Council, Washington, D.C., 1966.
267. C. C. Van Valin, B. J. Kallman, and J. J. O'Donnell, *Chemist-Analyst*, **52**, 73 (1963).
268. J. A. Burke and L. Giuffrida, *J. Assoc. Offic. Agr. Chemists*, **47**, 326 (1964).
269. J. A. Burke and W. Holswade, *J. Assoc. Offic. Agr. Chemists*, **47**, 845 (1964).
270. J. A. Burke and W. Holswade, *J. Assoc. Offic. Agr. Chemists*, **49**, 374 (1966).
271. A. B. Littlewood, in *Gas Chromatography*, Academic, New York, 1962.
272. J. Robinson, A. Richardson, and K. E. Elgar, Division of Agricultural and Food Chemistry, 152nd Meeting, American Chemical Society, New York, September 1966.
273. R. A. Conkin, *Residue Rev.*, **6**, 136 (1964).
274. D. C. Abbott and J. Thomson, *Residue Rev.*, **11**, 1 (1965).
275. M. F. Kovacs, Jr., *J. Assoc. Offic. Agr. Chemists*, **46**, 884 (1963).
276. M. F. Kovacs, Jr., *J. Assoc. Offic. Agr. Chemists*, **47**, 1097 (1964).
277. M. F. Kovacs, Jr., *J. Assoc. Offic. Agr. Chemists*, **48**, 1018 (1965).
278. M. F. Kovacs, Jr., *J. Assoc. Offic. Anal. Chemists*, **49**, 365 (1966).
279. L. J. Faucheux, Jr., *J. Assoc. Offic. Agr. Chemists*, **48**, 955 (1965).
280. W. A. Moats, *J. Assoc. Offic. Anal. Chemists*, **49**, 795 (1966).
281. B. Bazzi, R. Santi, M. Radice, and R. Fabbrini, *J. Assoc. Offic. Agr. Chemists*, **48**, 1118 (1965).
282. J. M. Finocchiaro and W. R. Benson, *J. Assoc. Offic. Agr. Chemists*, **48**, 736 (1965).
283. W. P. McKinley, in *Analytical Methods for Pesticides, Plant Growth Regulators and Food Additives* (G. Zweig, ed.), Vol. 1, Academic, New York, 1963, pp. 227–252.
284. D. E. Coffin, *J. Assoc. Offic. Anal. Chemists*, **49**, 638 (1966).
285. H. F. Beckman, R. B. Bruce, and D. MacDougall, in *Analytical Methods for Pesticides, Plant Growth Regulators and Food Additives* (G. Zweig, ed.), Vol. 1, Academic, New York, 1963, pp. 131–188.
286. R. C. Blinn and F. A. Gunther, *Residue Rev.*, **2**, 99 (1963).
287. R. C. Blinn, *Residue Rev.*, **5**, 130 (1964).
288. R. C. Blinn and F. A. Gunther, *Pesticide Research Bulletin*, **2**, Nos. 3 and 4, Stanford Research Institute, Menlo Park, California, 1962.
289. R. C. Blinn and F. A. Gunther, *Pesticide Research Bulletin*, **3**, No. 1. Stanford Research Institute, Menlo Park, California, 1963.
290. H. Frehse, *Pflanzenschutz-Nachrichten "Bayer"*, **16**, 182 (1963–1964).
291. N. T. Crosby and E. Q. Laws, *Analyst*, **89**, 319 (1964).
292. H. W. Boyle, R. H. Burttschell, and A. A. Rosen, *Advan. Chem. Ser.*, **60**, 207 (1966).
293. P. A. Giang and M. S. Schechter, *J. Agr. Food Chem.*, **14**, 380 (1966).
294. E. J. Broderick, J. B. Bourke, L. R. Mattick, E. F. Taschenberg, and A. W. Avens, *J. Assoc. Offic. Anal. Chemists*, **49**, 982 (1966).
295. W. R. Payne, Jr., and W. S. Cox, *J. Assoc. Offic. Anal. Chemists*, **49**, 989 (1966).

296. J. T. Chen, *J. Assoc. Offic. Agr. Chemists*, **48**, 380 (1965).
297. L. Giuffrida, *J. Assoc. Offic. Agr. Chemists*, **48**, 354 (1965).
298. R. C. Blinn, *J. Assoc. Offic. Agr. Chemists*, **48**, 1009 (1965).
299. D. MacDougall, *Residue Rev.*, **1**, 24 (1962).
300. D. MacDougall, *Residue Rev.*, **5**, 119 (1964).
301. C. A. Anderson, J. M. Adams, and D. MacDougall, *J. Agr. Food Chem.*, **7**, 256 (1959).
302. R. J. Keirs, R. D. Britt, Jr., and W. E. Wentworth, *Anal. Chem.*, **29**, 202 (1957).
303. H. A. Moye and J. D. Winefordner, *J. Agr. Food Chem.*, **13**, 516 (1965).
304. J. D. Winefordner and H. A. Moye, *Anal. Chim. Acta.*, **32**, 278 (1965).
305. W. J. McCarthy and J. D. Winefordner, *J. Assoc. Offic. Agr. Chemists*, **48**, 915 (1965).
306. H. A. Moye, unpublished work, 1966.
307. T. R. Kantner and R. O. Mumma, in *Residue Rev.*, **16**, 138 (1966).
308. J. M. Amy, E. M. Chait, W. E. Baitinger, and F. W. McLafferty, *Anal. Chem.*, **37**, 1265 (1965).
309. R. O. Mumma and T. R. Kantner, *J. Econ. Entom.*, **59**, 491 (1966).
310. J. N. Damico and W. R. Benson, *J. Assoc. Offic. Agr. Chemists*, **48**, 344 (1965).
311. T. R. Fukuto, E. O. Hornig, and R. L. Metcalf, *J. Agr. Food. Chem.*, **12**, 169 (1964).
312. B. J. Gudzinowicz and V. J. Luciano, *J. Assoc. Offic. Anal. Chemists*, **49**, 1 (1966).
313. C. M. Menzie, *Metabolism of Pesticides*, Special Scientific Report—Wildlife No. 96, Fish and Wildlife Service, U.S. Department of the Interior, May 1966.
314. J. E. Casida, in *Radioisotopes and Radiation in Entomology*, International Atomic Energy Agency, Vienna, 1962.
315. P. A. Dahm, in *Advances in Pest Control Research* (R. L. Metcalf, ed.), Vol. 1, Wiley (Interscience), New York, 1957, pp. 81–146.
316. B. W. Arthur, in *Radioisotopes and Radiation in Entomology*, International Atomic Energy Agency, Vienna, 1962.
317. R. D. O'Brien and L. S. Wolfe, *Radiation, Radioactivity and Insects*, Academic, New York, 1964.
318. G. N. Smith, in *Analytical Methods for Pesticides, Plant Growth Regulators and Food Additives* (G. Zweig, ed.), Vol. 1, Academic, New York, 1963, pp. 325–372.
319. W. C. Kaiser, *Anal. Chem.*, **38**, No. 11, 27A (1966).
320. F. P. W. Winteringham, in *Advances in Pest Control Research* (R. L. Metcalf, ed.), Vol. 3, Wiley (Interscience), New York, 1960, pp. 75–127.
321. R. L. Bogner, *Instrumental Methods for the Analysis of Food Additives* (W. H. Butz and H. J. Noebels, eds.), Chap. XIII, Wiley (Interscience), New York, 1961, pp. 171–181.
322. C. T. Redemann and R. W. Meikle, in *Advances in Pest Control Research* (R. L. Metcalf, ed.), Vol. 2, Wiley (Interscience), New York, 1958, pp. 183–206.
323. C. L. Dunn, D. J. Lisk, H. F. Beckman, and C. E. Castro, in *Analytical Methods for Pesticides, Plant Growth Regulators and Food Additives* (G. Zweig, ed.), Vol. 1, Academic, New York, 1963, pp. 253–280.
324. F. A. Gunther and J. H. Barkley, *Bull. Environ. Contam. and Toxicol.*, **1**, 39 (1966).
325. D. J. Lisk, *J. Agr. Food Chem.*, **8**, 119 (1960).
326. L. E. St. John, Jr., and D. J. Lisk, *J. Agr. Food Chem.*, **9**, 468 (1961).
327. R. C. Blinn, *J. Agr. Food Chem.*, **12**, 337 (1964).
328. C. A. Anderson, *Report No. 8544*, Chemagro Corp., Kansas City, Mo., February 1962.
329. W. W. Loeffler, *Report No. 11,476*, Chemagro Corp., Kansas City, Mo., June 1963.
330. J. B. Martin and D. M. Doty, *Anal. Chem.*, **21**, 965 (1949).
331. R. Anliker and R. E. Menzer, *J. Agr. Food Chem.*, **11**, 291 (1963).

332. P. A. Saliman, *Anal. Chem.*, **36**, 112 (1964).

333. C. E. Castro and R. A. Schmitt, *J. Agr. Food Chem.*, **10**, 236 (1962).

334. V. P. Guinn and J. C. Potter, *J. Agr. Food Chem.*, **10**, 232 (1962).

335. T. E. Archer, in *Analytical Methods for Pesticides, Plant Growth Regulators and Food Additives* (G. Zweig, ed.), Vol. 1, Academic, New York, 1963, pp. 373–397.

336. G. D. Winter and A. Ferrari, *Residue Rev.*, **5**, 139 (1964).

337. W. F. Phillips, M. C. Bowman, and R. J. Schultheisz, *J. Agr. Food Chem.*, **10**, 486 (1962).

338. W. F. Phillips, in *Analytical Methods for Pesticides, Plant Growth Regulators, and Food Additives* (G. Zweig, ed.), Vol. 1, Academic, New York, 1963, pp. 471–490.

339. Y. P. Sun, in *Analytical Methods for Pesticides, Plant Growth Regulators and Food Additives* (G. Zweig, ed.), Vol. 1, Academic, New York, 1963, pp. 399–423.

340. G. G. Patchett and G. H. Batchelder, *J. Agr. Food Chem.*, **8**, 45 (1960).

341. R. R. Watts, *J. Assoc. Offic. Agr. Chemists*, **48**, 1161 (1965).

342. W. A. Moats, *J. Assoc. Offic. Anal. Chemists*, **49**, 795 (1966).

343. R. C. Nelson, *J. Assoc. Offic. Agr. Chemists*, **48**, 752 (1965).

344. R. C. Nelson, *J. Assoc. Offic. Anal. Chemists*, **49**, 763 (1966).

345. B. L. Samuel, *J. Assoc. Offic. Anal. Chemists*, **49**, 346 (1966).

346. F. A. Gunther and D. E. Ott, *Residue Rev.*, **14**, 12 (1966).

347. D. E. Ott and F. A. Gunther, *J. Assoc. Offic. Anal. Chemists*, **49**, 669 (1966).

348. D. E. Ott and F. A. Gunther, *Bull. Environ, Contam. and Toxicol.*, **1**, 90 (1966).

349. D. E. Ott and F. A. Gunther, *J. Assoc. Offic. Anal. Chemists*, **51**, 697 (1968).

350. R. J. Gajan, private communication, 1962.

351. R. J. Gajan, *J. Assoc. Offic. Agr. Chemists*, **45**, 401 (1962).

352. P. H. Martens and P. Nangniot, *Residue Rev.*, **2**, 26 (1963).

353. R. J. Gajan, *J. Assoc. Offic. Agr. Chemists*, **46**, 216 (1963).

354. R. J. Gajan, *Residue Rev.*, **5**, 80 (1964).

355. R. J. Gajan, *Residue Rev.*, **6**, 75 (1964).

356. R. J. Gajan and J. Link, *J. Assoc. Offic. Agr. Chemists*, **47**, 1119 (1964).

357. D. O. Eberle and F. A. Gunther, *J. Assoc. Offic. Agr. Chemists*, **48**, 927 (1965).

358. R. J. Gajan, W. R. Benson, and J. M. Finocchiaro, *J. Assoc. Offic. Agr. Chemists*, **48**, 958 (1965).

359. J. Engelhardt and W. P. McKinley, *J. Agr. Food Chem.*, **14**, 377 (1966).

360. F. E. Hearth, D. E. Ott, and F. A. Gunther, *J. Assoc. Offic. Anal. Chemists*, **49**, 774 (1966).

361. R. A. Schmitt, in *Analytical Methods for Pesticides, Plant Growth Regulators and Food Additives* (G. Zweig, ed.), Vol. 1, Academic, New York, 1963, pp. 281–324.

362. V. P. Guinn and R. A. Schmitt, *Residue Rev.*, **5**, 148 (1964).

363. R. A. Schmitt and G. Zweig, *J. Agr. Food Chem.*, **10**, 481 (1962).

364. V. P. Guinn and J. C. Potter, *J. Agr. Food Chem.*, **10**, 232 (1962).

365. M. Szkolnik, K. D. Hickey, E. J. Broderick, and D. J. Lisk, *Plant Disease Reptr.*, **49**, 568 (1965).

366. R. W. Risebrough and V. Brodine, *Environment*, **12**, 17 (1970).

367. D. B. Peakall and J. L. Lincer, *BioScience*, **20**, 958 (1970).

368. C. G. Gustafson, *Environ. Sci & Technol.*, **4**, 814 (1970).

Chapter 6

TOXICOLOGY OF PESTICIDES
TO ESTABLISH PROOF OF SAFETY

Bernard L. Oser

FOOD & DRUG RESEARCH LABORATORIES INC.
MASPETH, NEW YORK

6-1   INTRODUCTION

The investigation of the potential hazards of pesticides† to man under conditions of normal or intended use is accomplished principally by means of toxicological tests in animals. No attempt is made in this chapter to

†The term pesticides is used herein in the sense of economic poisons as defined in the regulations of the Federal Insecticide, Fungicide, and Rodenticide Act, with particular reference to the use of these substances in agricultural practice. "The term 'economic poison'

discuss the toxicity of pesticidal agents toward insects, fungi, rodents, or other pests which they are intended to mitigate, inhibit, or destroy. Although not specifically designed for this purpose, the toxicological procedures to be discussed are to some extent applicable to the estimation of potential hazards to fish and wildlife. These species are themselves often used for the purpose of evaluating the dangers to which they may inadvertently be exposed from the manufacture and use of pesticides.

Human beings may be exposed to toxic chemicals in many ways. In laboratories where these substances are synthesized and examined for their chemical, physical, and biological properties, scientists and technicians are the first to be exposed. At this stage the risks may not be known but proper precautions are generally taken. In chemical plants where pesticides are manufactured, first on a pilot scale, then on a commercial basis, and subsequently in the preparation of formulations (sprays, dusts, etc.), human exposure may reach considerable proportions over prolonged periods. Similarly, application to vegetation in fields and orchards, from direct spraying or dusting equipment, or from airplanes, provides opportunity for contact, inhalation, and even ingestion of these toxic agents. In addition to occupational exposures, entry into the food channels of man and animals may occur in the form of residues in or on raw agricultural commodities or in processed or packaged foods produced therefrom. Thus traces of pesticides may be ingested by the young and old, the sick and well, and for long or short periods, at various stages or even throughout the life cycle.

Toxicological studies designed to ensure the safety of occupational exposure generally involve topical applications to the skin or mucus membranes and inhalation by experimental animals. Feeding studies are also conducted to assess the risks of chronic ingestion with the possibility of cumulative or subtle systemic effects that are not as readily detectable as are the effects of acute or topical exposure.

Toxicological studies of drugs have a somewhat different objective inasmuch as they are usually directed toward eliciting effects resulting from short-term administration at dosages not far removed from the therapeutic or use levels. It is the purpose of drugs to manifest pharmacological responses at these use levels. However as wide a difference as possible is desired between the therapeutic and toxic doses. The toxicity

---

means(1) any substance or mixture of substances intended for preventing, destroying, repelling, or mitigating any insects, rodents, nematodes, fungi, weeds, and other forms of plant or animal life or viruses, except viruses on or in living man or other animals, which the Secretary shall declare to be a pest, and(2) any substance or mixture of substances intended for use as a plant regulator, defoliant or desiccant."

of drugs may take the form of an exaggerated degree of pharmacological reaction of the type that that drug is functionally intended to produce (as in the case of prolonged or irreversible narcosis induced by barbiturates), or they may be designed to reveal unwanted or adverse effects of a different kind. The therapeutic:toxic ratio of drug dosage should be as high as possible compatible with effective use. In the case of a pesticidal agent, a major purpose of the toxicological investigation is to provide a basis for estimating the maximum dose that may be tolerated by animals (ultimately by man) throughout their lifetime without manifesting any adverse effects. It is necessary, however, to administer experimental dosages at levels high enough to ascertain the nature of the effects that might be induced by excessive dosage. The susceptibility of different species of animals to toxic agents may and often does vary. Since the observations in animals must be evaluated in terms of potential responses in man, several different species are employed, and man is assumed to be at least as sensitive as the most sensitive species tested. Moreover, in view of the uncertainties of applying the results of animal experiments to large human populations, the conditions of toxicological testing grossly exaggerate those of possible human exposure in respect to dosage and duration.

The estimation of the acceptable daily intake of a given pesticide (or other chemical) by man is usually based on the no-effect level in the most susceptible of the animal species tested, except where evidence for the metabolism of the substance indicates a closer relation between some other species and man.

The terms "toxicity" and "hazard" have been defined by the Food Protection Committee of the Food and Nutrition Board, National Academy of Sciences—National Research Council, and in regulations of the Food and Drug Administration, viz.: *toxicity*, a determinable value, is the capacity of a substance to produce injury; *hazard*, an indeterminable value, is the probability that injury will result from the use of a substance in a proposed quantity and manner. Any substance may be toxic if one ingests it, or is otherwise exposed to it, to an excessive degree. Whether it is hazardous depends more on how a substance is used than on how toxic it may be. See discussion under "Safety Factors and Tolerances" below.

Concomitant with the introduction of a large number of new synthetic pesticides during the past few decades, toxicological procedures have been considerably improved with the view toward making them more critical and meaningful. Toxicologists in the United States and abroad are in general agreement on the basic principles involved in safety evaluation. Reference may be made particularly to Refs. *1* and *2*. These have served as guidelines for the methods recommended by other national bodies such as the British Ministry of Agriculture, Fisheries and Foods, the German Research Association (Deutsche Forschungsgemeinschaft) and several international bodies such as the Council of Europe and the European Economic Community. Whereas these procedures vary in certain details, they are essentially quite similar and form the basis of the safety evaluation procedures discussed in this chapter.

6-2 OCCUPATIONAL EXPOSURE

The hazards of occupational exposure to pesticides result principally from direct physical contact with external body surfaces of the skin and eyes or with the nasopharyngeal and bronchopulmonary membranes by reason of the inhalation of sprays, mists, or dusts. Despite all efforts to reduce these hazards to a minimum by precautionary labeling and careful operational practice, unintentional exposure by these routes is inevitable and its possible effects must be considered before the use of a given pesticide or formulation can be permitted. Exposure under these conditions is generally of limited duration but it may be frequent and the effect may be cumulative. Furthermore the degree of exposure under unusual or extreme conditions, whether intentional or unintentional, may be rather severe. This would apply to pest control operators, spray pilots, farmers, orchardists, etc., as well as to the manufacturers of the basic chemicals and pesticide formulations. The consequences of accidental, suicidal, or homicidal overexposure should be evaluated whenever possible, not only to establish the nature of the effects but to provide a basis for developing and testing antidotal procedures.

The dangers of toxic reactions from drugs are mitigated somewhat by the fact that individuals are generally aware that they are being administered and physicians are expected to have knowledge of the potential risks involved; in contrast, the hazards that may ensue from the chronic ingestion of food additives or of pesticide residues, are not recognized or known to consumers. They are generally unaware or unconscious of their presence or effects in food. Fortunately there is little evidence of actual harm resulting from the traces of these chemical substances contained in our diets. This may be attributed in part to the careful precautions taken with regard to assessment of safety before the use of such chemicals is permitted. It must also be recognized that our bodies have the capacity to make biochemical and physiological adaptations to accommodate to small concentrations of most foreign substances, resulting in their elimination or detoxication.

6-3 WHAT TO TEST

The evaluation of pesticides from the standpoint of the safety of residues in foods must take into account the actual substances present, whether they be the pesticides themselves or reaction products resulting from their use. Chemical alterations of an oxidative, hydrolytic, or degradative nature may occur either after application to the surface of vegetation or in the soil with resultant entry to food crops by way of translocation. When plant materials such as grains, leaves, legumes, etc., or their by-products are used as components of feed for livestock or poultry, pesticide residues may be present either unchanged or in the form of metabolites in meat, milk, or eggs. It is these end-products whose safety must be assessed as pesticide residues. Certain pesticides are converted to single derivatives, e.g., heptachlor to its epoxide, aldrin to dieldrin, or parathion to paraoxon. In some cases the derivatives are multiple in nature, as in the case of DDT whose end products are DDE and DDD. When the transformation products are multiple or of unknown composition, the evaluation of safety of the residues may become highly complicated, especially since climatic conditions may vary their composition or relative proportions.

It is essential to have an approximate idea of the amounts of residue remaining in or on the edible portion of food crops in order to establish their quantitative relation to the graded dose response of test animals.

Legally, pesticide residue tolerances are permitted at levels no higher than necessary for their effective use as determined by good agricultural practice. This implies adequate intervals between the last application and the harvest, and the use of such procedures as washing, brushing, etc., to minimize residue levels in "raw agricultural commodities," which are defined to include "any food in its raw or natural state, including all fruits that are washed, colored, or otherwise treated in their unpeeled natural form prior to marketing." Not included in this definition are foods that have been processed, fabricated, or manufactured by means involving milling, cooking, freezing, or dehydration.

The factors that influence the amount of residue have been summarized in a bulletin of the National Academy of Sciences – National Research Council(3) as follows:

(1) Rate of application, i.e., magnitude of initial deposit
(2) Time of application relative to:
    (a) Development of edible plant parts
    (b) Exposure of edible parts to treatment
    (c) Time elapsed between last application and crop harvest

(3) Rate of loss of a pesticide deposit from the plant
    (a) Rate of decomposition or degradation of active ingredient as affected by:
       (i) Fluctuations in temperature, moisture, sunlight
       (ii) Plant secretions
    (b) Rate of evaporation of volatile materials as affected by environmental conditions
    (c) Rate of erosion of residual deposits as affected by:
       (i) Rainfall
       (ii) Wind
(4) Dilution due to growth of plant
(5) Adherence to or absorption by plant parts
(6) Efficiency of residue removal methods and extent of their use
(7) Miscellaneous practices in application
    (a) Effect of changes in formulation on any of the listed factors
    (b) Number of applications, and particularly, the date of the last application.

It is important not only to know the composition and identity of the substance or substances to be tested, but to have information with respect to its volatility (vapor pressure at ambient temperature), chemical stability, and reactivity in diet mixtures. These data are necessary in order to avoid the possibility of qualitative or quantitive changes in the test materials after incorporation in the diets of experimental animals.

6-4   WHEN TO TEST

Assessment of the potential hazard resulting from skin contact, inhalation, or, other physical exposure should be made as early as possible in the sequence of stages from research and development in the laboratory to production in the plant. Careful observation of plant operators should be made in order to elicit possible effects that were not demonstrable, or not previously revealed, by animal tests. Once the conditions for safe handling of a pesticide are determined and field tests are under way, the toxicological evaluation of the safety of the potential residues should be undertaken. This is a time-consuming process and often can effectively be carried out during the period that field tests to establish the functional efficacy of the pesticide are in progress. Unless the level of pesticide residue that might remain on the food crop is known, the initial toxicological work should be limited to short term tests. The chronic studies should not be undertaken until there is reasonable assurance as to the potential residue levels, since the selection of the critical dosage range in chronic studies is extremely

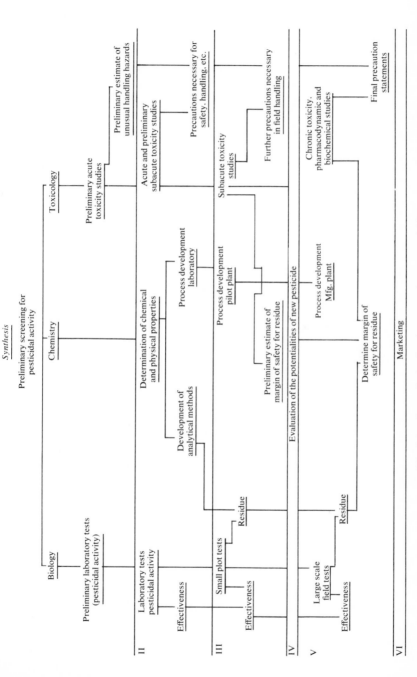

Fig. 6-1. Stages in the development of a new pesticide.

important. It is possible to overestimate these doses and thus obtain effects entirely unrelated to any potential hazard that might result from the use of a substance in a proper quantity and manner.

The timing of toxicological studies in relation to the sequence of steps in the synthesis, development, industrial production, and functional evaluation of a pesticide may vary in different companies. An example of such a program, from conception to marketing, is illustrated in Fig. 6-1 (*3*). Rather than following the screening for potential functional properties as shown in this chart, exploration of the chemical and physical properties of a candidate pesticide generally precedes it.

6-5 TOXICOLOGICAL ASSESSMENT OF OCCUPATIONAL HAZARDS

Exposure to toxic substances in the form of dust, mists, gases, vapors, or aerosols may occur through contact with the skin or mucus membrane (particularly of the eyes) and through inhalation into the respiratory pathways. It is thus necessary to test such substances by applying them to the skin and eyes of animals and by exposing animals to atmospheres in which the test materials are diffused in high concentrations.

A. Dermal Toxicity Tests

Rabbits are employed more often than any other species for studies of skin toxicity, although guinea pigs, rats, or mice are occasionally used. In tests designed to elicit the effect of single acute exposures, the results are generally expressed in terms of the LD_{50}, the dose which, under the conditions stated, will cause death in 50% of a group of test animals. Dermal toxicity tests are also conducted to show the effects of repeated applications or exposures. The test material is applied to the partially or completely depilated surface of the skin in graded concentrations. It is essential that the diluent or vehicle for the test material be tested simultaneously in a control group of animals.

A typical dermal toxicity test is described in the regulations governing the labeling of hazardous substances (*4*) as quoted verbatim below:

"§191.10 *Method of Testing Toxic Substances*

"(a) **Acute dermal Toxicity (single exposure).** In the acute exposures the agent is held in contact with the skin by means of a sleeve for periods varying up to 24 hours. The sleeve, made of rubber dam or other impervious material, is so constructed that the ends are reinforced with additional strips and should fit snugly around the trunk of the animal. The ends of the sleeve are tucked, permitting the central portion to "balloon" and furnish a reservoir for the dose. The reservoir must have sufficient capacity to contain the dose without pressure. In the following table are given the dimensions of sleeves and the approximate body surface exposed to the test substance. The sleeves may vary in size to accommodate smaller or larger subjects. In the testing of unctuous materials that adhere readily to the skin, mesh wire screen may be employed instead of the sleeve. The screen is padded and raised approximately 2 centimeters from the exposed skin. In the case of dry powder preparations, the skin and substance are moistened with physiological saline prior to exposure. The sleeve or screen is then slipped over the gauze which holds the dose applied to the skin. In the case of finely divided powders, the measured dose is evenly distributed on cotton gauze, which is then secured to the area of exposure.

DIMENSIONS OF SLEEVES FOR ACUTE DERMAL TOXICITY TEST
IN RABBITS[a]

Measurements in centimeters		Range of weight of animals (grams)	Average area of exposure (cm²)	Average percentage of total body surface
Diameter at ends	Overall length			
7.0	12.5	2500–3500	240	10.7

[a]Quoted verbatim from Ref.4.

"(b) **Preparation of Test Animals.** The animals are prepared by clipping the skin of the trunk free of hair. Approximately one-half of the animals are further prepared by making epidermal abrasions every 2 centimeters or 3 centimeters longitudinally over the area of exposure. The abrasions are sufficiently deep to penetrate the stratum corneum (horny layer of the epidermis), but not to disturb the derma—that is, not to obtain bleeding.

"(c) **Procedures for Testing.** The sleeve is slipped onto the animal, which is then placed in a comfortable but immobilized position in a multiple animal holder. Selected doses of liquids and solutions are introduced under the sleeve. If there is slight leakage from the sleeve, which may occur during the first few hours of exposure, it is collected and reapplied. Dosage levels are adjusted in subsequent exposures (if necessary) to enable a calculation of a dose that would be fatal to 50 percent of the animals. This can be determined from mortality ratios obtained at various doses employed. At the end of 24 hours the sleeves or screens are removed, the volume of unabsorbed material, if any, is measured, and the skin reactions are noted. The subjects are cleaned by thorough wiping, observed for gross symptoms of poisoning, and then observed for 2 weeks.

"§191.11   *Method of Testing Primary Irritant Substances*

"Primary irritation to the skin is measured by a patch-test technique on the abraded and intact skin of the albino rabbit, clipped free of hair. A minimum of six subjects are used in abraded and intact skin tests. Introduce under a square patch such as surgical gauze measuring 1 inch × 1 inch, two single layers thick, 0.5 milliliter (in case of liquids) or 0.5 gram (in case of solids and semisolids) of the test substance. Dissolve solids in an appropriate solvent and apply the solution as for liquids. The animals are immobilized with patches secured in place by adhesive tape. The entire trunk of the animal is then wrapped with an impervious material such as rubberized cloth for the 24-hour period of exposure. This material aids in maintaining the test patches in position and retards the evaporation of volatile substances. After 24 hours of exposure, the patches are removed and the resulting reactions are evaluated on the basis of the designated values in the following table:

EVALUATION OF SKIN REACTIONS[a]

Erythema and eschar formation:	*Value*[b]
No erythema	0
Very slight erythema (barely perceptible)	1
Well-defined erythema	2
Moderate to severe erythema	3
Severe erythema (beet redness) to slight eschar formation (injuries in depth)	4
Edema formation:	
No edema	0
Very slight edema (barely perceptible)	1
Slight edema (edges of area well defined by definite raising)	2
Moderate edema (raised approximately 1 millimeter)	3
Severe edema (raised more than 1 millimeter and extending beyond the area of exposure)	4

[a]Quoted verbatim from Ref. *4*.
[b]The "value" recorded for each reading is the average value of the six or more animals subject to the test.

"Readings are again made at the end of a total of 72 hours (48 hours after the first reading). An equal number of exposures are made on areas of skin that have been previously abraded. The abrasions are minor incisions through the stratum corneum, but not sufficiently deep to disturb the derma or to produce bleeding. Evaluate the reactions of the abraded skin at 24 hours and 72 hours, as described in this paragraph. Add the values for erythema and eschar formation at 24 hours and at 72 hours for intact skin to the values on abraded skin at 24 hours and at 72 hours (four values). Similarly, add the values for edema formation at 24 hours and at 72 hours for intact and abraded skin (four values). The total of the eight values is divided by four to give the primary irritation score. Example:

	Exposure time	Exposure unit
Erythema and eschar formation:	*Hours*	*Value*
Intact skin	24	2
Do	72	1
Abraded skin	24	3
Do	72	2
Subtotal		8
Edema formation:		
Intact skin	24	0
Do	72	1
Abraded skin	24	1
Do	72	2
Subtotal		4
Total		12

Primary irritation score is $12 \div 4 = 3$."
The data above concludes the extract from the Federal Hazardous Substances Act.

Under the conditions of this test, deaths may result from the systemic absorption rather than the purely local action of the toxic agent. A substance is described as a "primary irritant" if the empirical score, when tested by this method, is five or more. It should be noted that deaths may result from systemic absorption rather than purely local action. A substance is considered to be "highly toxic to the skin" if it "produces death within 14 days in half or more than half of a group of rabbits... tested in a dosage of 200 milligrams, or less, per kilogram of body weight" when tested by the method described in §191.10, above (4).

Tests of dermal toxicity involving repeated application are generally based on the use of three or four groups of rabbits with three or four in each group. Skin reactions are evaluated, in the regulatory procedure, by the Draize scoring system, as described above, which is the most popular in this country. The basic procedure quoted above may be extended by the inclusion of further observations, particularly necropsies, with gross and microscopic examination of the organs and tissues. In addition to the determination of the irritant properties of a substance when applied to the intact or abraded skin, it is often important to establish whether, and to what extent, a substance may penetrate the skin and be absorbed systemically. The outstanding example of such a case among pesticides is the organophosphorus compound parathion, which is more toxic when applied to the skin than when ingested. Percutaneous absorption from contaminated clothing has been known to cause deaths. The possibility of systemic poisoning through penetration of the intact cutis must be considered especially in the case of substances soluble in fat or fat solvents. Percutaneous toxicity is generally investigated by means of repeated applications to the depilated skin of rabbits held in stocks or provided with special collars to prevent oral ingestion.

The test material is applied daily for a period of 5 to 6 hr and is carefully removed by washing before the animals are returned to their cages. The applications are repeated for a period of 3 weeks during which observations are made for organic or behavioral effects. At the termination of the test the animals are subjected to post-mortem examination for evidence of gross or histomorphological alterations, not only at the site of application but throughout the various organs and tissues of the body.

Histochemical procedures have been developed for the determination of the extent to which test materials may penetrate the skin. These are generally based on the use of radioactive isotope derivatives of the test materials and the counting of microscopically thin sections of the cutaneous tissue prepared by slicing from the inner aspects outward in order to avoid contamination(4a).

B. Mucus Membrane and Eye Toxicity

The conjunctiva of the eye, the vaginal vault, and the tip of the penis of experimental animals have been employed in tests of the toxic or irritant effects of chemical substances on mucus membranes. By far the most commonly used test site is the eye of the rabbit, despite the fact that in this species the eye is both functionally and anatomically different from the human eye. Some investigators have preferred to use the monkey since, in contrast with the rabbit, its eye resembles the human eye at least in respect to the presence of lacrimating glands. The most frequently employed tests for eye irritation are those described by Draize which have been incorporated into the federal regulations for the evaluation of the toxicity of hazardous household substances(4). The procedure is as follows:

"§191.12   Test for eye irritants.

"(a) (1) Six albino rabbits are used for each test substance. Animal facilities for such procedures shall be so designed and maintained as to exclude sawdust, wood chips, or other extraneous materials that might produce eye irritation. Both eyes of each animal in the test group shall be examined before testing, and only those animals without eye defects or irritation shall be used. The animal is held firmly but gently until quiet. The test material is placed in one eye of each animal by gently pulling the lower lid away from the eyeball to form a cup into which the test substance is dropped. The lids are then gently held together for one second and the animal is released. The other eye, remaining untreated, serves as a control. For testing liquids, 0.1 milliliter is used. For solids or pastes, 100 milligrams of the test substance is used, except that for substances in flake, granule, powder, or other particulate form the amount that has a volume of 0.1 milliliter (after compacting as much as possible without crushing or altering the individual particles, such as by tapping the measuring container) shall be used whenever this volume weighs less than 100 milligrams. In such a case, the weight of the 0.1 milliliter test dose should be recorded. The eyes are not washed following instillation of test material except as noted below.

"(2) The eyes are examined and the grade of ocular reaction is recorded at 24, 48, and 72 hours. Reading of reactions is facilitated by use of a binocular loupe, hand slit-lamp, or other expert means. After the recording of observations at 24 hours, any or all eyes may be further examined after applying fluorescein. For this optional test, one drop of fluorescein sodium ophthalmic solution U.S.P. or equivalent is dropped directly on the cornea. After flushing out the excess fluorescein with sodium chloride solution U.S.P. or equivalent, injured areas of the cornea appear yellow; this is best visualized in a darkened room under ultraviolet illumination. Any or all eyes may be washed with sodium chloride solution U.S.P. or equivalent after the 24-hour reading.

(b) (1) An animal shall be considered as exhibiting a positive reaction if the test substance produces at any of the readings ulceration of the cornea (other than a fine stippling), or opacity of the cornea (other than a slight dulling of the normal luster), or inflammation of the iris (other than a slight deepening of the folds (or rugae) or a slight circumcorneal injection of the blood vessels), or if such substance produces in the conjunctivae (excluding the cornea and iris) an obvious swelling with partial eversion of the lids or a diffuse crimson-red with individual vessels not easily discernible.

"(3) The test shall be considered positive if four or more of the animals in the test group exhibit a positive reaction. If only one animal exhibits a positive reaction, the test shall be regarded as negative. If two or three animals exhibit a positive reaction, the test is repeated using a different group of six animals. The second test shall be considered positive if three or more of the animals exhibit a positive reaction. If only one or two animals in the second test exhibit a positive reaction, the test shall be repeated with a different group of six animals. Should a third test be needed, the substance will be regarded as an irritant if any animal exhibits a positive response.

"(c) To assist testing laboratories and other interested persons in interpreting the results obtained when a substance is tested in accordance with the method described in paragraph (a) of this section, an "Illustrated Guide for Grading Eye Irritation by Hazardous Substances" will be sold by the Superintendent of Documents, Government Printing Office, Washington, D. C. The guide will contain color plates depicting responses of varying intensity to specific test solutions. The grade of response and the substance used to produce the response will be indicated."

In order to predict more closely the sequelae of accidental contact of a chemical irritant in the eye, it has been recommended(5) that the size of the test group be increased to allow for additional observations, viz. the effects of (a) rinsing treated eyes with water at intervals of 2 and 4 seconds after application of the test material, while an equal number remain unwashed; and (b) extending the post-treatment observations over a period of one week to permit a better evaluation of reversibility of the reactions.

C. Inhalation Toxicity

The procedures for the evaluation of potential hazards of gases, dusts, mists, or vapors via the inhalation route vary considerably depending on the physical nature (solubility, particle size, etc.) of the atmospheric contaminant. In general, animals are exposed under conditions whereby they breathe the atmosphere containing known concentrations (expressed in milligrams per cubic meter or in parts per million) for various periods of time depending upon the possibilities of practical exposure. In order to avoid confusing oral or cutaneous toxicity with inhalation toxicity, the experimental conditions of exposure should be such that the animals inspire, but do not otherwise absorb or ingest, the test materials. The procedures for the evaluation of oral toxicity are considered in the next section of this chapter.

The regulations under the Federal Hazardous Substances Act describe as toxic, "any substance that produces death within 14 days in one-half of a group of white rats each weighing between 200 grams and 300 grams, when inhaled continuously for a period of 1 hour or less at an atmospheric concentration of more than 200 parts per million but not more than 20,000 parts per million by volume of gas or vapor or more than 2 milligrams but not more than 200 milligrams per liter by volume of mist or dust, provided such concentration is likely to be encountered by man when the substance is used in any reasonably foreseeable manner."

Depending upon the nature and degree of possible occupational exposure, particularly with reference to levels, frequency, and duration, various modifications of the procedure are employed.

The design of test chambers and of the dispersing equipment has become quite sophisticated since it was realized that not only concentration but particle or droplet size is extremely critical from the point of view of passage through the mucociliary barrier of the nasopharynx and trachea to the bronchopulmonary surfaces. It has been demonstrated that particles greater than $10\,\mu$ diameter are effectively prevented from reaching the lungs whereas those of 3 μ diameter or less may be inhaled into the alveolar region, and those of intermediate size do so with greater difficulty(6).

Inhalation test chambers that contain whole cages of animals have the disadvantage of permitting condensation of test materials on the surface of the animals themselves and of their cages so that the possibility of oral ingestion is not entirely excluded. Head-port chambers have been designed to overcome this difficulty by providing for immobilization of the animal with only its head inserted into the chamber in which the aerosol concentration is controlled(7). Equipment of this type has been designed for dogs, rabbits, and small rodents. In the latter case provision can be made for whole body plethysmography and recording of the various parameters of cardiopulmonary function.

D. Evaluation of Oral Toxicity

The oral route of administration is mandatory in studies designed to estimate the toxicity of dietary components, whether they be pesticide residues, food additives, or foods themselves.

The procedures employed entail the administration of the test materials either as components of the diet or intragastrically, by gavage. The latter route has the advantage of permitting precise measurement and adjustment of daily dosage to body weight. However, it is unsuitable for long-term daily administration. When reliable quantitative information is available with respect to the food consumption of the species and strain of animals used, incorporation of the test materials in the diet, with measurement of the daily food intake over a reasonably protracted period from weaning to maturity, will provide a reliable measure of doses administered via the diet(8). Experience has shown that observed dosage under these conditions is generally within 10% of the theoretical or expected dosage computed on a body weight basis. Table 6-1 shows typical data for food consumption and compound dosage expressed both as parts per million incorporated in the diet and as observed dosage in milligrams per kilogram body weight.

TABLE 6-1
LEVELS OF TOXICANT ADDED TO DIET OF RATS IN SUCCESSIVE
BIWEEKLY PERIODS[a]

Period, weeks	Normal food intake, g/kg body weight	Diets estimated to furnish		
		50	125	250
		mg/kg body weight/day (% in diet)		
1–2	120	0.042	0.103	0.210
3–4	100	0.050	0.125	0.250
5–6	75	0.067	0.167	0.333
7–8	65	0.077	0.183	0.383
9–10	60	0.083	0.208	0.415
11–12 +	50	0.100	0.250	0.500
		Observed dosage levels		
		45.7	131.5	227.5
		mg/kg body weight/day		

[a]Typical 12-wk experiment.

Some investigators, particularly in England and Germany, have for years employed the subcutaneous route in toxicological studies of food colors, in the belief that the induction of sarcomas by this means warrants the prohibition of a substance as a dietary component. The appropriateness of this route of administration as an index of oral toxicity is questionable, however. Many simple substances such as glucose, salt, distilled water, and indeed the trauma of repeated subcutaneous dosage have been known to induce sarcomas at the sites of injection. Moreover it is well recognized that substances that induce tumorigenic lesions in one organ by one route of administration in a given species may be quite innocuous when administered via another route in another species(9, 10). This is discussed below under the heading of "Carcinogenesis."

In evaluating oral toxicity, consideration must be given to all aspects of biochemical, physiological, pharmacological, and morphological change, in specific organs and in the body as a whole. When the nature of a chemical substance suggests the possibility of functional disturbance in a target organ, test procedures should be designed with particular reference to revealing such effects. The choice of test animals may depend upon existing knowledge with regard to the possible route by which the substance may be metabolized or excreted. Biotransformations may be in the direction of either detoxication or potentiation of toxicants. For example, in the case of aromatic amines, where acetylation is the usual detoxication process, emphasis should be placed on the rat which, like man, can acetylate, rather than on the dog which cannot. On the other hand, if detoxication occurs via the glucuronide pathway the response of the dog may be more significant in terms of human predictability.

Investigations for assessing oral toxicity generally take place in three stages, viz., acute, short-term, and long-term studies, the purpose being to determine the nature of toxic effects induced by massive dosage, and gradually to work down to the dosage range of minimal-to-negative effects. In order to compensate for the uncertainties involved in extrapolating results from one species to another (to wit, man) it is advisable to use at least two different species of animals in long-term toxicological evaluattions, and it is generally specified that at least one be a nonrodent.

In practice, the ultimate objectives of chronic oral toxicity studies are (a) to determine the nature of the toxic effects induced by the maximum dosage that rodent species can tolerate without substantially reducing normal longevity, and (b) to ascertain the maximum dosage of the substance that the test species can ingest without adverse effect under all conditions of normal physiological stress.

In the case of nonrodent species, these objectives are modified only with respect to reducing the duration of the test to not more than two years.

1. Acute Oral Toxicity. The point of departure for the evaluation of the potential hazard of repeatedly ingested traces of a pesticide in food, is the acute oral toxicity test, inasmuch as this provides a basis for estimating the critical dosage range to be employed in the subsequent feeding studies.

The first approximation of the oral toxicity of a substance is by way of the administration of massive doses sufficient to cause early death. This type of test not only permits comparison among various substances, both new and old, within a given chemical category, but allows some indication of the hazards of accidental ingestion and of the mode of action of the chemical in the animal organism. When such tests are conducted in several different species, an indication is obtained as to the assurance with which animal data may be applied in predicting safety to man.

It is important to remember, however, that not only the degree but the nature of the effects induced by massive doses may be entirely different from the responses to small doses ingested over a long period. In the latter situation, adaptive, metabolic, or excretory processes are often sufficient to prevent the accumulation of toxic substances in the tissues or organs of the host, and none of the manifestations characteristic of the massive dose response may occur.

Acute oral toxicity studies are designed to permit estimation of the lethal dose for 50% of a group of test animals, expressed per unit of body weight. This dose, the so-called LD_{50}, of a substance may be so close to the potential level of human intake in the form of a residue in food, that any proposed use of the substance should be abandoned. This is seldom the case, however, and acute oral toxicity tests are generally the forerunner of more protracted studies. Acute oral toxicity tests are conducted initially in the rat or mouse. After the critical dosage range is established in a rodent species, a sufficient number of doses within this range (or even the LD_{50} alone) is given to small groups of nonrodents, such as dogs or monkeys, to ascertain whether a species difference in susceptibility may exist. Test doses are administered either by mouth or by intubation after an overnight (16–18 hr) fast and are followed by the offer of food.

Test substances are administered in an inert vehicle in either solution or suspension depending on their nature. Water, physiological saline, vegetable oil, propylene glycol, and solutions of gum tragacanth or carboxymethylcellulose are among the commonly used diluents or vehicles. Whatever the vehicle, it is administered to a control group of animals by the same route as the test material to demonstrate that local irritant effects do not result from this source. The use of a surfactant to facilitate dispersion or solution is permissible provided that it does not potentiate intestinal absorption of a substance otherwise absorbed slowly or with difficulty. It is sometimes inexpedient to give the entire lethal dose at one time, and under these circumstances it may be administered in divided doses in the course of a day.

In the absence of information as to the critical dosage range, the doses may be spread over a wide range at the beginning (e.g., using 0.3 log intervals) with only one or two rats per group, followed by further tests at more closely spaced intervals (e.g., 0.1 log intervals) with larger groups in the critical range of 16 to 84% mortality. Four to 6 groups, of 6 to 20 rats each, within this range will generally give a reasonably reliable estimate of the LD_{50}, its standard error, and the slope of the dose: response curve.

Young adult animals are generally used in acute toxicity testing, in the case of rats, in the weight range of 200 to 300 g, and in mice 15 to 20 g. However age may have a determinant effect on toxic reactions. An LD_{50} value should always be accompanied by reference not only to the species of animals employed and the route of administration, but also to their age or maturity, and the duration of the post-dosage observation period.

The methods most commonly used for calculation of the LD_{50} are those of Miller and Tainter(11) and Litchfield and Wilcoxon(12). They are predicated upon a normal frequency distribution of the mortality response. The larger the number and size of groups of rats employed in the acute toxicity test the greater the probability of actually demonstrating such a frequency distribution. Cost and space requirements generally preclude the use of a sufficient number of dogs or monkeys for determining a statistically valid LD_{50} in these species. Significant information as to possible species differences may be obtained by administering to only one, or a few groups of two or three animals each, the LD_{50} dose (on a mg per kg body weight basis) found in the rodent species.

In any toxicological assay for safety evaluation, a gradual slope of the dose: response curve is a less desirable finding than a steep slope. In the latter case it is easier to delineate the threshold between effective (e.g, lethal) and noneffective doses, but when the slope is gradual it is difficult to establish a cut-off point and the estimation of the "no-effect" dose (e.g., LD_0) is therefore less reliable.

Whereas acute toxicity tests are often limited to observations of lethality, much valuable information can be obtained by noting the nature, severity, and duration of the premortal signs with respect to effects on posture, mobility, reflexes, salivation, laxation, diuresis, color of eyes, ears, and extremities, pupillary reaction, and water and food intake. Gross autopsies of animals that die or are sacrificed and histomorphological examinations also, of course, yield useful information, but inasmuch as acute tests are conducted principally for range-finding purposes, extensive (and expensive) pathological examinations are usually reserved for the longer feeding studies.

2. Short-Term ("Subacute") Oral Toxicity. Probably the most significant phase of the toxicological assessment of potential components of the diet is the short-term feeding study. Involving, as it does, the daily ingestion of the test substance for a period of one to six months, at levels ranging from the maximum tolerable to the minimum effective, a properly designed short-term study is capable of revealing virtually all of the toxic potential of an orally ingested substance. It is now recognized that such a comprehensive short-term investigation may indeed obviate the need for a chronic or lifetime study. A 90-day period in the young rat covers the span from infancy to full sexual maturity and only rarely is the toxicologist concerned with the evaluation of substances likely to be consumed daily throughout the life of man. The only information provided by tests of longer duration is reproductive performance (when included in the protocol) and cumulative or latent effects such as carcinogenesis, where the induction period in rodents may extend for a year or more. Current research efforts in the direction of discovering early signs of carcinogenic potential may avoid the necessity for lifetime studies altogether. Insofar as other aspects of toxicity are concerned, evidence has been assembled to show the relation between so-called "no-effect" doses based on 90-day tests in rats versus those based on 2-yr tests(*13*), the latter being one-half or less of the former in 50% of a series of 33 such comparisons(13).

The major objectives of a toxicological feeding study is to estimate the minimum daily intake of a substance that will induce an adverse affect in animals, and the maximum daily intake that they will tolerate throughout a normal lifetime. In some instances this information will suffice for the purpose of evaluating the safety for man of any expected level of pesticide residue intake, for example, when a sound judgment can be made of the metabolic fate of the compound based either on experimental data or on a knowledge of its chemical or pharmacological relationship to other substances. Short-term studies are considered by many qualified experts to be adequate in assessing the safety of food additives, or pesticide residues where the levels of exposure may be extremely low or infrequent. The present trend is toward intensifying short-term investigations by increasing the number of biochemical parameters measured, exploring more deeply the cellular changes, particularly in the liver, and following the metabolic pathways by which the test substance is distributed, transformed, and excreted.

The choice of species for toxicological feeding studies is determined by the expected predictability of effects (or lack thereof) in man, and by the desire to compare responses with those induced by similar or related test substances. For these reasons the animals most often used in evaluating the safety of pesticide residues are the rat and the dog, with the monkey coming into increasing prominence.

The rat is the rodent species in which most oral toxicity studies have been conducted and is preferred because it is omnivorous (like man), it grows and reproduces rapidly, it is economical of space and maintenance cost, it is docile, and is physiologically and anatomically similar to man in most respects. So much work has been done with the rat in the field of toxicological evaluation that results obtained with this species permit the comparison with many other compounds. Nutritionally the rat differs in its natural capacity to synthesize vitamin C, and anatomically it differs in that it lacks a gall bladder. Insofar as they have been investigated(14), the metabolic pathways by which chemical substances are transformed or detoxicated are generally alike in both the rat and man. The susceptibility of the rat to bronchopulmonary infection and its relatively high incidence of "spontaneous" mammary tumors militate against the choice of this species, although these disadvantages are not unique with the rat and, in any case, may be minimized with proper genetic selection of the strain.

The preferred nonrodent species for toxicological feeding studies is the dog, not only for physiological reasons but because of its availability, ease of handling, relative freedom from intercurrent infection, and size, which permits the collection of sufficient biological material for investigating the metabolic disposition of the test substance.

Nevertheless dogs must receive immunological treatment against leptospirosis and hepatitis, and must be deparasitized, not only prior to use but at occasional intervals thereafter. Seldom are dogs found to be completely free of intestinal parasites at autopsy even when bred and maintained under laboratory conditions. Among the various strains of dogs, the beagle is the one most used by toxicologists. Under no circumstances should mongrels of unknown history be employed in either short- or long-term feeding studies.

The availability of adequate numbers of disease-free primates is too recent to have permitted much experience in their use in chronic feeding studies. Supplies, however, are increasing and, as knowledge is acquired of the optimum conditions for maintenance and handling of these species and of their normal responses in terms of physiological and biochemical parameters, it is possible that the monkey will challenge the dog as the preferred nonrodent test animal. Currently investigations are being carried on with primates, including rhesus (*Macaca mulata*), stumptail (*Macaca speciosa*), cynamolgus (*Macaca irus*), squirrel (*Samiri*), and several other species of New- and Old-World monkeys.

In short-term toxicity tests with rats, groups are fed a series of three to five graded doses of the test material and a control group is fed the basal diet alone. Each group consists of 20 animals, 10 of each sex, although the control group should be somewhat larger, theoretically $N\sqrt{n}$ times the size of a test group, where N is the number of animals in a test group and n is the number of test groups. A common practice, however, is to use control and test groups of equal size. Statistically there is no sound reason why the group size of the nonrodent animals should be less than that of the rodent, but the factors of cost and space make this prohibitive. Since the purpose in using dogs or monkeys is to ascertain whether species differences exist, it is considered reasonable to use only three test groups of two to four animals of each sex.

Rats are started on test at or shortly after weaning, preferably in the 55 ± 5 g average weight range, to take advantage of their initial period of normally rapid growth. They are assigned to groups either randomly or, better, according to litter. Dogs or monkeys are placed on test at about six months of age, but only after preliminary conditioning and acclimatization have been completed.

The doses of test materials, whether incorporated in the diet or administered directly by mouth or intubation, are generally reckoned on a body weight basis, though they may be expressed in terms of per cent or parts per million of diet. The diet is fed ad libitum and a record of food consumption is kept as an index of effect on appetite. In the case of young rats, at least, it is also necessary to make frequent adjustments of dietary dose levels to compensate for the marked diminution of food intake relative to body weight as the rat approaches maturity. Table 1 illustrates how the food consumption of the weanling rat of the FDRL strain drops from the post-weaning level of 120 g/kg body weight to the adult level of 50 g, where it begins to plateau.

Accurate records of food intake are essential for establishing the actual dose of test supplement ingested via the diet. Except in cases where palatability is markedly impaired by the presence of the test material, agreement between actual and expected dosage is quite good. When this is not the case, additional "paired-feeding" tests are sometimes performed. This involves the feeding of each animal in a control group a quantity of diet equal to that consumed at libitum on the preceding day by its pair mate in the test group. Thus differences in growth response due to differing levels of food intake may be distinguished from differences attributable to toxicity per se.

The establishment of a critical dosage range for a short-term feeding study demands experience, judgment, and intuition on the part of the investigator. The highest dose level is expected to induce toxic response short of lethality, and is either estimated from the acute toxicity curve or is gradually approached by a preliminary range finding test. This may be performed by administering to, say, two rats some large fraction of the LD_{50}, and diminishing the doses to similar pairs by one-half or one-quarter to find the level which permits the animals to survive for two to four weeks. The dosage range employed in this preliminary test will depend somewhat upon the steepness of the acute toxicity curve.

The lowest dose in a short-range feeding study should induce no adverse affect in the test animals. It is theoretically impossible to establish the minimum "effective" dose since the animal organism is too complex a biological system to permit drawing a sharp line between positive and negative responses.

The Food and Drug Administration has recommended that five dosage levels be employed in short-term feeding studies, the doses intermediate between the highest and lowest dosage levels being established at equal logarithmic intervals. Many investigators use only three levels in these studies, but the greater the number the better is the likelihood of revealing a "no-adverse-effect" dosage level.

Incorporation of a test pesticide into the diet at graded levels requires considerable care to ensure uniform distribution. This, as well as the stability of the test material in the diet, should be checked by analysis. Dosing via the diet should be avoided in cases where the test material is likely to react with, or inhibit the utilization of, an essential nutrient constituent.

3. Observations. As indicated previously the objectives of toxicological feeding studies are to reveal not only the manifestations of effective doses of the test material, but also the threshold dose at which minimal or no adverse effects are induced. For the latter purpose, at least, much effort has to be expended to look for very little. The frank signs of organ damage or malfunction seen at high dosages may occur in lesser degree at lower dosages, but the more subtle effects of marginal doses are revealed only through the application of a battery of clinical, biochemical, and physiological tests. The greater the number and variety of such tests, and the more exhaustive the post-mortem examinations; the more meaningful is the toxicological appraisal. The search for dose-related responses includes observations in the intact living animal of growth, appearance, and behavior; chemical examinations of body fluids, tissues, and excreta; physiological tests of organ function (especially of the liver and kidneys); and biopsies of accessible target organs. These are followed post-mortem by gross and histomorphological examinations of all major organs and tissues. Modern sophisticated techniques of electron microscopy and histochemistry are currently being explored for their potential in disclosing early signs of injury at the cellular and subcellular levels.

In short-term feeding studies the clinical tests are conducted on representative animals in each group initially, terminally, and at one or more intervals in between. The length of the test period may not be sufficient to permit the entire gamut of examinations, hence only those considered to be more relevant may be included in the protocol. The following are among the major observations made in the course of short-term and subsequent chronic feeding studies.

(1) *Physical appearance.* Condition of fur, skin, eyes, ears, extremities, genitals, anal region, tail.

(2) *Behavior.* Signs of lethargy, posture, erratic movements, etc.

(3) *Growth response.* Body weight and possibly also body length in rats. Calculation of body surface area may be pertinent in certain types of metabolic intoxication.

(4) *Food consumption.* This is determined not only to observe effects on appetite but to check the dose actually ingested when incorporated in the diet. Efficiency of food utilization, i.e., grams gain in body weight per 100 g food consumed, is also calculated.

(5) *Water consumption.* Sometimes relevant in relation to urine volume.

(6) *Hemocytology.* (a) Erythrocyte count or hematocrit, leukocyte

count, and differential leukocyte count; (b) erythrocyte morphology; (c) reticulocyte count; (d) platelet count.

(7) *Hemochemistry.* (a) Blood glucose, urea nitrogen; (b) serum proteins (total protein, albumin/globulin ratio, and electrophoretic patterns); (c) triglycerides and cholesterol; (d) alkaline phosphatase, glutamic oxaloacetic and/or glutamic pyruvic transaminase, isocitric dehydrogenase, lactic dehydrogenase isoenzymes; (e) red cell cholinesterase; (f) prothrombin and/or clotting time; (g) sedimentation rate; (h) serum bilirubin.

(8) *Urinalysis.* Volume, specific gravity, reducing sugars, albumin, occult blood. Microscopic examination of sediment.

(9) *Special Metabolic and Functional Tests.* Carbohydrate metabolism–glucose tolerance; liver function–bromsulfalein retention; renal function–phenosulfonephthalein excretion, creatinine clearance.

(10) *Mortality.* This is best expressed in terms of the per cent mortality (or survival) of each group at quarterly periods throughout the test, and the median survival time, i.e., day or week in which more than 50% of each group have succumbed.

(11) *Necropsy.* Macroscopic inspection of all vital organs. Recording of absolute and relative weights of liver, kidneys, spleen, heart, thyroids, adrenals, pituitary, and gonads. Fixing in 10% formalin of these organs and the stomach, large and small intestines, pancreas, bladder, gonads, thymus, salivary gland, lymph nodes, lungs, marrow skin, muscle, spinal cord, and brain.

(12) *Histomorphology.* Section and stain (hematoxylin–eosin) major organs and tissues of animals in control and highest dosage groups, and livers and kidneys of all groups. Following microscopic examination, those organs showing significant pathological alteration are similarly examined in lower dosage groups.

In short-term rat feeding studies items (1) and (2) are examined daily, items (3), (4), and (5) weekly, or twice weekly during the first few weeks and weekly thereafter; items (6), (7a), and (8) initially (pretest) and at 1, 3, 6, and 12 weeks (and 24 weeks if the test continues for 6 months). The remaining determinations are made at one or more times (e.g., 6 and 12 weeks) during the course of the test period.

These parameters provide a broad spectrum of indices of physiological and morphological change in the animal species. Whereas the normal ranges for most of the biochemical values are to be found in the literature, it is highly desirable to establish for each strain of animal, in each laboratory, and for each analytical method, the standard error for each determination, in addition to computing the significance of differences between concurrent test and control groups (*15*). Standard reference works(*16*,

17) should be consulted for acceptable procedures for carrying out these clinical, biochemical, and pathological examinations.

From the foregoing outline it should not be assumed that all of these observations need to be carried out in cook-book fashion in every toxicological study of a pesticide. Some are pertinent in certain cases but not in others. For example, the cholinesterase determination is essential in tests of organophosphorus and carbamate insecticides but not in those of the organochlorine category, whereas the latter show virtually no effect on the levels of serum enzymes except in severe hepatic damage. In the last analysis the choice of procedures must be left to the discretion and judgment of the experienced investigator.

In the search for "early warning signs" of potential toxicity, efforts have been directed in recent years to the techniques of molecular biology. The mode of action of the gross toxic effects of chemicals is in most cases unknown, even though these effects are reflected in functional or biochemical deviations from the norm. How these changes are triggered at the cellular or subcellular level may be discovered through the application of electron microscopy and the correlation of histomorphological with histochemical tests. The effect of certain agents on the reticulated endothelium of hepatic cells and the relation of this effect to the elaboration of drug-metabolizing enzymes(*18, 19*) give promise of useful developments in this field of research. The possibility of demonstrating similarities or differences in the ultrastructure of human and animal tissues such as the liver, that vital organ of metabolic transformation and detoxication, suggests that electron microscopy may furnish a much needed tool for the selection of appropriate test species. However, research in this area has not yet reached the stage where this technique is a sine qua non in safety assessments.

4. Long-Term ("Chronic") Oral Toxicity. The term "chronic" in this context should, strictly speaking, be inferred to mean the major fraction of the lifetime of the species in question; in the case of the rat, two years. However, it has sometimes been applied to 2-year tests in the dog, monkey, and other longer-lived species.

Except for the greater duration and numbers of animals employed, long-term feeding studies are in most respects simply an extension of the short-term studies. In the interest of saving time, the latter are, in fact, often started with larger groups, for example with 35 rats (or 12 dogs) of each sex, so that 10 rats (or 2 dogs) may be sacrificed at the termination of the short-term test, leaving the desired number of animals per group for the long-term study.

At the conclusion of the short-term test, a reasonably reliable judgment can be made of the proper dosage range to employ for continued investigation. These are usually limited to three dose levels which occasionally may extend above or below those used in the short-term test. Following the initial 12 weeks of rapid growth, when the rodents have reached adulthood, body weight records are taken at less frequent intervals, e.g., monthly, and food intake is recorded and food efficiency calculated only for 1-week periods during the 6th, 12th, and 18th months. At these periods, and just before the termination of the 2-year test, clinical examinations are repeated, particularly those which earlier observations indicated to be significant in terms of revealing adverse responses.

The long-term investigation provides the opportunity for more observations and determinations than are possible in the shorter test period. Included among these are tests of hepatic or renal function, tissue biopsies (especially of the liver in dogs), metabolic balance studies, investigations of the route of distribution or excretion of the test material, effects on fertility, reproduction, and lactation, teratological examination of the newborn, and tumorigenesis.

At the conclusion of a chronic study the responses are carefully evaluated in relation to the graded dosage levels employed in the test groups, and the responses of the untreated or control group. It is here that distinctions must be drawn between "incidental" observations, such as occur in colonies of "normal" animals, and relevant or dose-related findings.

The significance of differences between groups is generally assessed according to some statistical criterion, as illustrated in Table 6-2, based on a 95% probability factor. Such interpretations must be made with discretion, inasmuch as a marginal, statistically significant difference may not necessarily be biologically or toxicologically significant. Sporadic observations, which are not graduated to increasing dosage, or which are found in most animal colonies under "normal" conditions, may not be dose-related. These factors are especially important in establishing the "no-effect" level, which is perhaps the main objective of the study. Problems involved in this estimation are discussed below.

E. Reproduction and Teratology

The endocrinological changes associated with the reproductive cycle in the female and the anabolic systems involved in embryologic and fetal growth constitute a challenging background against which to test the toxic potentiality of a drug or other foreign substance. It is conceivable that under these conditions, enzyme systems are called into play that are otherwise dormant, if indeed existent. Behavioral phenomena such as loss of sexual interest or maternal neglect, and defective prenatal

TABLE 6-2
DIFFERENCES BETWEEN TWO GROUPS NECESSARY FOR SIGNIFICANCE AT A p VALUE OF 0.05 (9)

50 animals per group		40 animals per group		30 animals per group		20 animals per group		10 animals per group	
Number less affected[a]	Number more affected	Number less affected	Number more affected	Number less affected	Number more affected	Number less affected	Number more affected	Number less affected	Number more affected
0	6	0	6	0	6	0	6	0	5
2	8	2	8	1	8	1	7	1	7
5	13	4	11	3	11	2	9	2	8
10	19	8	17	6	14	4	11	3	9
15	25	12	21	9	17	6	13	4	10
20	30	16	25	12	20	8	15	5	10
25	35	20	29	15	23	10	17	6	—
30	40	24	33	18	26	12	18	7	—
35	44	28	36	21	28	14	20	8	—
40	47	32	38	24	30	16	—	9	—
45	—	36	—	27	—	18	—		

[a] "Less affected" generally refers to control group.

437

development are manifestations of toxic effects that can be seen only in appropriately planned experiments.

Effects on fertility, reproduction, and lactation are generally determined in mating studies with rats. After about 12 weeks on their respective test and control diets (or at the termination of a 90-day test period), a male and one or more female rats from the same experimental groups are placed in mating cages. In teratological studies it is necessary to establish the day of impregnation by the presence of a vaginal plug or of sperm in smear. Pregnancy is determined either visually and by palpation or by the marked weight increment. One male may be housed with as many as three females. In any case, if pregnancy is not established within two estrus cycles or in two weeks, the male rats may be transferred to other cages and if again no response is obtained, the females should be mated with males from a breeding colony of known fertility. Pregnant rats are preferably isolated for the remainder of the gestation period and are kept with their litters until they are weaned. After a brief rest period, the mating process is repeated with the same females until a second litter is weaned. Some investigators consider it sufficient to terminate the reproduction study at this point; others, however, prefer to continue the reproductive cycles as a measure of the cumulative effect on the duration of fertility in the females in the initial (F_0) generation. More often, however, multigeneration studies are conducted by selecting mating pairs from the second litters of each generation, maintaining them on the same experimental rations as their parents, and repeating the reproduction procedure. The design of such a study is illustrated in Fig. 6-2 (20).

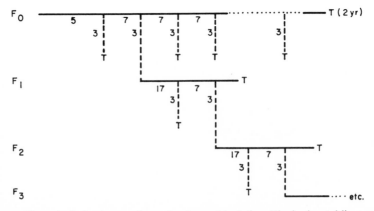

Fig. 6-2.   Chronological scheme of reproduction and lactation. The horizontal lines represent the generations of rats through their successive matings and the dotted vertical lines indicate litters; termination of a litter or a generation is shown by the letter T; the figures indicate the number of weeks elapsed at each stage, beginning with the first mating in F_0 (20).

The pups in each litter are counted and weighed on the 4th and 12th day and on the 21st day, when they may be weaned. With care they may also be weighed at birth, but at the risk of arousing the mother to the point of destroying her litter. In addition to these criteria, the following indexes of reproductive and lactational performance are determined:

Fertility Index — the percentage of matings resulting in pregnancies

Gestation Index — the percentage of pregnancies resulting in the birth of live litters

Viability Index — the percentage of rats born that survive for 4 days

Lactation Index — the percentage of rats alive at 4 days that survive for the 21 days of lactation

Failure of the newborn to survive may be a result of impaired prenatal development and/or cannibalism, or of lack of maternal interest, any of which may be a toxic manifestation.

Prior to the thalidomide episode, it was neither customary nor expected that drugs or pesticides be investigated for the possibility that they might be teratogenic. In fact no generally satisfactory methods were available whereby this potentiality could be reliably predicted. Since then procedures employing various species of rodents, primates, and the developing chick embryo have been studied, certain of which for want of better methods, are currently recommended for the evaluation of chemical substances that may be ingested during pregnancy. In the absence of human experience, it is not possible to gauge the predictive value of these procedures, though prudence would certainly dictate the proscription of any substances as a component of food if it were capable of inducing congenital malformations in experimental animals.

Present recommendations require that the usual reproduction studies in rats be supplemented by three types of experiments especially designed to elicit effects on the fetus resulting from the ingestion of the toxicant before or during gestation. The first involves daily dosing of male and female rats at least 60 days prior to mating, and examining the males for any effect on spermatogenesis and the females for signs of copulation. This may be indicated by the presence of spermatozoa or a plug in the vagina. Dosing is continued during pregnancy and on the 13th day after copulation, half the females in the group are sacrificed and examined for the number of embryos and implantation sites and for evidence of resorption. The other half are allowed to go to term. Grossly abnormal pups are examined for skeletal defects, while those apparently normal are carried to weaning, their number and weight being recorded on the 4th and 21st days.

The second type of teratogenic experiment may be conducted either with rats or rabbits.Occasionally mice have also been used. After mating with normal, untreated males, females are dosed from the 6th to the 15th (in the case of rabbits the 18th) days of pregnancy only. On the day preceding expected parturition, delivery is effected by Caesarian section and the uterine contents inspected as described above. A half or a third of the fetuses are examined by dissection or by serial sectioning(21) and low-power microscopic examination for abnormalities of the soft tissues and organs, while the remainder are "cleared" and their skeletal structure stained and examined for the presence of missing, misshapen, or fused bones by techniques described in the proceedings of the second Workshop in Teratology(22) published as a supplement to the first [cited in(21)].

In the third type of teratogenic study emphasis is directed to the effect of administering the test substance to rats during the last trimester of pregnancy and through the weaning of the litter. The duration of pregnancy is recorded and observations are made of prolonged or difficult labor, maternal attention to the young, and their growth and survival through the lactation period. In the event of a manifest toxic response of the does, such as neglect or cannibalism, litters may be transferred to foster mothers.

It is apparent that these procedures are of an exploratory nature, but when conducted on a sufficient number of animals of each species (at least 20 female rats or 10 rabbits per group and at least two dosage levels plus an untreated control group) the data obtained provide the most practicable approach presently available for the assessment of teratogenic potential of drugs and food chemicals.

As a screening procedure for possible teratologic potency, a test has been proposed based on the effect of injecting the compound into the yolk or air sac of the developing chick embryo(23). This is a rather severe test, since it does not permit excretion of the potential toxicant as in the living animal, but a negative response would appear to have some predictive value. However, many substances known to be innocuous have been shown to produce structural defects in chicks by this method, therefore a positive response must be interpreted with caution.

F. Carcinogenesis

The experimental approach to evaluating the carcinogenic potential of a pesticide is beset with more complications than the usual chronic toxicity study. Differences in susceptibility between species and between genetic strains within species, the "normal" incidence of spontaneous tumor formation, and the long latent period before the signs of neoplasms become evident, present difficulties not only in the execution of carcinogenetic investigations but in the interpretation and practical application of the findings.

The presence of a potent carcinogen in the diet of the rat is generally manifested in the course of the 2-year feeding study and may also be detected during this period in dogs. It is when these studies reveal a weak or borderline tumorigenic response that experienced judgment is required and additional, more critical tests may become necessary. These are fully described in publications of the Food Protection Committee of the National Academy of Sciences/National Research Council(9) and the Joint FAO/WHO Expert Committee on Food Additives(10). Despite the extensive search for short, yet definitive approaches to the estimation of the carcinogenic potential of pesticides and other environmental chemicals, no satisfactory substitute has yet been found for chronic tests in relatively large numbers of experimental animals.

Rats and mice have been the species of choice not only because of their short life cycle but because more is known about the predisposition to tumorigenesis of various strains of these animals than of any other species(24). Dogs are the nonrodent mammals most often used in studies of oral carcinogenesis. Practical limitations in the numbers needed to given meaningful results and the minimum of 5 years required to reach a negative probability as to tumorigenic activity are serious obstacles to the use of the canine species.

Owing to the fact that a great deal of the early work in experimental carcinogenesis was based on subcutaneous injection or on topical application to the skin of rodents, much consideration has been given to the appropriate route of administration for the evaluation of dietary carcinogenesis. It is generally agreed that "carcinogenesis... by one route of administration does not imply carcinogenesis... by another route of administration"(25). Whereas the induction of local sarcomas at the site of subcutaneous injection is regarded by some oncologists as sufficient cause for proscribing the use of food additives, the view of the Joint FAO/WHO Expert Committee is that "when large doses of material must be given repeatedly, the subcutaneous route of testing substances for their carcinogenic activity is of limited value and of dubious interpretation. Despite this criticism of the test, the subcutaneous route should not be disregarded and might prove of value in the routine evaluation of the carcinogenicity of some substances proposed for use in foods"(9).

To compensate in some degree for the variations in genetic susceptibility among strains of rats and mice, it is recommended that at least two strains of each species be used, one of which should have a definite predisposition to the development of spontaneous tumors. More information concerning this tendency is known about mice (of which a greater number of pure strains is available) than about rats (24).

Two or 3 dosage levels are usually employed in carcinogenesis studies, the highest dose being as much as the animals can tolerate without materially reducing their life span. Because of the normal rate of mortality and the necessity that a sufficient number of survivors remain to permit valid statistical assessment of low-incidence responses, larger numbers of animals per group must be used than for the usual chronic toxicity test, at least twice as many of each sex in both test and control groups as in the usual chronic studies.

The duration of a feeding study for cancer-inducing potential is determined by the fact that this is most frequently a disease associated with advanced age, whether because of its long induction period or because of a latent susceptibility of the host. Some investigators recommend that the test animals should continue on treatment until they become moribund or die rather than that the experiment should be terminated at any fixed period. The dosage period, however, is generally limited to two years, or even to one, according to British investigators (26). During the course of the test, observations are made at weekly or more frequent intervals for the time of appearance, characteristics, and rate of development of visible or palpable growths. The most critical phase of these studies is the necropsy and the subsequent histological examinations of not only the grossly affected or suspect organs but also all major organs and tissues in at least the highest dosage group of animals. Here, too, the detection of frank carcinogenesis presents little problem, whereas distinguishing between a benign and a malignant process involves the cautious judgment of the experienced pathologist well versed in histomorphology of the species in question.

At this point it should be recalled that the law prohibits the presence in any amount of a pesticide residue or food additive in food if at any level of oral ingestion it induces cancer in man or animal (27). Notwithstanding the accumulating evidence that the carcinogenic response, like most pharmacologic or toxic effects, is graded to dose, and that at some dose no such response is manifested, regulatory agencies have construed the statutory language of the Federal Food, Drug and Cosmetic Act literally, and refuse to recognize a safe level of a substance if at any dose it is demonstrated to induce cancer upon ingestion.

There is little hope of a change in administrative policy so long as it is guided by the view that "No one at this time can tell how much or how little of a carcinogen will be required to produce cancer in any human being" (28). It can seriously be doubted that this objective will ever be capable of achievement short of quantitatively controlled, lifetime studies in man.

For obvious reasons, clinical pathologists attempt to distinguish between benign tumors, which are generally reversible and nonlethal, and malignant neoplasms, which are irreversible, often metastatic, and lethal. The term "cancer" is applied to the latter type of lesion, but despite the fact that it is used in the legal prohibition, the tendency is to regard any orally induced tumor as suspicious and possibly "precancerous" since the benign lesion may progress to the malignant stage, given sufficient time or repeated insult.

The liver is the most frequent target organ of carcinogenic pesticides, such as the chlorinated hydrocarbons, most of which are hepatotoxic in some degree. The stages of transition in the rat liver from localized protoplasmic hyperplasia to frank, invasive hepatocarcinoma, have been described by Popper(29). This degree of expertise is essential for the definitive interpretation of experimentally induced neoplastic lesions.

6-6 LIMITATIONS OF SAFETY EVALUATION IN ANIMALS

Aside from their high cost in terms of time, facilities, and manpower, chronic toxicological investigations have certain inherent limitations which are implicit in the oft-expressed statement that the prediction of safety can never be guaranteed on an absolute basis. There is, first of all, the problem of translating conclusions based on experiments with a few species of laboratory animals to the human species. In addition to the rat, mouse, dog, and monkey, investigations have been carried out in recent years with chickens, ducks, turkeys, Japanese quails, pigs, rabbits, ferrets, gerbils, hamsters, and several other species. The greater the similarity in inter-species response, the more reliance is placed on the validity of extrapolation to man. The mammalian species used in toxicological studies for safety assessment (principally the rat and the dog) resemble the human in most physiological and anatomical respects, and judgments based on findings obtained with these species have provided a high degree of assurance of safety under the practical conditions of use of pesticides, food additives, etc.

The common assumption that monkeys bear a close resemblance to man and hence should be the preferred species for safety evaluation studies may be questioned on the ground of the known physiological differences among the numerous species of primates.

Variations among species and strains of experimental animals with respect to the "normal" incidence of characteristic pathological conditions, such as pyelonephritis and pneumonitis in rats and certain viral conditions and parasitism in dogs, add to the difficulties of interpretation of chronic toxicity data. In safety evaluations emphasis is directed toward the relative incidence of marginal as well as more pronounced effects. Thus the small size of test groups compared to the entire population of any given species becomes a limiting factor from a purely statistical standpoint. In this respect negative responses in animals are less meaningful than positive or adverse responses.

The wide range of dosages employed in animal studies presents the possibility of another problem of interpretation, namely that of the relation of dosage to metabolic disposition of the toxicant. The pathways of biotransformation (and hence the toxicity) may be quite different at elevated doses from those at the minimal levels of ingestion of a chemical compound.

Toxicological experiments are conducted under well-defined conditions not only as to species and numbers of animals, but also with regard to their initially normal state of health and the nutritional adequacy of their basal diets. The signs of injurious effect or impaired function are either grossly evident or are observed only upon careful inspection. Aberrations in growth, appearance, behavior, reproduction, etc., are manifestations of toxicity that generally become apparent at high dosage levels, but disclosure of the more subtle or latent signs of biochemical or histological change requires a specifically directed search. Hence irrespective of how numerous and varied such examinations may be, the finding of a "no-effect" dose level can only mean that no adverse effect was seen as a result of the criteria employed by the particular investigator. The application of tests or determinations not previously included in an experimental program, or possibly not yet available, might show a response not hitherto seen. This is considered further below.

No satisfactory animal tests are available today for uncovering the capacity of a drug for inducing certain types of blood dyscrasia, such as agranulocytosis, or allergic reactions of moderate, yet not uncommon incidence. Methods for estimating blood lipid fractions or serum enzyme levels were nonexistent until relatively recently and, therefore, were not employed in the evaluation of pesticides in wide use today. Notwithstanding the animal investigations that preceded the use of chloromycetin and thalidomide, their toxic potential was not demonstrable by the methods then available and did not become apparent until after clinical experience had been built up with these drugs.

Promising as they may be, observations by such sophisticated techniques as those of histochemistry, electron microscopy, and psychopharmacology have not thus far been sufficiently correlated with data for the more conventional parameters of physiological response to warrant their general adoption in protocols for safety appraisal.

In any case, the experimenter's armamentarium lacks suitable means for observing or measuring subjective effects in animals such as fatigue, headache, nausea, or minor central nervous system abnormalities. The latter emphasizes a direction that a considerable amount of pharmacological and toxicological research is presently taking, namely the study of behavioral effects of drugs and chemicals in the animal organism. Dogs, monkeys, rodents, and pigeons are the principal species that have been used in these investigations, and thus far, at least, tests for behavioral reactions have not been officially proposed as criteria for the appraisal of safety of pesticide residues.

A. The "No-Effect" Level

In no phase of toxicological appraisal is the application of statistical criteria more important than in the interpretation of responses at the level of minimal effect. Whether a real or suspected adverse response occurs with greater frequency in test groups than in the untreated controls is determined by means of Student's t test, the chi-square test, or modifications thereof (cf. Table 2). The use of modern computer techniques promises to eliminate much of the drudgery of statistical calculations.

The "no-effect" level is the dose (presumably the highest attainable dose) in a chronic toxicity study at which no adverse effects are observed in a group of test animals compared with the untreated control group. It is in this connection that statistical procedures come into play inasmuch as aberrations seen in test animals may also occur with only slightly lower frequency in the controls. This applies especially to tumors of the type that occur spontaneously.

Estimation of the "no-effect" level of a pesticide is a critical step in arriving at the maximum acceptable dietary intake. Acceptable daily intake has been defined(30) as the daily dosage of a chemical which, during an entire lifetime, appears to be without appreciable risk on the basis of all the facts known at the time. "'Without appreciable risk' is taken to mean the practical certainty that injury will not result even after a lifetime of exposure." From the "no-effect" level, the acceptable daily intake is derived by the application of an appropriate safety factor. Both the "no-effect" level in test species and the acceptable daily intake for man are expressed in terms of mg per kg body weight per day.

Certain of the difficulties involved in determination of the "no-effect" level have been discussed in the foregoing section. The safety factor is intended, in part, to compensate for these difficulties and the uncertainties inherent in extrapolating from animals to man.

"No effect" implies no toxic or adverse effect. A clear-cut decision as to whether a deviation from the "normal" response or from the response of control animals is indeed an adverse effect, is not always possible. An effect may be real, though not relevant. A slight, though statistically significant, increase in the relative weight of an organ, such as the liver, may not be an adverse effect, particularly if it is unaccompanied by histomorphological alteration or change in the functional efficiency of the organ. The concentration of a residue of the pesticide in the fat depots of the body (such as occurs with the chlorinated hydrocarbons) unaccompanied by manifestations of injury may also be regarded as not an adverse effect.

Regardless of how many parameters of response may be measured in toxicological studies, it is conceivable that one or more tests not employed, or perhaps not yet devised, might later reveal an effect where none was previously observed. The failure to detect a given effect under one set of experimental conditions does not preclude the possibility of such an effect under other conditions.

Despite these limitations the threshold or no-effect dose in animals has served satisfactorily as the basis for arriving at acceptable daily intakes, and has permitted the evaluation of safety of food additives and pesticide residues with a degree of assurance sufficient to satisfy existing public health requirements.

B. Safety Factors and Tolerances

Paracelsus taught that "no thing is without poison. The dosage makes it either a poison or a remedy." Modern toxicologists have abandoned the "poison per se" concept in favor of evaluating hazard, or conversely, safety, under conditions of intended use.

The injury which a food chemical is capable of producing in man may be determined either (a) by observing the effects of accidental or homicidal ingestion, or of occupational exposure by chemical operators, formulators, and applicators, or (b) by conducting controlled studies with intact animals, or with living tissues. Deliberately planned experiments, of course, provide the more reliable criteria upon which to base assessments of toxic potential. Evaluation of the hazard or risk associated with the use of a toxic substance must go beyond simply studying the toxic effects of unintentionally excessive exposure or of intentionally high test dosages. It must take into account the nature and extent of probable exposure in respect to route, dosage, frequency, and duration, as well as the metabolic capabilities of the host species. These factors are implicit in the term "conditions of use" which must be considered in predicting whether a poisonous substance is likely to be harmful.

Prediction of safety under conditions of use by man from toxicological data obtained with laboratory animals involves both qualitative and quantitative elements of uncertainty. In the absence of complete and reliable information as to the metabolic disposition of a foreign chemical, the transition from species to species cannot be made with absolute assurance. Furthermore, such factors as variation in individual response within populations, irrespective of species, and the relatively small number of test animals of a given strain as compared with the entire population of that species, much less the entire human population "at risk" with respect to the substance under test, militate against absolute assurance of predictability.

To compensate for the uncertainties in applying experimental animal data to man, a liberal margin of safety is allowed between the observed maximum no-effect dosage and the acceptable dietary intake for man. The more-or-less arbitrary safety factor of 1/100 has been generally employed "except where evidence is submitted which justifies use of a different safety factor," to quote an interpretative regulation of the Food and Drug Administration. Such evidence may relate to the nature of the chemical substance, experience in prior use, controlled tests with volunteers or examination of workers occupationally exposed to it.

Estimation of the acceptable dietary intake (ADI) is based on dosage expressed on a body weight basis, e.g., milligrams per kilogram body weight per day. The ADI is thus the limiting daily quantity of the substance that may be safely ingested from whatever source or sources e.g., as permitted pesticide residues and as unintentional food components resulting indirectly from persistence in soil, drift from use on non-food crops, etc.

It is important to recognize the distinction between the ADI for a pesticide and a tolerance governing its use. The former is based solely on toxicological evidence, with allowance of a presumably adequate margin of safety; the latter is based on a showing of practical need to accomplish effective use in the production or protection of agricultural commodities. A legal tolerance, or maximum permitted residue of a pesticide is established at no higher level than that necessary for its intended purpose. In granting tolerances for any specific substance, the regulatory agency must be certain that the sum of all tolerances for that substance does not, in actual practice, exceed the ADI. Experience has shown that "tolerances are not additive." Permission to use a pesticide up to its tolerance levels on various crops does not imply that it is in fact employed on all such crops, at all seasons, and in all geographic areas.

Surveillance studies are conducted regularly and on an increasing scale by the Food and Drug Administration. By means of analytical procedures capable of detecting the presence of 54 pesticide chemicals, it has been shown that out of 49,000 samples tested over a three-year period, the most commonly found pesticides and their frequency of occurrence were as shown in Table 6-3.

TABLE 6-3
RESIDUE LEVELS OF VARIOUS PESTICIDES
FOUND IN DOMESTIC AND IMPORTED FOOD
SAMPLES

	Frequency, per cent	
Pesticide chemical	Domestic samples	Foreign samples
DDT	26.7	30.4
DDE	25.5	26.2
Dieldrin	17.8	13.2
TDE	10.0	11.8
Heptachlor epoxide	7.5	1.5
Lindane	4.8	4.5
BHC	2.8	6.7
Endrin	2.5	1.4
Aldrin	2.0	2.9
Toxaphene	1.5	0.7
Kelthane	0.7	1.8

In addition to these monitoring surveys, the Food and Drug Administration conducts regular "market basket" studies in which 82 foods, representing the total diet of a young adult male for two weeks, are prepared for consumption, composited in classes, and analyzed for the presence of pesticide residues. As illustrated in Table 6-4, the data indicate that the actual levels of intake of pesticides in the U.S. are considerably below the ADI's proposed by a Joint Committee of the Food and Agriculture Organization and the World Health Organization.

Enforcement of any legal proscription of the presence of a toxic sub-

TABLE 6-4
ESTIMATED vs. ACCEPTABLE INTAKE OF PESTICIDES

Pesticide	Estimated from "market basket" studies, mg/kg	ADI, Body weight/day
DDT	0.0005	0.01
Lindane	0.00006	0.0125
Malathion	0.009	0.02
Carbaryl	0.0012	0.02

stance, such as a pesticide, in food, demands the application of procedures capable of proving the complete absence of the substance. Since all chemical analytical methods have limits of sensitivity beyond which they are incapable of detecting the presence, much less the amount, of a trace contaminant, it is apparent that so-called "no-residue" or "zero-tolerance" regulations cannot be enforced literally. Certain pesticide uses have been permitted by the Department of Agriculture on the ground that no apparent or demonstrable residues resulted therefrom. Subsequent development of analytical techniques of exquisite sensitivity, however, revealed the presence of small though finite traces. This resulted in reconsideration and later abandonment of the no-residue concept and its replacement by the more realistic "negligible residue" tolerances, which recognize the existence of detectable, though toxicologically insignificant, concentrations of pesticides.

Since the adoption of the Federal laws regulating the use of pesticides in agriculture, many registrations were promulgated allowing the use of certain pesticides on a "no-residue" or "zero-tolerance" basis. In some cases, this was because the substances in question were considered to be unsafe at any concentration, as, for example, those found to be carcinogenic; in other cases it was because their use resulted in no detectable residues when tested by the currently acceptable analytical methods.

However, in the course of time, analytical methods of improved sensitivity were developed, and it was possible to detect amounts of residue where none had previously been thought to exist. Whereas this was not believed to pose a health hazard, it became advisable for administrative purposes to substitute the "no-residue" or "zero-tolerance" concept for the more rational and realistic "negligible residues" standard. A tolerance established on this basis will, generally speaking, require less toxicological supporting data (e.g., a short-term rather than a two-year feeding study) and will be controlled by an analytical method for residue content which would not be changed "without notice and opportunity for comment by interested parties."

The ad hoc advisory committee of the National Academy of Science — National Research Council, in recommending the adoption of the negligible residue policy, suggested that the limit be set at some small fraction of the ADI, and as an example mentioned 5%. Assuming 1/100 to be the commonly used safety factor in establishing ADI's, it follows that 1/2000 would be the suggested factor for converting the no-effect level in test animals to an acceptable limit for negligible levels. Like any safety factor, it should be used only as a guideline, with at least the same degree of flexibility as applied to the determination of the ADI for permitted pesticide chemicals.

6-7 TOXICOLOGICAL TESTS OF PESTICIDES IN MAN

The grossly exaggerated conditions imposed on animal tests for pesticide toxicity, together with the wide margin of safety between minimal effective doses in animals and maximum acceptable intake by man, are fundamental factors that are considered in setting the ceiling on permissible residue levels ("tolerances") with the degree of "reasonable certainty that no injury will result from the proposed use" sufficient to satisfy regulatory authorities. The evaluation of the safety of a drug takes into account the relatively small margin between its pharmacologically effective dose and its toxic dose. The prudent pharmaceutical manufacturer has therefore traditionally sponsored (and the law now requires) cautious clinical trials of any new drug before it is approved for general use.

Notwithstanding the difference between the intentional use of a therapeutic agent and the unintentional, though unavoidable, presence of a pesticide residue in food, it has been urged by some toxicologists that pesticides be subjected to human tests prior to granting approval of residue tolerances. It is argued that drugs are generally used voluntarily and under medical control, that adequate precautionary labeling is required on preparations used for self-medication, and that toxic reactions or side-effects are recognizable and traceable to their cause. On the other hand, the presence of pesticide residues in foods is not known to the consumer, and it is extremely difficult, if not impossible, to establish an etiological relationship between a subtle or delayed toxic reaction and a residual component of a food.

Because it would help fill the gap of uncertainty in the transition from the experimental animal to the human species, controlled observations of the effects of the ingestion of pesticides by man are believed to be important. There can be little doubt, however, that they add to the fund of useful, if not essential, information involved in the prediction of safety. Subjects occupationally exposed to pesticides, either in chemical plants or in the preparation or application of pesticide formulations, would seem to offer a valuable opportunity for clinical appraisal since their exposure level is often rather high and continuous. The difficulties of limiting industrial exposure to a single substance and of regulating it from the quantitative standpoint, not to mention the problem of labor relations that may arise, are among the obstacles to the wider use of this type of clinical material for toxicological appraisals.

From time to time patients brought into hospitals following accidental or suicidal ingestion of pesticides are studied for toxic (sometime premortal) reactions and for the response to antidotal treatment. However the effects observed in these cases are of an acute nature and give little indication of the cumulative response that may be expected from repeated ingestion of trace amounts of the toxicant.

More reliable sources of human subjects for this purpose are prisons and other institutions with "captive" populations. A valid question may be raised as to whether such subjects are typical of the general populations to which the experimental conclusions are to be applied. Nevertheless studies with inmates of these institutions are capable of a greater degree of control from the standpoint of dosage, diet, activity, and clinical observation, than is possible with ambulatory volunteers. The opposite situation has also been cited of the drug found to be free of side effects when tested in prisoners but toxic when administered to patients following the consumption of moderate amounts of alcoholic beverages.

A committee of the National Academy of Sciences/National Research Council, reporting on "Some Considerations in The Use of Human Subjects in the Safety Evaluation of Pesticides and Food Additives" (31), has recommended that such studies be conducted only after adequate animal experiments have been completed and when the information sought can be obtained in no other way. Apart from the subjective or behavioral effects of moderate doses, the principal data that might be revealed by human experimentation relate to the metabolic disposition of the toxicant, the biotransformations it may undergo, and its effect on organ (particularly liver) function and morphology at the cellular and subcellular levels. The necessity to restrict dosage and the limitations of sampling and of analytical detection in body tissues and excreta (even by means of radio-isotope techniques) are obvious handicaps to the achievement of these desirable objectives. Nonetheless, any conclusions or suggestions derived from such investigations can be fed back for further, more critical animal studies in the species most closely resembling man in respect to these responses.

Sight must not be lost of the fact that even under optimal conditions, human experiments are severely limited by their short duration relative to the normal life span of the species; by lack of superimposed physiological stresses such as pubescense, pregnancy, lactation, and senescence; by the magnitude of the doses that can be administered with impunity, and by the absence of exhaustive macroscopic and histopathological examination of the viscera after treatment.

These limitations preclude the use of human subjects for assessing the risks of carcinogenicity and teratogenicity, as well as of irreversible impairment of growth or development of vital organs.

Experience with toxicological tests of pesticides in human subjects is too recent to have revealed any cases where pathological changes, not manifested in experimental animals, should preclude the use of these chemicals.

In the pharmacological and clinical evaluation of drugs for safety and efficacy it is required by law that voluntary and informed consent be obtained prior to the use of a human subject. The question has been raised as to how "voluntary" such consent is in the case of prison inmates offering to be subjects for pesticide tests in exchange for special dispensations. But this, as well as the moral aspects and legal responsibilities implicit in such investigations, is beyond the scope of this chapter.

Suffice it to say that the major criteria to consider before embarking on toxicological tests of pesticides with human volunteers have been well summarized by the late Professor Alastair Frazer(32), as follows:

(a) The information sought should be needed in the interests of the community, or for the benefit of the individual receiving the substance, or those with similar medical problems.

(b) The information should not be readily obtainable by any other means.

(c) The potential risks arising from the administration of the substance, or from the methods used for investigating the effects of the substance, should have been adequately defined by previous animal studies.

(d) The potential risks should have been fully explained and adequately understood by the human subjects taking part in the investigation and, where appropriate, by legal guardians or dependents.

(e) The effects to be studied should be clearly defined, whether they be wanted or unwanted effects, and appropriate methods for the assessment of these effects should be available.

(f) Only those effects that have been shown to be reversible in animal studies and that would give rise to no serious permanent damage should be studied.

(g) The human subject under investigation should retain a continued power of veto throughout the study and a similar power of veto should be accorded to an objective supervisory body.

6-8 ESTIMATING HAZARD TO FISH AND WILDLIFE

This chapter has been concerned primarily with the assessment of toxicity of pesticides for man via tests in experimental mammals. The increased agricultural and environmental use of these agents for insect, weed, and rodent control and the demonstration of the ubiquitous presence of traces in soil, streams, and air, have given rise to serious concern on the part of conservationists and ecologists, lest our populations of birds, wildlife, and fish be extirpated. There are legitimate grounds for exploring the possible effects of pest control on these natural resources. Aerial applications of pesticide formulations have been responsible in large measure for unintentional dispersion over woods, fields, and orchards. Dissemination through drift is affected by particle size, height of application, volatility, and wind. The nature of soil type and the run-off waters therefrom clearly influence the extent of stream pollution with pesticides, to which must be added the occasional direct contamination by chemical-plant effluents.

The records show that destruction of birds and fish has occurred upon occasion in localized areas. Forebodings of widespread or permanent eradication of natural life, as expressed in Rachel Carson's *Silent Spring*, however, are not supported by statistics over the past quarter-century, during which the number and uses of synthetic organic pesticide chemicals have increased to a substantial degree. Nevertheless, the U.S. Department of Interior, through its Fish and Wildlife Service, has shown increased interest in the situation and has joined with the Department of Agriculture and the Department of Health, Education, and Welfare in requiring safety evaluation of pesticides in avian and marine species. A Federal Committee on Pest Control has been established representing these three agencies.

Little concern is felt for compounds which volatilize or decompose readily, leaving innocuous end products, as compared with the more stable and persistent compounds, such as the organochlorine derivatives. The latter, as well as substances applied repeatedly over long stretches of time, pose the more serious possibility of chronic toxicity to fish and wildlife.

The procedures suggested by the U.S. Fish and Wildlife Service for evaluating new economic poisons are principally acute toxicity tests. The species proposed for these studies obviously cannot cover all forms of life "at risk," but are believed to be sufficiently representative to warrant a general appraisal of potential toxicity. They are shown in Table 6-5.

In addition to tests with these species, the Department of Interior requires full information on the chemical and physical properties of the proposed pesticide, the formulations to be employed, the manner and pattern of application, and data on persistence in soil and water.

The acute oral toxicity test in birds involves the administration of graded doses to groups of three to six birds, five to seven days of age. Dosing is continued for five days, but observations are made over an eight-day period. The LC_{50} (the concentration in ppm estimated to kill half a group) is then calculated. The acute toxicity is compared with that of DDT tested concurrently.

TABLE 6-5

ANIMAL SPECIES PROPOSED FOR TOXICOLOGICAL EVALUATION
OF PESTICIDES IN WILDLIFE

Tests	Species
Acute toxicity	Mammals: rats, dogs
	Waterfowl: Mallard duck
	Upland birds: coturnix quail, bobwhite quail, ring-necked pheasant
	Cold-water fish: rainbow trout
	Warm-waterfish: bluegill, goldfish, channel catfish
	Salt-water fish: bluegill
Subacute toxicity	Marine-mollusks: eastern oyster
	Birds (may be requested)

Acute toxicity is estimated in each species of fish by first acclimatizing groups of ten fish, each weighing 1 to 2 g, for a period of 10 days, then exposing them to grade concentrations of the test compound. Aquarium conditions for aeration, water composition, temperature control, etc., are described. The LC_{50} is computed after 96-hr of exposure. DDT is tested concurrently as the reference pesticide.

A similar procedure is employed in testing the effect on a marine mollusk. It is based on observing the extent of linear deposition of new shell in groups of young (1 to 2-in.) oysters exposed for 96 hr to graded concentrations of the toxicant in flowing sea water. Shell growth is measured and compared with that of the untreated controls. It is expressed as EC_{50}, i.e., the concentration of test material calculated to induce 50% depression. Here, too, DDT is used as the reference pesticide. (It should be mentioned, however, that pesticides differ in their mode of action and metabolic disposition, and the choice of any specific compound, even one as widely used as DDT, leaves something to be desired from a toxicological standpoint.)

These procedures have obvious limitations in terms of the desired objectives, but appear to be all that can reasonably be justified in the present state of knowledge and capabilities.

REFERENCES

1. Division of Pharmacology, Food and Drug Administration, Department of Health, Education, and Welfare, *Appraisal of the Safety of Chemicals in Foods, Drugs, and Cosmetics*, Association of Food and Drug Officials of the United States, Austin, Texas, 1959.

2. Second Report of the Joint FAO/WHO Expert Committee on Food Additives, *Procedures for the Testing of Intentional Food Additives to Establish Their Safety For Use*, Food and Agriculture Organization of the United Nations, Rome, 1958.

3. Food Protection Committee, *Safe Use of Pesticides in Food Production*, National Academy of Sciences–National Research Council Publication 470, Washington, D.C., 1956.

4. Federal Hazardous Substances Act Public Law 86-613, 1960 and Code of Federal Regulations Part 191, Chapter 1, Title 21.

4a. S. Carson and R. Goldhamer, *Proc. Sci. Sect. Toilet Goods Assoc.*, **38** (1962).

5. S. Carson, *Test Methods and Toxicity Considerations under the Federal Hazardous Substances Labeling Act*, Proceedings of the 49th Annual Meeting, Chemical Specialties Manufacturers Association, New York, 1962.

6. R. I. Mitchell, *Am. Rev. Respirat. Diseases*, **82**, 5 (1960).

7. *Respiratory Challenge of Animals with Microorganisms by Means of the Henderson Apparatus*, U.S. Army Chem. Corps Res. & Dev. Comm., U.S. Army Biological Warfare Laboratories Ft. Detrick, Frederick, Maryland, 1957.

8. B. L. Oser, M. Oser, and H. C. Spencer, *Toxicol. Appl. Pharmacol.*, **5**, 142 (1963).

9. Food Protection Committee, *Problems in the Evaluation of Carcinogenic Hazard from Use of Food Additives*, National Academy of Sciences – National Research Council Publication 749, Washington, D.C., 1959.

10. Fifth Report of the Joint FAO/WHO Expert Committee on Food Additives, *Evaluation of the Carcinogenic Hazards of Food Additives*, World Health Organization, Geneva, 1961.

11. L. C. Miller and M. L. Tainter, *Proc. Soc. Exp. Biol. Med.*, **57**, 261 (1944).

12. J. T. Litchfield, Jr., and F. Wilcoxon, *J. Pharmacol. Exp. Therap.*, **95**, 99 (1949).

13. C. S. Weil and D. D. McCollister, *J. Agr. Food Chem.*, **11**, 486 (1963).

14. R. T. Williams, *Detoxication Mechanisms*, 2nd ed., Wiley (Interscience), New York, 1959.

15. M. S. Weinberg, S. Carson, and B. L. Oser, *Toxicol. Appl. Pharmacol.*, **7**, 480 (1965).

16. B. L. Oser (ed.) *Hawk's Physiological Chemistry*, 14th ed., McGraw-Hill, New York, 1965.

17. *Manual of Histologic and Special Staining Technics*, 2nd ed., McGraw-Hill, New York, 1960.

18. J. J. Burns, S. A. Cucinell, R. Koster, and A. H. Conney, *Ann. N.Y. Acad. Sci.*, **123**, 273 (1965).

19. H. Remmer, *Ann. Rev. Pharmacol.*, **5**, 405 (1965).

20. B. L. Oser and M. Oser, *J. Nutrition* **60**, 489 (1956).

21. J. G. Wilson in *Teratology Principles and Techniques*, (J. G. Wilson and J. Warkany, eds.), University of Chicago Press, Chicago, 1964.

22. I. W. Monie, in *Supplement to Teratology Workshop Manual*, Pharmaceutical Manufacturers Association, Washington, D.C., 1965.

23. M. J. Verrett, J.-P. Marliac, and J. McLaughlin, Jr., *J. Assoc. Offic. Agr. Chemists*, **47**, 1003 (1964).

24. *Handbook of Laboratory Animals*, National Academy of Sciences–National Research Council Publication 317, Washington, D.C. 1954.

25. R. E. Zwickey and K. J. Davis, in *Appraisal of the Safety of Chemicals in Foods, Drugs, and Cosmetics*, Association of Food and Drug Officials of the United States, Austin, Texas, 1959.

26. *Monthly Bulletin*, British Ministry of Health, **19**, 108 (1960).

27. Federal Food Drug and Cosmetic Act, Section 409 (c) (3).

28. G. B. Mider, in testimony before the Committee on Interstate and Foreign Commerce, House of Representatives, January 26, 1960.
29. H. Popper, S. S. Sternberg, B. L. Oser, and M. Oser, *Cancer*, **13**, 1035 (1960).
30. World Health Organization Technical Report Series 240, 1962, cited in *Evaluation of the Toxicity of Pesticide Residues in Food*, Report of a Joint Meeting of the FAO Committee on Pesticides in Agriculture and the WHO Expert Committee on Pesticide Residues, 1964.
31. National Academy of Sciences – National Research Council Publication 1270, Washington, D.C. 1965.
32. A. C. Frazer, Presented before the Food Protection Committee – Food and Drug Law Institute Conference on the Use of Human Subjects in Safety Evaluation, Washington, D.C., November 19, 1966.

Chapter 7

PEST RESISTANCE TO PESTICIDES

A. W. A. Brown

DEPARTMENT OF ZOOLOGY
UNIVERSITY OF WESTERN ONTARIO
LONDON, CANADA

7-1   NATURE AND DISTRIBUTION OF RESISTANCE

A. Toxicants in General

The use of chemicals to kill large proportions of pest organisms has frequently resulted in the development of resistant strains or populations, as has been the experience with many antibiotics against bacteria(397) and with some antimalarials and trypanocides applied against parasitic protozoa(74,654). The fire-blight bacterium, *Erwinia amylovora*, and the pepper bacterium, *Xanthomonas vesicatoria*, are now resisting control by streptomycin sprays in Delaware(236). Among the fungicides, biphenyl has induced resistance to itself in the green mold *Penicillium digitatum* on lemons in California(185a, 315). Resistance to chlorinated nitrobenzene fungicides has been reported to have developed in *Fusarium solani* on cucurbits in North America and in *Sclerotium voltsi* on tomatoes in Greece (236, 271b). Among the herbicides, 2,4-D can select out resistant strains of chickweed within two generations, while certain strains of flax and Canada thistle(350) are naturally tolerant. Out of 33 instances of increasing difficulty in control reported from the Americas to the Food and Agricultural Organization (FAO) in 1967, 19 of them involved 2,4-D, nine involved 2,4,5-T, and five involved dalapon(236). Dalapon can select out resistant strains of barley within four years(45), and the variation between different strains of Johnson grass in their response to dalapon may account for the survival of resistant plants in the field(307). Corn (*Zea mays*) is normally resistant to the herbicides atrazine and simazine, detoxifying them by hydroxylation; but a line susceptible to them has been produced by inbreeding, and the atrazine susceptibility has proven to be due to a single recessive gene allele(208,296).

The molluscicide NaPCP (sodium pentachlorophenate) has induced resistant strains of the snail *Australorbis glabratus*, a vector of schistosomiasis; after ten generations of selection at the 90% mortality level, the LC_{50} for NaPCP increased by 50–60% (*576*). In the field, increases in tolerance to NaPCP were developed in some populations of the Japanese snail *Oncomelania nosophora*, while the use of DNOCHP (dinex) for six years induced increased tolerance to this molluscicide (*805*).

Among the rodenticides, the anticoagulants warfarin and diphacinone have induced the development of a resistant population of the Norway rat on a farm in western Scotland (*83, 171*). Resistance of *Rattus norvegicus* to warfarin has also been found in Jutland, Denmark, where the rats came to show 75% survival to a dose which normally killed 95% (*484*). More recently it has been encountered along the border between England and Wales (*832*) and in Drenthe Province, Netherlands (*591*). The development of warfarin resistance has also been confirmed in the house mouse, *Mus musculus*, in English cities and in the paddyfield mole rat, *Gunomys gracilis*, in northwestern Ceylon (*832*). As for the piscicides or fish poisons, it has been noted that the occasional bluegill sunfish (*Lepomis macrochirus*) was especially tolerant to rotenone (*771*).

B. Insecticides and Nonpest Organisms

It is the insecticides and acaricides that have induced so many resistant strains of insects and mites (no less than 194 species by 1966), and this is the main theme of the present chapter. It should, however, be observed at the outset that insecticide resistance has also been induced in certain vertebrate groups, such as fish, frogs, and mice. Populations of the mosquito fish, *Gambusia affinis*, in an area of Mississippi State (Sidon), that had been heavily treated with insecticides proved to have four times the normal tolerance to DDT (*795*) and TDE, and showed as much as a 120-fold resistance to dieldrin and other cyclodiene derivatives (*81*); they were not crossresistant to the organophosphorus (OP) insecticides, except for methylparathion, to which they showed a fivefold cross-resistance (*221*). Populations of the golden shiner, *Notemigonus crysoleucas*, the green sunfish, *Lepomis cyanellus*, and the bluegill sunfish, *L. macrochirus*, collected at Twin Bayou in this state showed tolerances some 50 times above normal to the cyclodiene insecticides, while remaining susceptible to DDT, despite its having been widely applied in the area (*224*). These resistant fish populations originate from a few resistant genotypes (*220*); postadaptation to truly sublethal doses has not been noted with toxaphene, though it has been noted with dieldrin and DDT (*478*). Samples of the northern cricket frog, *Acris crepitans*, taken from the Sidon area, showed only 35% mortality to DDT dosages that killed 100% of specimens from untreated areas (*82*), and they were just as resistant to aldrin, which had not been used in the area (*796*); the southern cricket frog, *A.*

gryllus, was DDT resistant to about the same extent(*82*). A similar response was obtained in warm-blooded animals; when a colony of white mice (*Mus musculus*) was selected by injecting them with DDT for nine generations, its LC_{50} was nearly doubled(*605*). These mice showed no respiratory increase when injected with a dose of DDT that produced it in normal mice(*607*); they were cross-resistant to lindane and dieldrin when injected, but not when orally administered(*606*). A population of the pine mouse, *Pitymys pinetorum*, in a Virginia apple orchard, treated with endrin or a rodenticide for 11 successive years, developed a 12-fold tolerance to endrin(*809*). However, when field populations of the cotton rat, *Sigmodon hispidulus*, in the cotton-growing areas of Mississippi were tested, they were found to be no more DDT-tolerant than those from untreated areas(*222*).

Insecticide resistance has also developed in beneficial arthropods. Colonies of the honey bee, *Apis mellifera*, maintained at Riverside, California, were found to have developed as much as eight-fold resistance to DDT in a period of eight years(*35*). Similar DDT resistance levels were found recently in several Louisiana colonies and two imported from England(*288*). The mayflies, *Heptagenia hebe* and *Stenonema fuscum*, have developed a 15-fold DDT resistance in New Brunswick forests that have been repeatedly sprayed for spruce budworm control(*285*). Although the freshwater shrimp, *Palaemonetes kadakiensis*, has developed only a two- to three-fold tolerance to chlorinated hydrocarbons in Mississippi(*223*), the parasitic copepod, *Argulus*, in fish ponds in Israel, developed a significant resistance to gamma-BHC(*453*), and the paddyfield crab, *Paratelphusa ceylonensis* in northwestern Ceylon(*236*) developed resistance to endrin. The predacious mites *Typhlodromus occidentalis* and *Amblyseius hibisci* have developed parathion resistance in British Columbia(*550*) and California(*415a*), respectively. It has been suspected that *Macrocentrus ancylivorus*, a braconid parasite of the oriental fruit moth, has developed DDT resistance in the peach orchards of south-western Ontario since it survives well in sprayed orchards; certainly this species became more tolerant after laboratory selection with DDT(*639*), but the peak 12-fold level could not be maintained and the resistance disappeared completely in 14 subsequent generations without DDT selection(*682*). *Bracon mellitor*, a parasite of the cotton boll weevil, has developed a fourfold resistance to DDT, carbaryl, or methyl parathion after selection with these compounds in the laboratory for 5 generations(*5*).

Whether insects can develop resistance to insecticide substitutes has been investigated by laboratory selection experiments. When a colony of *Aedes aegypti* was exposed in the larval stage to the chemosterilant apholate for generation after generation, it became more tolerant to the sterilizing effect of the chemical(*333*). By the 43rd generation it had developed a 20-fold resistance, with a three-fold cross-resistance to metepa and almost none to tepa(*618*). Selection with hempa has also induced resistance(*256*), while the resistance induced by metepa selection has been associated with increased detoxication(*433*). Initial experiments in Florida indicated that houseflies could not develop resistance to chemosterilants; an Italian strain did develop considerable resistance to metepa by the ninth generation but lost it in the subsequent two generations(*694*), perhaps because of inherited chromosome damage. With bacterial toxins, exposure of *Musca domestica* for 27 generations to a selection pressure at about 40% mortality from the spores of *Bacillus thuringiensis* did not result in any resistance(*218*). However, selection at levels well above 50% mortality induced an 8- to 14-fold resistance after 27–50 generations, and this was fairly stable when selection was discontinued(*329*).

C. Types of Resistance in Pest Arthropods

Insecticide resistance has now been reported and adequately documented in 225 species of insects and acarines, nearly all of them destructive; of these, 119 species are found on crops, in forests, or on stored products, and 106 species are of public health or veterinary importance (Table 7-1). Resistance in the public health field has been reviewed frequently and is well documented(*102,128,532*), so that this chapter omits reference to most of the original publications except the most recent, citing only the reviews that contain the pertinent literature on the particular topic. Resistance in plant-feeding insects has been fully summarized in the past(*38,40,101,526*) and there are some more recent reviews(*111,219,249*), but since basic research on biochemistry and genetics has been infrequent, due to the difficulty of rearing these insects in quantity, an effort is made in this chapter to cite all the original papers on resistance in the agricultural field. The FAO Working Party on Resistance to Pesticides(*236*) in 1967 published questionnaire returns indicating reported or suspected resistance; those instances which were confirmed by laboratory-controlled tests are included in the following tables, although their dates cannot be given.

In these species, populations have developed resistance to an insecticide group to which they were formerly susceptible. Thus, it is convenient to reserve the word "resistance" for cases in which the characteristic has been developed within a normally susceptible species and not apply it to forms such as grasshoppers or the Mexican bean beetle, which could scarcely be killed by DDT in the first place. A population may be judged

TABLE 7-1

NUMBER OF SPECIES OF THE VARIOUS ORDERS OF ARTHROPODS IN WHICH POPULA-
TIONS HAVE DEVELOPED RESISTANCE TO VARIOUS TYPES OF INSECTICIDES

	DDT	Cyclo-diene	Organo-phosphate	Other	Agr.[a]	Public[b] health	Total[c]
Diptera	49	71	16	3	11	81	92
Lepidoptera	14	14	6	6	29	0	29
Hemiptera	8	15	14	4	33	2	35
Acarina	3	7	16	4	16	8	24
Coleoptera	6	19	1	1	23	0	23
Siphonaptera	5	5	0	0	0	5	5
Thysanoptera	3	1	0	2	4	0	4
Phthiraptera	5	6	1	0	0	7	7
Ephemeroptera	2	0	0	0	2	0	2
Orthoptera	2	2	1	0	0	2	2
Hymenoptera	1	0	0	0	1	0	1
Total	98	140	54	20	119	105	224

[a]Including species attacking forests and stored products.
[b]Including species of veterinary importance.
[c]Less than the sum of Columns 1–4, since some species have developed resistance
to three, or even four insecticide groups.

to have become resistant when this change has been confirmed by some
set test in which it is compared to populations from untreated areas or to
normal laboratory colonies, and when in the field it has come to resist
control by the insecticide as currently employed. A population that is
significantly less susceptible by laboratory test but is nevertheless still
controlled by that insecticide in the field is not called resistant but is, for
convenience, described by the word "tolerant."

Whereas tolerance is often nonspecific, resistance is usually specific to
one insecticide or group of insecticides. The chlorinated hydrocarbon
insecticides pose the greatest problem; in this area there are two different
types of resistance, one to DDT and its relatives and the other to the
cyclodiene derivatives and to gamma-BHC(126). Houseflies and mosquitoes made DDT resistant by selection pressure with DDT are cross-
resistant to TDE and methoxychlor but not to the cyclodiene derivatives
and only very occasionally to gamma-BHC(59, 540); those made cyclodiene resistant by selection with dieldrin or related compounds are cross-
resistant to gamma-BHC but not to DDT(127, 526). These rules have
been found to apply to mosquitoes, bed bugs, fleas, cockroaches, ticks,
and the sheep blowfly(136). Cyclodiene resistance in the housefly, sheep

blowfly, and mosquito is most strongly shown to dieldrin, aldrin, and chlordane(*126, 136*) and most weakly to isobenzan(*528*). Among agricultural insects, DDT-resistant cabbageworms (*Pieris rapae*) are cross-resistant to PERTHANE but not to the cyclodienes, while cyclodiene-resistant onion maggots (*Hylemya antiqua*) and boll weevils (*Anthonomus grandis*) are not cross-resistant to DDT(*93, 517*) nor are BHC-resistant strains of the cocoa capsid, *Distantiella theobroma*(*199*). Neither type of resistance to the chlorinated hydrocarbons confers any appreciable cross-resistance to the organophosphorus insecticides. But it is curious that selection of houseflies or mosquitoes with OP compounds results in a high DDT resistance and high cyclodiene resistance in houseflies(*114, 825*).

The known cases of developed resistance are therefore tabulated below under DDT resistance, cyclodiene resistance, OP resistance, and other types (Table 7-1). It will be seen that there are 98 species that have developed DDT resistance, 141 species that have cases of developed cyclodiene resistance, 54 species that have developed OP resistance, and 20 species that have developed resistance to other types of insecticides (Table 7-1). The actual number of species involved is more than 225, since many species resemble the bed bug, German cockroach, or housefly in that they are resistant to two, three, or even four groups of insecticides. Nearly half of the species belong to the Diptera, especially mosquitoes, and it is in this order of insects that cyclodiene resistance is particularly frequent. The Hemiptera, Lepidoptera, and Acarina follow in importance, OP resistance being frequent among the mites; the Coleoptera are conspicuous for their developing resistance to the cyclodiene insecticides used against soil insects.

D. Insecticide Resistance in Agriculture

The earliest case of resistance occurred in 1908, when the San Jose scale became no longer controllable by lime sulfur in an area of Washington State (Table 7-2). Subsequently, three species of scale insects in California became resistant to hydrogen cyanide fumigation. The codling moth progressively lost susceptibility to acid lead arsenate sprays in many parts of the world, while two species of cattle ticks became resistant to sodium arsenite dips and the peach twig borer became resistant to basic lead arsenate. To these four cases of arsenic resistance were added two species that became resistant to tartar emetic, and one each that became resistant to cryolite† and to selenium, making a total of 14 species showing

---

†This cryolite resistence in the walnut husk fly for which the evidence was deterioration of control [Quayle, J. E. E., **36**, 493 (1943)], is considered invalid by Boyce [*Herb. Commun.* (1967)].

TABLE 7-2
GEOGRAPHICAL AND CHRONOLOGICAL INCIDENCE OF OBSERVED RESISTANCE TO NON-
ORGANOCHLORINE AND NONORGANOPHOSPHATE PESTICIDES AMONG VARIOUS ECONOMIC
INSECT SPECIES

Resistant species	Pesticide	Geographic and chronological incidence
Aspidiotus perniciosus (San Jose scale)	Lime-sulfur	Wash. 1908(*521*); Ill. 1920(*235*); W. Australia(*236*)
Saissetia oleae (Black scale)	Hydrogen cyanide	Cal. 1912(*826*)
Aonidiella aurantii (California red scale)	Hydrogen cyanide	Cal. 1913(*190,848*)
Coccus pseudomagnoliarum (Citricola scale)	Hydrogen cyanide	Cal. 1925(*657*)
Carpocapsa pomonella (Codling moth)	Lead arsenate	Col. 1928(*364*); N.Y. 1930(*611*); Wash. 1931 (*810*); Va. 1932(*367*); S. Africa 1940(*749*); Cal. 1943(*658*); Syria 1958(*377*); Ohio, N.S. (*236*)
Anarsia lineatella (Peach twig borer)	Lead arsenate	Cal. 1944(*750*)
Boophilus microplus (Cattle tick)	Sodium arsenite	Argentina 1935(*226*); Queensland 1937(*349*); Brazil(*36*); Colombia, Jamaica, NSW 1954 (*457*)
Boophilus decoloratus (Blue tick)	Sodium arsenite	Cape Prov. 1938(*819*); Natal 1945(*820*)
Scirtothrips citri (Citrus thrips)	Tartar emetic	Cal. 1939(*80*); Transvaal 1945(*721*)
Taeniothrips simplex (Gladiolus thrips)	Tartar emetic	Cal. 1943(*658*)
Tetranychus telarius (Two-spotted mite)	Selenium	Eastern U.S.A. 1943 (*166,658*)
Tetranychus urticae (Greenhouse mite)	Azobenzene	England 1965(*378*)
Rhagoletis completa (Walnut husk fly)	Cryolite	Cal. 1943(*658*)
Epilachna varivestris (Mexican bean beetle)	Rotenone	N.Y. 1949(*611*); N.C. 1951(*94*); Conn. 1952 (*784*)
Ephestia cautella (Tobacco moth)	Pyrethrins	Ga., Fla. 1960(*339*)
Musca domestica (Housefly)	Pyrethrins	Sweden 1958(*183*)
Megaselia halterata (Mushroom phorid)	Pyrethrins	England 1965(*378*)
Epiphyas postvittana (Light brown apple moth)	Carbaryl	New Zealand 1963(*720*)
Heliothis virescens (Tobacco budworm)	Carbaryl	Tex. 1964(*650*)
Tetranychus mcdanieli (McDaniel mite)	Binapacryl	Wash. 1965(*373*)
Spodoptera littoralis (Cotton leafworm)	Carbaryl	Egypt 1966(*769*)

resistance by 1944. These were later followed by one species that developed resistance to rotenone, three to pyrethrins, and three to carbaryl, although the pyrethrin resistance of the housefly had been enhanced by previous selection with chlorinated hydrocarbons, and the carbaryl resistance of *Epiphyas* had been strengthened by the use of azinphosmethyl, of *Spodoptera* by chlorinated and organophosphate insecticides, and of *Heliothis virescens* by both types of chlorinated insecticides(6); in addition, five other species are known from laboratory experiments to have the potential to develop pyrethrin resistance(228). Recently, the McDaniel mite on apple has developed resistance to the dinitro compound, binapacryl, in Washington State(373).

Resistance to DDT has been developed by 27 species of plant pests, 14 of them being caterpillars (Table 7-3) including two that became resistant to TDE. By the year 1951, about five years after the introduction of DDT, resistance appeared in the codling moth, cabbageworm, cabbage looper, tomato hornworm, diamondback moth, and two species of grape leafhoppers. DDT-resistant codling moth populations are now present in all apple-growing regions of the world, although they often remain quite local. Resistance in the spruce budworm is becoming important in areas that have been resprayed more than twice. Indications of DDT resistance in the elm bark beetle, *Scolytus multistriatus*, reported from Connecticut in 1951, were not borne out by subsequent experience. Also not included in the tables and totals are the garden leafhopper, *Empoasca solana*, in the U.S.(249) and the cotton jassid, *E. lybica*, in Sudan(106), which is not really DDT-resistant at present, and three species of *Lygus* (*L. elisus*, *desertus*, and *lineolaris*) suspected of having developed DDT resistance on sugar beets in the Salt River Valley, Arizona(346).

No less than 53 species of plant pests and three species affecting stored products have developed resistance to cyclodiene derivatives or gamma-BHC (Table 7-4). Of these, 10 are pests of cotton, including *Estigmene acraea* and *Trichoplusia ni*, and the resistance resulted mainly from the use of toxaphene and endrin. Among the soil-inhabiting pests, resistance to aldrin, dieldrin, or heptachlor has been developed in three species of wireworms and four species of *Diabrotica* rootworms,† as well as in five species of root maggots of the general *Hylemya*, *Psila*, and *Euxesta*; this type of resistance has recently developed in a sixth, namely *Hylemya cilicrura*. Developed resistance to control by dieldrin or toxaphene has been observed in three species of thrips, namely *Frankliniella tritici* in California, *F. occidentalis* in New Mexico, and *Chaetaphanothrips*

---

†Tolerance to the OP compounds diazinon and phorate, used as substitutes, increased in *Diabrotica virgifera* by 60% in the period from 1963 to 1967(47).

TABLE 7-3
GEOGRAPHICAL AND CHRONOLOGICAL INCIDENCE OF RESISTANCE TO DDT AND DDE
AMONG VARIOUS SPECIES OF PLANT-FEEDING INSECTS

Resistant species	Geographical and chronological incidence
Pieris rapae (Imported cabbageworm)	Wis. 1951(*517*); N.Y. 1952(*343*); Fla.(*332*); Conn. 1953(*814*); S.C., Ill. 1954(*145*); Japan 1956(*440*); Australia 1959(*418*); Ind. 1960(*281*)
Trichoplusia ni (Cabbage looper)	N.Y. 1952(*343*); Cal., Ala. 1954(*249*); Ont.(*314*); Quebec 1956(*313*); La., Okla., Ariz., Tex. 1958(*395*); Tenn., Ark., 1959 (*785*); Ind. 1960(*281*); S.C., La., N.C., Cal.
Protoparce quinquemaculata (Tomato hornworm)	Fla. 1951(*39*)
Carpocapsa pomonella (Codling moth)	Ohio 1951(*174*); N.Y. 1952(*278*); Wash. 1953(*303*); Australia(*475,725,726*); Ky. 1955(*304*); S. Africa 1956(*566*); N.S.W.(*418*); Cal. 1957(*54,486,487*); B.C., Ont., 1958(*233*); Victoria(*551*); Syria(*377*); Tasmania(*418*); Va., W.Va., Ill., Del., Utah 1959(*277*); Poland, Turkey, Greece(*235*)
Grapholitha molesta (Oriental fruit moth)	Victoria 1958; Mich., Va. 1958(*277*)
Argyrotaenia velutinana (Red-banded leafroller)	N.Y. 1954(*276*); Ont. 1958(*345*); Va.(*683*), W.Va. 1959(*785*) (TDE)
Epiphyas postvittana (Light brown apple moth)	Tasmania 1958(*418*); New Zealand 1960 (*720*) (TDE)
Choristoneura fumiferana (Spruce budworm)	New Brunswick 1963(*670*)
Plutella maculipennis (Diamondback moth)	Java 1951(*17*); Ceylon(*236*)
Phthorimaea operculella (Potato tuber moth)	Queensland 1965(*143*); N.S.W., Transvaal, S. Rhodesia(*236*)
Pectinophora gossypiella (Pink bollworm)	Mexico 1959(*480*); Tex.(*236*)
Spodoptera exigua (Beet armyworm)	Cal.(*681*); Ariz. 1958(*785*)
Heliothis zea (Corn earworm)	La.(*290*), Ark., Tex. 1962(*91*); Ga., Colombia(*236*)
Heliothis virescens (Tobacco budworm)	Tex. 1962(*90*); Miss. 1964(*92*); La. 1965 (*289*); Colombia, Peru(*236*)
Leptinotarsa decemlineata (Potato beetle)	N.Y. 1949(*659*); N.D. 1952(*653*); Minn, 1954(*173*); Spain 1955(*448*); Czechoslovakia 1965(*376*); Pa., Va., Alta, Poland, Switzerland, Germany, Italy, Spain, Portugal(*236*)
Epitrix cucumeris (Potato flea beetle)	Ind.(*282*); Conn. 1951(*443,444*); Ohio 1955(*717*); Va. 1956(*351*); R.I.(*298*); Wis. 1957(*232*)
Costelytra zealandica (Grass grub)	New Zealand(*236*)
Miccotrogus picirostris (Clover seed weevil)	Idaho 1964(*394*)

TABLE 7-3 *(cont.)*

Resistant species	Geographical and chronological incidence
Sitophilus oryzae (Rice weevil)	Queensland 1963 *(142)*
Euxesta notata (Spotted root maggot)	Ont. 1960 *(763)*
Thrips tabaci (Onion thrips)	Tex. 1957 *(146)*; British Columbia *(236)*
Scirtothrips citri (Citrus thrips)	Cal. 1949 *(215)*
Diarthrothrips coffeae (Coffee thrips)	Tanganyika 1956 *(761)*
Lygus hesperus (Alfalfa plant bug)	Wash. 1952 *(523)*; Cal. 1953 *(16)*; Ariz. 1954 *(346)*
Erythroneura variabilis (Grape leafhopper)	Ariz., Cal. 1951 *(56)*
Erythroneura elegantula (Grape leafhopper)	Cal. 1951 *(737)*
Erythroneura lawsoniana (Apple leafhopper)	Ky. 1953 *(305)*
Typhlocyba pomaria (White apple leafhopper)	Eastern U.S.A. 1959 *(277)*
Myzus persicae (Green peach aphid)	England *(841)*
Epitrix tuberis (Tuber flea beetle)	B.C. 1965 *(498,50)*.

orchidii in Costa Rica *(236)*, but no resistance tests have been reported. Cyclodiene resistance can be seen to involve the principal pests of the principal crops of at least three important regions of the world, namely the boll weevil in all the cotton-growing states of the U.S., the cotton leaf-worm in the Egyptian province of the UAR, and the cocoa capsid in Ghana.

E. Mites and OP Resistance

Organophosphorus compounds have induced resistance in 31 species of insects and mites attacking plants, in three species attacking stored products, and in the northern fowl mite (Table 7-5). Organophosphorus resistance is also suspected in the serpentine leaf-miner (*Liriomyza pusilla*) in Florida *(827)* and in the mushroom phorid (*Megaselia halterata*) in the south of England *(236)*. Certain species that have developed resistance to chlorinated hydrocarbons have subsequently developed OP resistance, namely the German cockroach *(293)*, the bedbug *(51)*, the Australian cattle tick *(712)*, and the cattle-sucking louse in Alberta *(330)*. Of the plant pests listed, four species are caterpillars and 12 are Homoptera, with *Myzus persicae* being most important. This aphid developed resistance to dimethoate in England *(569)* and to methyl demeton in Europe *(44)*. The remaining 14 species are all plant-feeding mites, all but one being *Tetranychidae*.

The OP resistance developed first in the two-spotted spider mite, *T. telarius*, on roses and other greenhouse plants a year after the substitution of parathion of TEPP, and had become general throughout the eastern U.S. by 1950. In this year the OP resistance became evident in the

TABLE 7-4

GEOGRAPHICAL AND CHRONOLOGICAL INCIDENCE OF BHC AND CYCLODIENE INSECTICIDE RESISTANCE AMONG VARIOUS SPECIES OF AGRICULTURAL INSECTS

Resistant species	Pesticide	Geographical and chronological incidence
Protoparce sexta (Southern tobacco hornworm)	Endrin	S.C. 1961(785); N.C. 1962(664)
Estigmene acraea (Salt marsh caterpillar)	Chlor. hydro.	Cal. 1957(395); Ariz. 1959(745,812)
Alabama argillacea (Cotton leafworm)	Toxaphene	Tex.(384); Venezuela 1951(575); Southeastern U.S.A. 1959(395); Colombia(236)
Spodoptera littoralis (Cotton leafworm)	Toxaphene	Egypt 1961(490); N.India(236)
Euxoa detersa (Sandhill cutworm)	Aldrin	Ont.(236)
Euxoa messoria (Dark-sided cutworm)	Dieldrin	Ont. 1960(322)
Agrotis ypsilon (Black cutworm)	Aldrin	Brazil, Taiwan(236)
Heliothis virescens (Tobacco budworm)	Endrin	Tex. 1964(248); Miss. 1964(92); La. 1965(289)
Trichoplusia ni (Cabbage looper)	Endrin	Ariz. 1958; S.C. 1959; Cal. Ark., Okla. 1960(785); N.C., Va.(236)
Earias insulana (Spiny bollworm)	Endrin	Israel 1956(26); Spain(236)
Phthorimaea operculella (Potato tuber moth)	Endrin	Queensland 1965(143)
Diatraea saccharalis (Sugar cane borer)	Endrin	La. 1963(842)
Chilo suppressalis (Rice stem borer)	BHC	Japan 1966(450,812); Taiwan(236)
Bucculatrix thurberiella (Cotton leaf perforator)	Chlor. hydro.	Cal. 1958(395)
Anthonomus grandis (Boll weevil)	Toxaphene	La. 1955(689); Tex., Ark(246); Miss.(690); S.C. 1957(249); Okla., Ala., N.C. 1959(687); Mexico, Venezuela(236)
Conoderus fallii (Southern potato wireworm)	Chlordane	S.C. 1955(677); Fla. 1957(581)
Conoderus vespertinus (Tobacco wireworm)	Dieldrin	S.C. 1960; N.C. 1961(301)
Limonius californicus (Sugar beet wireworm)	Aldrin	Wash. 1964(589)
Diabrotica virgifera (Western corn rootworm)	Aldrin	Neb. 1959(48); Kan.(121), S.D. 1962(368); Iowa, Mo., Minn. 1964(306)
Diabrotica undecimpunctata (Southern corn rootworm)	Aldrin	N.C., Va.
Diabrotica balteata (Banded cucumber beetle)	Aldrin	La. 1958; S.C. 1961(172)
Diabrotica longicornis (Northern corn rootworm)	Aldrin	S.D. (368); Ohio 1962(75); Ill. 1963(71); Iowa, Minn., Wis. 1964 (306)

Leptinotarsa decemlineata (Potato beetle)	BHC	France(*461*); Germany 1960(*335*); Italy, Spain, Portugal(*236*)
Galerucella birmanica (Singhara beetle)	BHC	N.India(*236*)
Lema oryzae (Rice leaf beetle)	BHC	Japan 1960(*715*)
Epitrix tuberis (Tuber flea beetle)	Dieldrin	B.C. 1965(*50*)
Hypera postica (Alfalfa weevil)	Heptachlor	Utah, Mont., Wyo.(*785*), Nev., Cal., Va., Md., N.Y. 1962(*7*); N.C., Pa., Del.
Lissorhoptrus oryzophilus (Rice water weevil)	Aldrin	Ark. 1964(*685*); La. 1966(*291*); Miss.
Graphognathus leucoloma (White-fringed beetle)	Dieldrin	Ala. 1964(*713*)
Sitophilus oryzae (Rice weevil)	BHC	England 1960(*132*); Queensland 1963 (*142*); Trinidad(*236*)
Sitophilus granarius (Granary weevil)	BHC	S.Africa(*236*)
Sitophilus zeamais (Maize weevil)	BHC	Kenya(*236*)
Tribolium castaneum (Red flour beetle)	BHC	Kenya 1960(*132*)
Cosmopolites sordidus (Banana tree weevil)	Dieldrin	Guinea, Ivory Coast, Cameroun 1967(*792*); N.S.W. 1969(*706*)
Hylemya antiqua (Onion maggot)	Dieldrin	Wis. 1957(*146*); Mich.(*302*); Ont.(*514*); Wash.(*369*); Ore., B.C. 1958(*231*); Ill., N.Y. 1959(*116*); Man., Que. 1961(*322*); Minn. 1962(*637*)
Hylemya florilega (Bean seed maggot)	Dieldrin	Ont.(*61*,*324*); Conn. 1960; Que., Nfld.(*236*)
Hylemya platura (Seed corn maggot)	Dieldrin	B.C. 1964(*230*); Ont. 1965(*320*); Japan, England 1966(*236*)
Hylemya brassicae (Cabbage maggot)	Dieldrin	Ill., Wis.. Wash.(*106*); B.C. 1960(*322*); Que. 1961(*322*); Nfld.(*552*); PEI(*673*); Ont. 1962(*578*); England 1963 (*156*); N.S., Me., Pa., Ohio, Belgium, Germany, Netherlands, Sweden(*236*)
Hylemya floralis (Turnip maggot)	Heptachlor	Saskatchewan 1963(*746*); Germany, Norway(*236*)
Psila rosae (Carrot rust-fly)	Dieldrin	Ore. 1958(*370*); B.C. 1960(*579*); Ont. 1961(*579*); Wash., Netherlands, France(*236*)
Euxesta notata (Spotted root maggot)	Dieldrin	Ont. 1961(*763*)
Merodon equestris (Large bulb fly)	Aldrin	England 1965(*833*)
Liriomyza archboldi (Serpentine leaf miner)	Aldrin	Fla.[a] 1951 (*827*); S.C. 1963(*785*)
Scirtothrips citri (Citrus thrips)	Dieldrin	Cal. 1959(*215*)
Dysdercus peruvianus (Cotton stainer)	BHC	Peru(*236*)
Distantiella theobroma (Cocoa capsid)	BHC	Ghana 1961 (*199*); Nigeria(*236*)
Sahlbergiella singularis (Brown cocoa capsid)	BHC	Nigeria 1964(*213*)
Blissus pulchellus (A chinch bug)	BHC	Panama 1958(*13*)
Leptocoris varicornis (Rice paddy bug)	BHC, Endrin	Ceylon, Thailand(*236*)

TABLE 7-4 (cont.)

Resistant species	Pesticide	Geographical and chronological incidence
Aeneolamia varia (Sugar cane froghopper)	BHC	Trinidad 1957 (568)
Scotinophora lurida (Black rice bug)	BHC	Taiwan (236)
Delphacodes striatella (Smaller brown leafhopper)	BHC	Japan 1961 (715)
Psallus seriatus (Cotton fleahopper)	Chlor. hydro.	Tex. 1958 (612,613)
Psylla pyricola (Pear psylla)	Dieldrin	Wash. 1958 (318)
Eriosoma lanigerum (Woolly apple aphid)	BHC	Queensland (236)
Aphis gossypii (Cotton aphid)	Endrin	Southeastern U.S.A. 1956 (435); Peru (236)
Chaetosiphon fragariaefolii (Strawberry aphid)	Endosulfan	Wash. 1965 (707)

[a] Described as Liriomyza pusilla.

TABLE 7-5
GEOGRAPHICAL AND CHRONOLOGICAL INCIDENCE OF ORGANOPHOSPHORUS-RESISTANCE AMONG MITES AND PLANT-FEEDING INSECTS

Resistant species	Geographical and chronological incidence
Tetranychus urticae (European spider mite)	Germany *(196)*; Holland 1950*(88, 337)*; Norway*(234)*; Denmark*(236)*; France 1951*(254)*; Italy 1955*(336)*; Switzerland 1958*(336)*; Poland 1964*(454)*; Czechoslovakia, Austria, Italy*(236)*
Tetranychus telarius[a] (Two-spotted spider mite)	(Greenhouses) Conn.*(250)*: N.J., Pa.*(724)*; N.Y.*(611)*; Cal. 1949*(393)*; Md., Ind.*(236)*; Ont. 1957*(141)*; (Apple) S. Africa 1957*(818)*; Australia 1958*(489)*; Argentina*(236)*; (Peach) Va., Pa., Ind., Mich., Ohio 1954 *(277)*; Wash. *(236)*; (Beets) Israel 1958*(176)*; (Cotton) Miss., La., Colombia*(236)*
Tetranychus cinnabarinus (Carmine mite)	Arizona 1958*(272)*; Cal., Ala. 1960*(785)*; Wis., Germany*(236)*.
Tetranychus tumidus (Tumid spider mite)	Texas 1952*(384)*
Tetranychus mcdanieli (McDaniel mite)	Wash. 1952*(573)*; Utah 1958*(184)*; B.C.*(236)*
Tetranychus schoenei (Schoene spider mite)	Va. 1957*(277)*
Tetranychus canadensis (Four-spotted spider mite)	Ill. 1958*(277)*
Tetranychus atlanticus (Strawberry spider mite)	Cal. 1956*(15)*; France*(236)*
Tetranychus pacificus (Pacific spider mite)	(Apple) Wash. *(573)*; Japan 1952*(277)*; (Cotton) Cal. 1956*(15)*
Panonychus ulmi (European red mite)	Ind. *(785)*; Wash. 1950*(572)*; N.Y.*(466)*; B.C. 1952*(197)*; Cal., Miss., S.C. 1957*(249)*; Ont. *(656)*; Holland*(790)*; England 1958*(164, 807)*; Japan*(353)*; Syria*(377)*; Tasmania 1959*(418)*; Australia*(338)*; Lebanon 1960*(852)*; Que., N.S., Ill., Ohio, Del., Md., Pa., Va., W. Va., New Zealand, France, Denmark, Norway, Poland, Switzerland, Austria*(236)*
Panonychus citri (Citrus red mite)	Cal. 1956*(390)*; Japan 1961*(715)*; Fla.*(236)*
Aculus cornutus (Peach silver mite)	Wash.*(236)*
Vasates schlechtendali (Apple rust mite)	B.C. 1952*(550)*; Wash.*(236)*
Bryobia praetiosa (Clover mite)	Poland*(236)*
Ornithonyssus sylbiarum (Northern fowl mite)	Tex. 1962*(684)*

471

TABLE 7-5 (cont.)

Resistant species	Geographical and chronological incidence
Myzus persicae (Green peach aphid)	Wash. 1953 (18); Fla. 1957 (828); Ore., Wis. 1959 (146); Cal. 1960(259); Germany, Norway 1962 (44); England 1963 (200, 569); Sweden. Ceylon, Taiwan (236); France 1967 (168)
Toxoptera graminum (Greenbug)	Okla. (236)
Chromaphis juglandicola (Walnut aphid)	Cal. 1953 (531)
Aphis pomi (Green apple aphid)	Wash., Switzerland(236)
Myzus cerasi, Sapaphis pyri & S. plantaginis	Switzerland 1954 (822)
Phorodon humuli (Hop aphid)	England (236); France 1967 (168); Czechoslovakia 1968(375)
Therioaphis maculata (Spotted alfalfa aphid)	Cal. 1956 (739, 740)
Psylla pyricola (Pear psylla)	Northwestern U.S.A. (319); B.C. 1959 (125); Israel(236)
Erythroneura elegantula (Grape leafhopper)	Cal. (236)
Nephotettix cincticeps (Green rice leafhopper)	S. Japan 1961 (439)
Blissus leucopterus (Lawn chinch bug)	Fla. (236)
Chilo suppressalis (Rice stem borer)	Japan 1960(604)
Spodoptera littoralis (Cotton leafworm)	Egypt (769); Israel 1967 (791)
Ephestia cautella (Tobacco moth)	Fla. Ga. (236)
Epiphyas postvittana (Light brown apple moth)	New Zealand 1961 (720); Tasmania 1966(236)
Tryporyza incertula (Paddy borer)	Taiwan (236)
Plodia interpunctella (Indian meal moth)	Ga. (236)
Dasyneura pyri (Pear midge)	Poland 1964(454)
Tribolium castaneum (Red flour beetle)	Nigeria 1964(635); Fla., Ga. (735)

[a]The two-spotted mite, formerly called T. bimaculatus, now synonymous with T. urticae (Boudreaux and Dosse, 1963).

472

greenish two-spotted mite of Europe (*T. urticae*) on greenhouse plants and on carnations grown outdoors in the south of France; OP resistance is now general in western Europe, this species being considered to be the same as the American greenhouse mite(*338*). By 1954 parathion resistance was apparent in *T. telarius* in American peach orchards and subsequently in the apple orchards of South Africa and Australia and the beet fields of Israel. By this time OP resistance was appearing in the carmine mite, *T. cinnabarinus*, on all types of crops in the south-western U.S. Parathion resistance has also developed in *T. pacificus*, infesting apple in Washington and Japan and cotton in California, as well as in *T. mcdanieli*, *T. schoenei*, and *T. canadensis* on apple and in *T. tumidus* and *T. atlanticus* on cotton.

Parathion resistance in *T. telarius* could first be corrected by sulfotepp or demeton in greenhouses(*724*) and by methyl demeton or azinphosmethyl in orchards(*277*), but these progressively failed so that the only OP compounds that remained effective against OP-resistant strains were ethion and carbophenothion. Parathion-resistant strains of the citrus mite, *P. citri*, were, however, even more resistant to carbophenothion, but parathion selection resulted in little cross-resistance to ethion(*378*) and none to the phosphoroamidothioate DOWCO 133(*388*). Parathion-resistant strains of *Tetranychus mcdanieli* may nevertheless still be controlled by demeton or azinphosmethyl(*184*). In the case of the European red mite, *P. ulmi*, control continues to be obtained by rotating treatments between carbophenothion, malathion, and parathion(*277*), but resistance to carbophenothion finally appeared in New York State in 1966(*465*).

The tetranychid mites have also developed at least five types of resistances to the chlorinated acaricides. The dates of their first reported appearances in the two-spotted mite and the European red mite are as follows:

	T. telarius	*P. ulmi*
Ovex (OVOTRAN)	East. U.S. 53	Cal. 54
Genite and fenson	Cal. 54	Cal. 54
Chlorobenzilate and Chlorbenside	Cal. 55	Cal. 54
Dicofol (KELTHANE)	Conn. 55	W. Va. 60
Tetradifon (TEDION)	U.S.A. 64	Pa. 61

The resistance of *P. ulmi* to ovex soon spread over the U.S.A. and by 1958 it had appeared in Ontario and Nova Scotia(*237*); resistance to fenson and chlorbenside appeared in England in 1957 and 1958(*164, 283*). The resistance to dicofol was developed by the McDaniel mite in Washington by 1958(*374*), the European red mite in England by 1963 (*188*), and the citrus mite in California by 1961(*392*). Ovex resistance and tetradifon resistance are readily developed by *P. citri*(*387*). Tetradifon resistance has been developed by the two-spotted mite throughout the U.S.A. and western Europe, but not in England(*378*). Resistance to the sulfur-containing acaricide ARAMITE first appeared in *Tetranychus telarius* in Cincinnati in 1955(*723*), so that by 1964 the only effective materials for spraying greenhouse roses were the phosphate BIDRIN and the cyclodiene derivative PENTAC(*10*). By 1965, the best control of *P. ulmi* in Ohio was given by the sulfur acaricide MORESTAN, the chlorinated acaricides tetradifon and dicofol, and the dinitro compounds binapacryl and dinocap(*240*). ARAMITE resistance has not yet appeared in *P. ulmi*, even in areas such as Ohio and Indiana where the red mite is resistant to all the chlorinated acaricides(*785*). ARAMITE has also been employed for many years against *P. citri* without the development of resistance, and its addition to demeton delays the onset of the demeton resistance to which this species is normally liable(*390*). Another possible type of resistance has never developed in tetranychid mites; the dinitro acaricide DNOCHP has remained effective against *P. citri*, despite many years of use(*392*), and selection of *Tetranychus telarius* for 60 generations with binapacryl induced no resistance to this dinitro compound(*11*). However *T. mcdanieli* has developed resistance to binapacryl as already noted(*373*).

The separate nature of the resistances to the chlorinated acaricides is best exemplified in *P. citri*(*388,565*); in this species, resistance to ovex (a chlorinated sulfone) involved no cross-resistance to dicofol (a chlorinated alcohol), and neither ovex resistance nor dicofol resistance involved any appreciable cross-tolerance to chlorobenzilate (a chlorinated ester), to tedradifon (a chlorinated sulfone), to ARAMITE (a chlorinated sulfite), or to ERADEX (a sulfur-containing heterocyclic). However, in *Tetranychus urticae* in Holland, chlorobenzilate resistance was closely linked with dicofol resistance(*89*). ARAMITE resistance induced in *T. pacificus* by laboratory selection (Table 7-6) carried little or no cross-resistance to

TABLE 7-6

RESISTANCE RATIOS IN *Tetranychus pacificus* AFTER SELECTION WITH VARIOUS ORGANO-PHOSPHORUS AND CHLORINATED ACARICIDES AT THE 70% MORTALITY LEVEL FOR THE NUMBER OF GENERATIONS STATED(*246*)

	Aramite	Chloro-benzilate	Dicofol	Parathion	Ethion
Aramite (15 generations)	8.0	1.6	1.0	600	460
Chlorobenzilate (23 generations)	3.0	1.0	1.0	30	17
Dicofol (22 generations)	1.7	1.0	1.6	100	24
Parathion (10 generations)	1.8	1.8	1.0	1000	1000
Ethion (9 generations)	1.8	1.3	1.0	90	1000

dicofol or chlorobenzilate(*389*). Although the induction of OP resistance by parathion or ethion pressure on this species involved no cross-resistance to the chlorinated acaricides, selection pressure with the chlorinated acaricides resulted in cross-resistance to the OP compounds. Similar results were obtained in *T. telarius* (Table 7-7), in which the lines

TABLE 7-7
RESISTANCE AND CROSS-RESISTANCE RATIOS IN *Tetranychus urticae* AFTER SELECTION
WITH VARIOUS INSECTICIDES FOR 8–12 GENERATIONS(*202*)

Selecting Agent	Gen- erations	Parathion	Me- demeton	Malathion	Aramite	Chloro- benzilate	Dicofol
Parathion	10	39	31	20	1.7	1.1	1.8
Methyl- demeton	10	71	65	39	1.2	1.2	2.6
Malathion	11	57	66	32	1.7	1.4	2.7
Aramite	8	63	42	22	0.7	1.6	1.6
Chloro- benzilate	12	45	37	25	1.0	0.5	1.8
Dicofol	12	55	80	35	1.5	1.0	1.6

selected with chlorinated acaricides became highly OP resistant, although in this case the unselected control line also became OP resistant, possibly because of vapor contamination(*312*); on the other hand, the cross-tolerance from OP compounds to the chlorinated acaricides did not exceed twofold(*804*). Most OP-resistant strains of mites are cross-resistant to ZECTRAN and some of the other carbamates(*388*). The phosphoroamidothioate DOWCO 133 (having O-isopropyl-O-trichlorophenyl substitution) is unusual among OP compounds in that selected strains of *T. pacificus* do not become cross-resistant to it, and it does not induce any appreciable resistance to itself or cross-resistance to other organo-phosphate compounds(*388*). The corresponding phosphoroamidate, in which P=S is converted to P=O, is even more effective against OP-resistant strains, not only of the Pacific mite, but also of the citrus mite (*391*).

It is interesting to note that of the two types of OP resistance that have developed in the Australian cattle tick, *Boophilus microplus*, as a result of dioxathion usage, one of them (the Ridgelands type) showed no cross resistance to ethion and malathion(*711*), while the other (the Biarra type) resisted all OP compounds investigated(*712a*); both strains were, however, highly cross resistant to carbophenothion and to carbaryl.

F. Resistance in Flies

Among the insects of public health and veterinary importance, the group that features in the resistance problem is the *Diptera*, with no less than 81 species involved. This is epitomized by the housefly (Table 7-8), in which DDT resistance first appeared in 1946, cyclodiene resistance in 1949, and OP resistance in 1955. The general experience in housefly control operations is that DDT resistance develops in two years, cyclodiene resistance one year after the substitution of BHC or dieldrin, and OP resistance becomes significant five years after the substitution of malathion or diazinon. At first, malathion resistance was restricted to North America and diazinon resistance to Europe, offering promise that there might be little cross-resistance between methyl phosphates and ethyl phosphates. But subsequently the use of a variety of OP insecticides for fly control blurred this difference, although a distinction between malathion resistance and resistance to diazinon and other organophosphate compounds was still detectable(*264*). Moreover, cross-resistance had already handicapped certain new organophosphate insecticides, such as fenthion. However, little cross-resistance is shown to dimethoate, and this OP compound induces little or no resistance to itself(*413*).

Other barn-inhabiting and synanthropic flies did not pose a resistance problem as the housefly does, although eventually resistant populations appeared in the stable fly, *Stomoxys*, the horn fly, *Haematobia*, the little housefly, *Fannia*, the blowfly, *Protophormia*, and the latrine fly, *Chrysomyia*. Most important is the cyclodiene resistance that has developed in the sheep blowfly (*Phaenicia*, formerly *Lucilia*) of South Africa, Australia, and New Zealand, and that has negated dieldrin as a cure for breech strike. Appreciable resistance to diazinon in *Phaenicia cuprina*, first observed at Dubbo in the interior of New South Wales(*706*), has become general in Australia, while quite strong organophosphate resistance has been induced by further selection with diazinon in the laboratory(*327*). Moderate dieldrin resistance has been developed by the Japanese flesh fly, *Sarcophaga peregrina*, in the laboratory(*513*). A number of midges and gnats have developed resistance to either one or both types of chlorohydrocarbons, namely the eye gnat, *Hippelates*, the Clear Lake gnat, *Chaoborus*, the Bodega Bay gnat, *Leptoconops*, the filter fly, *Psychoda*, the midges *Chironomus* and *Glyptotendipes*, the biting midge, *Culicoides*, and the borborid midge, *Leptocera*. Moreover, organophosphate resistance has been developed by the horn fly, *Haematobia*, the coprophilous fly, *Chrysomyia*, and the midge, *Glyptotendipes*. With the appearance in 1963 of DDT resistance in the blackfly, *Simulium aokii* and *S. ornatum* in Japan, and developing DDT resistance in *S. venustum* at

TABLE 7-8

GEOGRAPHICAL AND CHRONOLOGICAL INCIDENCE OF RESISTANCE TO THREE GROUPS OF INSECTICIDES BY NOXIOUS SPECIES OF DIPTERA (FLIES)

Species	DDT group	BHC-Dieldrin group	Organophosphorus group
Musca domestica	Sweden, Denmark 1946; U.S., Mediterranean 1947; N.Z., S. America 1948; W. Europe, Canada 1949; USSR, Africa 1950; Japan 1953; China 1956; Czechoslovakia, Poland 1958; India 1960(405)	Cal., Sardinia, Egypt 1949; U.S.A., Scandin. 1950; S. America 1951; Africa 1952; USSR 1953; Japan 1954; India 1957; Caribbean 1961; Romania 1962 (485)	Denmark, Fla. 1955; Switzerland, Italy, Ga. 1956; N.J. 1957; Cal., Ariz., La. 1958; Japan 1960; Germany, France 1961(835); Australia 1962(328)
Simulium ornatum	Japan 1968(22)		Japan 1968(22)
Simulium aokii	Japan 1963(755)		
Simulium venustum	Queb. 1967(813)		
Glyptotendipes paripes		Fla. 1953	Fla. 1955
Chironomus zealandicus		New Zealand(236)	
Chaoborus astictopus	Cal. 1961(835) (TDE)		
Psychoda alternata	Ill. 1949	England 1953	
Leptoconops kerteszii	Cal. 1961(835)		
Culicoides furens		Fla. 1958; Panama 1959	
Hippelates collusor		Cal. 1957(563)	
Leptocera hirtula	Malaya 1955	Malaya 1955	
Drosophila virilis[a]	Japan 1952		
Fannia canicularis	Spain 1953, Japan 1962 (752); England(236); Cal. 1967(263)	Cal. 1967(263)	
Fannia femoralis	Cal. 1967(263)	Cal. 1967(263)	
Stomoxys calcitrans	Sweden 1948; Norway 1958; Germany, Italy (236)	Norway 1958(733); Fla. 1965(560)	

477

TABLE 7-8 (*cont.*)

Species	DDT group	BHC-Dieldrin group	Organophosphorus group
Phaenicia cuprina		Australia 1957; S. Africa 1959	Australia 1966(*706*)
Phaenicia sericata		New Zealand 1959; Zanzibar 1961(*286*)	
Chrysomia putoria	Malagasy 1960	Congo 1949; Malagasy 1960 (*150*)	Congo 1954
Haematobia irritans		Tex. 1959(*615*)	La. 1962(*123*)
Protophormia terraenovae	European Russia 1964(*187*)		

[a] Also *D. melanogaster*; both field strains.

478

Petit Bras River, Quebec(*813*), the only important genera left in this group that have not yet shown resistant strains are *Phlebotomus*, sand-flies, and *Glossina*, tsetse flies.

G. Parasites of Man and Animals

DDT-resistant populations of the body louse, *Pediculus* (Table 7-9), have been found on human hosts in almost all parts of the world, especially in Japan and Korea and areas where mass delousing operations have been carried out. Resistance to lindane powders has also developed in many places, but malathion powders have not induced any organophosphate resistance in the body louse.† Resistance to DDT, gamma-BHC, and organophosphate compounds has been observed in the cattle-sucking louse, *H. eurysternus*, at Lethbridge, Alberta(*330*). DDT resistance has also developed in three species of long-nosed cattle louse (*Linognathus*) and toxaphene resistance in *Bovicola*, goat lice. Resistance of both types within the chlorinated hydrocarbons also occurs in five species of fleas and is particularly important in the oriental rat flea, *Xenopsylla cheopis*, although in this species the BHC resistance is not too far advanced. In the bedbug, *C. lectularius*, and its tropical congener, *C. hemipterus*, the BHC resistance is more intense than the DDT resistance. *C. lectularius* has developed malathion resistance in Israel, with a strong cross-resistance to fenthion(*51*). The German cockroach, *Blattella*, has developed resistance to chlordane in many parts of the world and most places in the U.S. and to DDT in Europe and the Caribbean region. They are cross-resistant to other cyclodienes but remain susceptible to the chlorinated polycyclic KEPONE, employed in baits(*383*). Unlike houseflies, chlordane-resistant *Blattella* are strongly cross-resistant to isobenzan and so are DDT-resistant strains(*160*). Diazinon-resistant cockroaches appeared in Kentucky and Indiana in 1960(*293*) and in Texas in 1964(*294*). Malathion selection does not induce cross-resistance to diazinon or other organophosphate compounds(*344*). Yet a survey made in Louisiana in 1966 found that all populations had at least a sixfold tolerance to all organophosphate compounds tested, the highest resistance ratios being 110 to malathion, 13 to diazinon, and 11 to fenthion, with a 14-fold resistance to propoxur in three populations(*63*). The two species of *Boophilus* — the blue tick of South Africa and the cattle tick of Australia and elsewhere — both became BHC resistant around 1950 and DDT resistant around 1957. Organophosphate resistance to remedial insecticides such as dioxathion (DELNAV has already developed at Rockhampton, Queensland(*712*), with an even higher cross-

---

†Although malathion-tolerant body lice have recently been encountered in Burundi and Egypt.

TABLE 7-9

GEOGRAPHIC DISTRIBUTION OF CERTAIN ARTHROPOD SPECIES OF PUBLIC HEALTH AND VETERINARY IMPORTANCE RESISTANT TO ORGANO-CHLORINE PESTICIDES

Arthropod species	Resistant to DDT group of pesticides	Resistant to BHC-Dieldrin group of pesticides
Pediculus corporis (Human body louse)	Korea, Japan 1951; Egypt Levant 1952; Iran, Turkey, Ethiopia, W. Africa, S. Africa, Peru, Chile 1955; France 1956; Jordan, Libya, Afghanistan, India 1958; Mexico, Uganda, 1959; Sudan 1961*(622)*; Romania 1964*(202)*	France, Japan 1955; W. Africa, S. Africa 1956; Iran 1957; India, Korea 1958; Tanganyika 1959; Sudan 1961*(662)*; Egypt 1962*(839)*; Turkey 1966*(836)*
Linognathus vituli (Long-nosed cattle louse)	Va. 1957*(19)*; Alberta 1966*(330)*	
Linognathus africanus		
L. stenopsis (Goat lice)		S. Africa*(204)*
Haematopinus eurysternus (Short-nosed cattle louse)	Alberta 1964*(330)*	Alberta 1965*(236)*
Bovicola limbata		
B. caprae (Cattle louse)		Tex. 1957*(544)*
Blatta orientalis (Oriental cockroach)	Czechoslovakia 1964*(651)*	Germany 1958*(835)*; Czechoslovakia 1964*(651)*
Blattella germanica (German cockroach)	France, Germany*(808)*, Cuba, Bahamas, Puerto Rico 1958; Trinidad, Poland 1959*(20)*; England 1960*(284)*	Tex. 1951; Southeastern U.S.A. 1955; Northeastern U.S.A. 1956, Cal., Panama, Cuba, Puerto Rico 1958; Canada*(116)*; Trinidad, Japan*(347)*; Poland 1959*(20)*; England*(284)*; Germany*(808)*; Central U.S.A. 1960 *(738)*; Denmark 1963*(410)*; Hawaii 1964*(383)*; New Guinea, N.S.W., N.Z.*(236)*
Periplaneta brunnea (Brown cockroach)	Hawaii 1947; Ohio, Ill., Ind., Utah, Congo, Israel, Korea, Greece 1952; Japan, Italy, 1953;	Florida 1965*(495)*
Cimex lectularius (Bedbug)	Iran, Colo., Pa., Tex. 1955; French Guiana 1956;	Italy 1954; Israel 1956; Indonesia, Zambia, Borneo 1960; S. India 1961*(710)*; S. Africa 1962*(816,279)*; N. India 1965

Species		
Pulex irritans (Human flea)	Trinidad, Turkey 1957; Hungary, Poland 1959; Borneo, Indonesia, Colombia, Rhodesia 1960 *(676)*; S. India 1961*(700)*; S. Africa 1962*(816)*; N. India 1965*(400)*	Tanganyika 1959; Turkey 1965*(280)*
Ctenocephalides canis C.felis (Dog and cat fleas)	Peru 1949; Ecuador 1950; Greece 1951; Brazil, Palestine 1952; Turkey 1965*(280)* Ga. 1952, French and British Guiana, Colombia 1953; Various points U.S.A. 1956; Hawaii 1958	U.S.A. 1956; Hong Kong 1957; Hawaii, Japan 1958
Xenopsylla cheopis (Oriental rat flea)	W. India 1959*(616)*; N. India 1960*(446)*; Southeast India 1963*(669)*; S. Vietnam 1964*(153)*; Thailand 1966*(836)*	W. India*(616)*; Southeast India 1962*(541)*; Thailand 1966*(836)*
Xenopsylla astia (Rat flea)	N. India 1961*(708)*; S. India 1962*(541,669)*	Southeast India 1962*(541)*
Boophilus decoloratus (Tick)	Cape Province, S. Africa 1956	Cape Province, S. Africa 1958; Transvaal 1952; N. Rhodesia 1956
Boophilus microplus (Tick)	Queensland 1954; Brazil 1956	Queensland 1950; Brazil 1952; N. India 1960*(148)*; Malagasy 1963*(786)*; Guadeloupe 1961*(548)* Oklahoma 1954
Amblyomma americanum (Lone star tick)		N.J. 1954; Panama, Tex. 1958
Rhipicephalus sanguineus (Brown dog tick)		S. Africa 1964*(46)*
Rhipicephalus appendiculatus (Tick)		
Rhipicephalus evertsi (Tick)		S. Africa 1960*(817)*
Dermacentor variabilis (American dog tick)	Mass. 1959*(806)*	Mass. 1959*(806)*

resistance to carbophenothion (TRITHION). Resistance to cyclodiene derivatives has appeared in the lone star tick, *Amblyomma*, the brown dog tick, *R. sanguineus*, the relapsing fever tick, *R. appendiculatus*, and the African red tick, *R. evertsi*, while the wood tick, *D. variabilis*, in Massachusetts has developed resistance to both types of chlorinated hydrocarbons.

A notably high pyrethrin tolerance coupled with resistance to chlorinated hydrocarbons has been observed in populations of *Blattella* in Alabama(*415*), of *C. hemipterus* in Kenya(*130*), and of *Boophilus decoloratus* in South Africa(*815*). The Swedish pyrethrin-resistant strain of houseflies, when further selected with pyrethrins, developed a high cross-resistance to DDT(*227*), while pyrethrin pressure on *Pediculus corporis* resulted in moderate pyrethrin tolerance but high cross-resistance to DDT(*162*); pyrethrin-resistant populations have been discovered on people from Oran, Algeria, and Shimonoseki, Japan(*102*). Selection with carbaryl (SEVIN) failed to induce resistance until the 70th generation was reached(*155*). Whereas DDT-resistant strains of *Pediculus* are slightly more carbaryl-susceptible than normal(*121*), on the other hand, a DDT-selected strain of *Blattella* has been found to have gained a 75-fold carbaryl resistance, with a threefold cross-tolerance to arprocarb(*515*). Increased tolerance to some carbamate insecticides is shown by OP-resistant strains of houseflies and by strains resistant to chlorinated hydrocarbons; on the other hand, housefly strains made resistant to carbaryl or other carbamates often become cross-tolerant and occasionally highly cross-resistant to chlorinated hydrocarbons(*257,271, 362*). A strain selected by the OP insecticide ronnel was definitely cross-resistant to carbaryl, the molecules of both compounds possessing regions of high electron density ortho to each other(*258*).

H. ·Resistance in Mosquitoes

Among the *Diptera*, 24 species of culicines have developed resistance, 10 of them to both classes of chlorinated hydrocarbons and eight to OP insecticides also (Table 7-10). The tropical house mosquito, *Culex fatigans*, normally somewhat DDT tolerant, develops DDT resistance readily and can become even more resistant to gamma-BHC and dieldrin. Unlike its northern counterpart, *C. pipiens*, certain populations of *C. fatigans* have shown a resistance to organophosphorus compounds in the field, but this disappears when the mosquitoes are colonized in the laboratory. On the other hand, *C. tarsalis*, the vector of western equine encephalitis, developed a specific malathion resistance that has been thoroughly studied in the laboratory.

Resistance to both classes of chlorinated hydrocarbons has been devel-

TABLE 7-10

INCIDENCE OF RESISTANCE AMONG CULICINE MOSQUITOES TO THREE MAJOR GROUPS OF INSECTICIDAL COMPOUNDS IN RELATION TO GEOGRAPHICAL AREA AND YEAR OF FIRST RECORDED OBSERVATION

Mosquito species	Geographical areas and year first observed		
	DDT group	BHC-Dieldrin	Organophosphate
C. fatigans (*quinquefasciatus*)	India 1952; Reunion 1953; Venequela, Taiwan 1956; Puerto Rico 1957; W. Africa, S. Australia, Panama 1958; Hawaii, Congo 1959(*455*); W. Pakistan, Peru 1960(*4*); Malagasy 1961(*149*); Cuba 1961(*835*); Brazil 1961(*167*); Ga. 1961(*680*); Trinidad 1961(*588*); Ryukus 1966(*836*); China 1964(*840*); Queensland 1966(*355*)	Cal. 1951; Malaya, India 1953; E. Asia 1954; S. America 1956; W. Africa 1957; Panama 1958; Zanzibar, Congo 1959; Tex. 1960(*533*); Mali 1961(*835*); Malagasy 1961(*149*); Brazil 1961(*167*); Trinidad 1961(*588*); Upper Volta, Togo, Ivory Coast, Senegal, Niger, Mauritania 1966(*383*)	Cameroun 1959(*309*); Calif. 1960(*564*); Sierra Leone 1963 (*766*); Ryukyus 1967 (*622*)
C. pipiens	Italy 1947; Mass., N.J. 1955; Israel 1959; Japan 1959(*753*); Cal. 1959; China 1960 (*847*); N.Y., Mass., Ill., Utah 1961(*835*); France 1963(*734*); Korea 1965(*379*); Turkey 1966(*838*)	Italy 1950; Israel 1955; France, Japan 1959(*845*); Korea 1965(*379*); Morocco 1966(*836*)	
C. tarsalis	Cal. 1951; Oreg. 1956; Wash. 1961; Utah 1961(*680*)	Cal. 1951; Oreg. 1961(*680*)	Cal. 1956
C. coronator	Panama 1958		
C. tritaeniorhynchus	Ryukyus 1958	Ryukyus 1958; Dahomey 1959; Korea 1965(*379*)	Ryukyus 1967(*622*)
C. peus	Oreg. 1961(*96*)		Cal. 1965(*829*)

TABLE 7-10 (cont.)

Mosquito species	Geographical areas and year first observed		
	DDT group	BHC-Dieldrin	Organophosphate
Aedes aegypti	Trinidad, Dominican Republic 1954; Venezuela 1955; Haiti 1956; Antigua; Columbia 1957; S. Vietnam 1958; Puerto Rico, Jamaica, Guadeloupe, French Guiana 1959; Fla. 1961(*214*); British Guiana 1963 (*124*); Thailand 1964 (*570*); E. India 1964; Japan 1965(*552*); Tex. 1965(*242*); St. Vincent, W. India 1966(*836*); Liberia, Ivory Coast, Dahomey, Cameroun 1968(*557*)	Puerto Rico 1959(*241*); Jamaica. Haiti 1962; Netherlands, Antilles, Grenadines 1962(*432*); Virgin Islands 1963 (*243*); Surinam, French Guiana 1964(*610*); Cambodia 1964(*558*); S. Vietnam 1964 (*663*); Tex. 1965 (*242*); Cameroun 1966 (*556*); Tahiti 1966(*559*); Thailand 1966 (*386*); Ivory Coast 1966(*555*); Senegal, Congo, Cuba, St. Vincent 1966(*836*); Liberia, Togo, Upper Volta, Nigeria 1968 (*557*)	
A. sollicitans	Fla. 1947; Del. 1951	Fla. 1951; Del. 1958	Fla. 1965(*247, 275*)
A. taeniorhynchus	Fla. 1949; Ga. 1959	Fla. 1951; Ga. 1959	Cal. 1958(*57, 460*)
A. nigromaculis	Cal. 1949	Cal. 1951	Cal. 1962(*680*)
A. melanimon (formerly *A. dorsalis*)	Cal. 1951	Cal. 1951	
A. dorsalis			N. Mex. 1966(*316*)
A. cantator	New Brunswick 1960(*636*)	New Brunswick 1960(*636*)	
A. detritus	France 1959(*434*)		
A. cantans	Germany 1958(*835*)		
A. albopictus	S. Vietnam 1964(*663*); S. India 1964(*554*)		
A. vittatus	W. India 1964(*697*)		
Psorophora confinis	Miss. 1954		
P. discolor	Miss. 1954		

oped by the salt-marsh mosquitoes *Aedes sollicitans* and *A. taeniorhynchus*, and the irrigation-water mosquitoes *A. nigromaculis* and *A. melanimon*. In western Florida *A. taeniorhynchus* has finally developed a significant malathion tolerance; in New Mexico *A. dorsalis* shows resistance to fenthion(*315*), and repeated larvicidal operations in California have induced resistance in *Ae. nigromaculus* and *Cu. tarsalis* to parathion and methyl parathion(*117, 263a*), and more recently to fenthion and even to DURSBAN(*271b*). Two species of rice-field mosquitoes of the genus *Psorophora* have responded to dieldrin treatments with cyclodiene resistance.

The yellow-fever mosquito, *Aedes aegypti*, has become DDT resistant in the Caribbean area, where it has survived the eradication program that has been so successful elsewhere in the Americas. Caribbean populations are also frequently dieldrin-resistant and have occasionally developed malathion tolerance(*416*). Diazinon resistance has recently been observed at Brazzaville, Congo(*836*). Populations of this species in the U.S. have either already developed DDT resistance or have the potentiality for it (*2*). *Aedes pseudoscutellaris* has developed a 100-fold DDT resistance in 13 generations of laboratory selection, while *A. fijiensis* is unusual in that the larva but not the adult has a high normal DDT tolerance(*122*). Indications of developed DDT resistance have been found in *Culex salinarius* at New Gretna, New Jersey(*751*), in *Aedes vexans* at Kamloops, British Columbia, in *A. atropalpus* at Fort Sill, Oklahoma (but this may be a normal species characteristic), in *Culex erythrothorax* at Fort Conkhite, California, and of dieldrin resistance in *C. restuans* at Camp Drum, New York(*610*).

No fewer than 37 species of anopheline mosquitoes have developed resistance (Table 7-11), 35 of them to dieldrin. Of the 15 species that have developed DDT resistance, only *Anopheles nunez-tovari* and *A. hyrcanus sinensis* have not shown dieldrin resistance. Many populations of *A. sacharovi*, *A. pharoensis*, *A. stephensi*, and *A. albimanus* are simultaneously doubly resistant, although they can still be controlled with high dosages of DDT; in most cases, this situation has been caused by the effect of agricultural insecticides on the larvae in rice fields. Insecticide-resistant anophelines(*107,251*) now occupy between 5 and 10% of the worldwide malaria eradication program. OP resistance has developed only in *Ae. albimanus*, with malathion-tolerant strains in El Salvador (*656a*), Guatemala and Nicaragua(*837*), and propoxur tolerance in eastern Nicaragua(*611a*).

I. Stored Products Insects

Resistance has not developed to any extent among the insects attacking stored products. The BHC resistance of *Tribolium castaneum* in Kenya

TABLE 7-11

INCIDENCE OF RESISTANCE AMONG ANOPHELINE MOSQUITOES TO DDT AND DIELDRIN
IN RELATION TO GEOGRAPHICAL AREA AND YEAR OF FIRST RECORDED OBSERVATION

Species resistant to DDT	Year	Geographical area
A. sacharovi	1951	Greece, Lebanon, Iran, Turkey (855)
A. sundaicus	1954	Java, Burma, Sumatra (787)
A. stephensi	1955	Arabia, Iraq, Iran, India, W. Pakistan (69, 671)
A. subpictus	1956	N. India, W. Pakistan (455); Nepal (492); Java (43)
A. albimanus	1958	Central America, Mexico, Cuba (836)
A. pharoensis	1959	Egypt (403); Sudan (493)
A. quadrimaculatus	1959	Ga., Md., Mexico (499)
A. annularis	1959	W. India, E. India, N. India (447)
A. culicifacies	1960	W. India (696); S. India (68); Nepal (609); W. Pakistan (787); N. India (447)
A. albitarsis	1961	Colombia (492)
A. nunez-tovari	1961	Venezuela (492)
A. aconitus	1962	Java (730)
A. fluviatilis	1963	W. India (493)
A. hyrcanus sinensis	1964	Ryukyus (836)
A. gambiae	1967	Upper Volta, Senegal (310)

Species resistant to Dieldrin		
A. sacharovi	1952	Greece (62)
A. quadrimaculatus	1953	Miss., Ga., Mexico (499)
A. gambiae	1955	Nigeria, Liberia, Ivory Coast, Dahomey, Upper Volta, Cameroun, Sierra Leone, Togo, Ghana, Mali, Congo (Braz.) (308); Madagascar, Sudan, Kenya (150)
A. subpictus	1957	Java, Ceylon, N. India, W. Pakistan (455)
A. coustani	1957	Arabia (619)
A. pulcherrimus	1957	Arabia (619)
A. albimanus	1958	Salvador, Guatemala, Nicaragua, Honduras, Jamaica, Ecuador, Mexico, British Honduras, Cuba, Dominican Republic, Haiti, Colombia (787)
A. pseudopunctipennis	1958	Mexico, Nicaragua, Peru, Venezuela (787); Ecuador (788)
A. aquasalis	1958	Trinidad, Venezuela, Brazil (665)
A. culicifacies	1958	W. India (709); Nepal (609)
A. vagus	1958	Java (152); Phillipines (492); Malaya (609)
A. barbirostris	1958	Java (152)
A. annularis	1958	Java (152)
A. sergenti	1958	Jordan (788)
A. fluviatilis	1958	Arabia (492)
A. splendidus	1958	N. India (608)

TABLE 7-11 (*cont.*)

Species resistant to Dieldrin	Year	Geographical area
A. stephensi	1958	Iran, Iraq, Arabia(787); S. India(69)
A. minimus flavirostris	1959	Philippines(787); Java(43)
A. pharoensis	1959	Egypt(401); Sudan(493); Israel(492)
A. albitarsis	1959	Colombia(491); Venezuela(492)
A. labranchiae	1959	Morocco, Algeria(467)
A. strodei	1959	Venezuela(491)
A. triannulatus	1959	Venezuela, Colombia(492)
A. sundaicus	1960	Java, Sumatra(730); Sabah(787)
A. aconitus	1960	Java(730); India(609)
A. neomaculipalpus	1960	Trinidad(587); Colombia(492)
A. crucians	1960	S. C.(680); Dominican Republic(492)
A. filipinae	1960	Philippines(492)
A. maculipennis messeae	1961	Romania(201); Bulgaria(492)
A. rangeli	1962	Venezuela(835)
A. philippinensis	1962	Sabah(788)
A. funestus	1962	Nigeria(203); Ghana(385); Kenya
A. nili	1966	Ghana(836)
A. rufipes	1968	Mali(168a)

(Table 7-4) is about 12 times the normal, while the BHC resistance of the rice weevil, *Sitophilus oryzae*, in Queensland is some 85 times the normal; BHC-resistant rice weevils, either *S. oryzae* or *S. sasakii*, have been encountered in England, BHC-resistant granary weevils (*S. granarius*) in South Africa, and BHC-resistant maize weevils, *S. zeamais*, in Kenya. The red flour beetle, *Tribolium castaneum*, is also showing decreased susceptibility to malathion in England(615), various degrees of malathion tolerance in the eastern U.S.(736,793), and a considerable malathion resistance at Kano, Nigeria (Table 7-5) that does not extend to fenthion or diazinon(635). Tolerance to malathion has also been developed by the caterpillars *Ephestia cautella* and *Plodia interpunctella* in Georgia (Table 7-5). Under laboratory selection with DDT, *Tribolium castaneum* has developed a 160-fold DDT resistance(205).

An English strain of *Sitophilus granarius*, already slightly pyrethrin tolerant, when submitted to laboratory selection with pyrethrins, developed a 35-fold resistance in 20 generations(614) and subsequently a 130-fold resistance to pyrethrins, but showed only a twofold increase in tolerance to synergized pyrethrins. A Canadian strain, however, developed a 2.2-fold tolerance to synergized allethrin in only six generations(704). Another Canadian strain developed an eightfold resistance to arprocarb after 14 generations of laboratory selection(450). The granary weevil has

not developed much resistance to the fumigant methyl bromide, 30 generations of selection producing only a fourfold increase in LC_{50}, with a
similar cross-tolerance to totally unrelated fumigants and a 20% increase
in body weight (543).

On the other hand, this increased tolerance did not disappear when
selection was discontinued (542). The selected strain was characterized
by a 20% decrease in respiratory rate, which may account for the 13-fold
cross-tolerance that it showed to hydrogen phosphide (660). In general,
however, North American millers have not yet noted any resistance
developing to their fumigants. Samples of *Tribolium confusum* and *T.
castaneum* collected from various parts of the U.S.A. and tested in 1964
showed no important increase in tolerance to HCN, methyl bromide, or
ethylene dibromide. Laboratory selection of *Sitophilus granarius* and
S. oryzae with these three fumigants did not produce any more than three-
fold tolerance in 25 generations (468).

J. Behavioristic Resistance

Whereas physiological resistance refers to the insects' withstanding a
lethal dose, behavioristic resistance refers to ways and means whereby
they have come to avoid taking a lethal dose. Populations of the housefly
have developed a characteristic of not coming readily to malathion in
sugar baits (426); this avoidance is not shown to trichlorfon or dichlorvos
and is intensified by further malathion selection. Strains of the spotted
root fly, *Euxesta notata*, have developed an avoidance of residual
deposits of malathion or parathion; they take off soon after contacting the
deposits, being more liable to be irritated by them since they are now less
active in detoxifying these compounds (358). DDT-resistant houseflies
on sprayed premises are characteristically found at low levels and on
untreated surfaces, presumably because it is in these places that they can
rest and detoxify the dose that would have killed their susceptible counter-
parts (553). Populations of the potato flea beetle, *Epitrix cucumeris*, that
had developed DDT resistance simultaneously lost their normal characteristic of avoiding DDT-sprayed leaves (445), presumably because the
deposit could no longer stimulate the nerves through the feet, as found
with DDT-resistant flies (811).

It is in the reaction of anopheline mosquitoes to DDT deposits that this
type of behavioristic resistance becomes important. DDT-resistant
populations of *Anopheles sacharovi* (855), *A. culicifacies* (66), and *A.
albimanus* (41) and DDT-resistant laboratory strains of *A. stephensi*, *A.
albimanus*, *Culex fatigans* (134), and *Aedes aegypti* (112) are found to be
less irritable than normal to the excitorepellent effect of DDT deposits.
Strains of *Anopheles atroparvus* hyperirritable to DDT deposits have

been developed by laboratory selection(*273*), but malathion-avoidance could not be selected so definitely in houseflies(*727*). The development of increased irritability to deposits of parathion and malathion by laboratory selection in *Euxesta notata* was found to be associated with a decrease in detoxifying ability(*358*).

In the world malaria-eradication program, it has been found that some species, notably *Anopheles gambiae* of Africa, are characteristically irritated by DDT deposits and for this reason escape being killed. In *A. maculipennis* this characteristic may aid in the reduction of malaria transmission in Italy, since it drives the mosquitoes out of the sprayed houses(*856*). In Mexico this characteristic has developed, after many years of the selective effect of DDT treatments, in populations of *A. pseudopunctipennis* and *A. albimanus*; in the former species, it has been quantitatively established that they now enter sprayed houses as readily as unsprayed and show a lower mortality after entry(*500*). The deterrency of DDT having been lost, these populations can enter sprayed houses, bite the inhabitants, and leave without having picked up a lethal dose, despite the fact that they show no greater physiological DDT resistance than the original norm(*857*). The same changed response to DDT deposits had been noted in a population of *A. albimanus* on the Chagres River in Panama(*773*) which, though of normal physiological DDT susceptibility, had become slightly more irritable to DDT deposits (*103,203*).

Another type of behavioristic resistance among anophelines could derive from the development of an increased exophily, the mosquitoes entering the houses less. Outside-resting has been found to characterize *A. nunez–tovari* in western Venezuela and *A. sundaicus* in southern Java, but this characteristic was not developed as a result of the DDT spraying program. Increased exophily has been reported to have been induced by several years of DDT treatment in populations of *A. punctimacula* in Columbia and of *A. cruzii* in southern Brazil(*835*), but proof that this behavioristic avoidance developed as a result of the DDT selection pressure is considered to be lacking(*562*). The increase in exophily observed in *A. gambiae* in Mauritius, Rhodesia and elsewhere may be due to the preferential survival of sexually isolated sibling species that are characteristically exophilic(*834*).

7-2 PHYSIOLOGICAL MECHANISMS OF RESISTANCE

A. DDT Resistance

Studies of DDT-resistant houseflies have provided a model for the understanding not only of DDT resistance in insects but insecticide resistance in general, which so frequently derives from increased detoxication. The first DDT-resistant flies to be studied, collected from Arnas in northern Sweden, were found in 1945 to have a thicker cuticle on their tarsi and pulvilli, which suggested that the DDT would be less readily absorbed(821). Subsequent studies of various DDT-resistant strains, however, failed to find any increase in cuticle thickness or any consistent decrease in cuticular absorption (Table 7-12). The flies of the Arnas strain were also larger and darker than normal, but again these differences were not characteristic of DDT-resistant but of Swedish flies(102).

TABLE 7-12

ABSORPTION AND DEHYDROCHLORINATION OF DDT BY SUSCEPTIBLE AND RESISTANT HOUSEFLIES AND THEIR HOMOGENATES

	LD$_{50}$ μg/fly	Absorption in % of the dose applied[a]	Dehydrochlorination	
			In vivo[a] % of amount absorbed	In vitro[b] % conversion
NAIDM	0.25	18	6	0
Beltsville	0.3	12	0	—
Savannah	0.3	13	4	—
Urbana	0.3	6	18	0
Pollard	1.85	—	—	6
Orlando	2.0	4	54	9
Roberds	25	18	54	12
Edgewood	40	5	56	8
DMC-5	50	5	70	23
DDT-45	55	11	67	30

[a]Per cent of 6.8 μg contact dose absorbed, and per cent of absorbed dose dehydrochlorinated in 2 hr(395).
[b]Per cent of 200 μg DDT converted in 4 hr by homogenate from six flies(466).

In certain DDT-resistant strains, the lipoid content was reported to be higher in the fly's body(823) and in its ganglia and tarsi(678), to differ qualitatively from the normal lipoid(65), and to be more resistant to dissociation from lipoprotein complexes by the action of DDT poisoning (679). When many DDT-resistant housefly strains were compared with susceptible ones, however, no consistent difference could be discovered in the quality and quantity of the lipoid(113) or sterols(212). In mosquitoes an increased lipoid content was found in the adults of a laboratory-selected DDT-resistant strain of Anopheles atroparvus(571), but not in the larvae of DDT-selected strains of Aedes aegypti(217). The differences in DDT tolerance of individuals within a strain or population are however positively correlated with lipoid content, as in the greater tolerance of female over male *Periplaneta* cockroaches(*105*) or in well-fed *Lygus* bugs(*42*).

Although an increased level of cytochrome oxidase was at first found in a DDT-resistant strain, comparison of many resistant and susceptible strains revealed no consistent difference in this respect(*629*). No consistent interstrain difference was shown with respect to the content of succinic dehydrogenase and the rate of oxygen consumption nor to the titer of cholinesterase or the content of phosphatase enzymes(*105,113*).

All of the DDT-resistant strains, however, were characterized by their ability to detoxify DDT to the nontoxic metabolite DDE(*625,628,741*) by a process of dehydrochlorination (Fig. 7-1). The amount of DDE

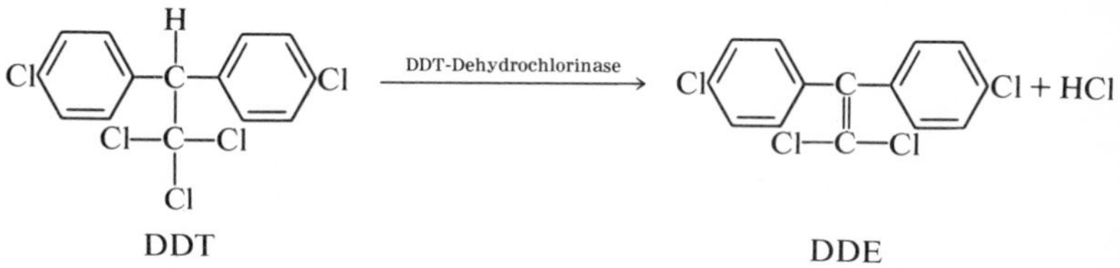

Fig. 7-1. Enzymic detoxication of DDT to DDE by dehydrochlorination.

produced proved to be proportional to the DDT resistance of the strains studied (Table 7-12). The detoxication was due to an enzyme requiring glutathione for activation *in vitro*(*744*); the titer of this enzyme, called DDT dehydrochlorinase(*743*), was also proportional to the DDT-resistance level of the strain (Table 7-12). Purification of this enzyme resulted in a protein with a molecular weight of 36,000 and an optimum pH of 7.4; it also dehydrochlorinated methoxychlor, and TDE even more rapidly than DDT(*472*). The housefly DDT-ase cannot dehydrochlorinate *o,p'*-DDT, nor the ortho-chloro derivative of DDT(*341*). It is inhibited by DMC (chlorfenethol), and this is the reason that DMC is a synergist for DDT against resistant houseflies. DDT is also synergized by WARF-ANTI-RESISTANT, which inhibits the enzyme *in vitro*, and by piperonylcyclonene, which inhibits dehydrochlorination *in vivo*. The enzyme is synthesized along with the protein synthesis during larval growth, so that the last larvae to pupate have the most DDT-dehydrochlorinase. Resistant larvae can completely dehydrochlorinate a DDT dose that is only one-third detoxified by susceptible individuals and that kills them as they emerge from the pupa(*698*). Houseflies can, in turn, develop resistance to the DMC–DDT mixtures, and these strains carry even higher DDT-ase levels and, in addition, some mechanism that has now made them resistant to the ortho-chloro derivative of DDT(*113,473*).

The ability of DDT-susceptible flies to break down the bromoethane analogs of DDT may derive from a nonspecific enzyme system(64). Houseflies in which the DDT resistance has been induced by selection with an organophosphate compound have acquired DDT-dehydrochlorinase(113) and break down the DDE produced to a greater extent than ordinary DDT-resistant flies(58). They also show reduced absorption of DDT(58), a characteristic also of strains selected with DMC–DDT mixtures(623). A Danish strain made DDT-resistant by diazinon selection converted about half of the applied dose of DDT to water-soluble metabolites(421); this conversion could be accomplished by microsomal preparation, producing four unknown metabolites(599).

The isolated nerves of DDT-resistant houseflies are resistant in themselves to DDT directly applied, the repetitive discharges of action potentials induced thereby being more transitory and requiring a higher threshold concentration to produce them(811,844). The taste receptors in the labella(118) and tarsi(728) of DDT-resistant flies are not stimulated by DDT in their action potentials and responses to the extent that are those of normal flies. This resistance of nerves may derive not only from their content of protective lipoprotein, which depots DDT, but also from their unusually high content of DDT-dehydrochlorinase, which destroys DDT before it finally reaches the axon sheaths(539). DDT-resistant flies have an additional defense mechanism in that they respond to DDT, unlike normal flies, by increased protein synthesis and glutathione turnover(9).

The enzymic dehydrochlorination of DDT is a resistance mechanism in many species of mosquitoes. Resistant strains of *Aedes aegypti*, *A. taeniorhynchus*, and *Culex fatigans* show a much higher larval and adult production of DDE than normal strains(3,105). Although in *Aedes aegypti* dehydrochlorination is the only metabolic path(357), it is not so important as a defense mechanism in *Culex fatigans*(402), which can also develop a reduced absorption(356). Whereas the DDT resistance of *Anopheles subpictus* is specific for p,p'-DDT, that of *Culex fatigans* extends to the o,p'-isomer also(399). The DDT-dehydrochlorinase of *Aedes aegypti* differs from the housefly enzyme in requiring more glutathione protection and in dehydrochlorinating o-chloro-DDT but not deutero-DDT(427), while the DDT-ase of *Culex fatigans* is different in being less active on TDE than DDT(428). Increased adult production of DDE has been found in DDT-resistant *Anopheles sacharovi*(624) and *A. subpictus*(401) but not in *A. stephensi*, *A. albimanus*, and *A. quadrimaculatus*, as compared to normal strains(471). DDT-resistant adults of *A. sacharovi* or *Aedes aegypti* are less susceptible to the excitorepellent effect of DDT and show less of this behavioristic resistance type of response(*112*), just as DDT-resistant houseflies are much less sensitive to the stimulating effect of DDT deposits on their tarsi(*728*).

In the body louse, *Pediculus h. humanus*, both the DDT-resistant and the susceptible strains contain a peculiar DDT-detoxifying enzyme that is heat-stable and resists destruction by ethanol or protease(*626*). It differs from the housefly DDT-ase in being activable not only by glutathione but also by cysteine or ascorbic acid, and in not being inhibited by DMC (*633*). Homogenates of both DDT-resistant and susceptible body lice detoxify DDT to a mixture of bis(*p*-chlorophenyl) acetic acid (DDA) and *p,p'*-dichlorobenzophenone (DBP), but *in vivo* the metabolism operates only in the resistant strain. Purified enzyme fractions were less heat-stable, one fraction producing a small amount of DDE and the other an equal mixture of DDA and DBP(*538*).

In the German cockroach, *Blattella germanica*, the detoxication of DDT (Fig. 7-2) produces not DDE but dicofol (KELTHANE) and three

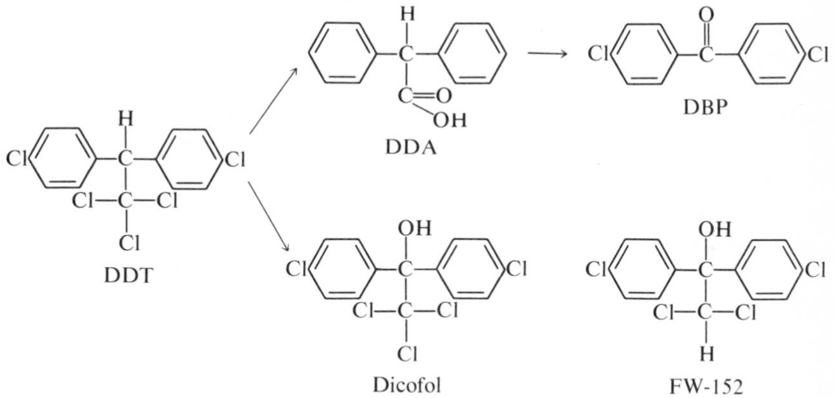

Fig. 7-2. Enzymatic detoxication of DDT to dicofol by oxidation, or to DDA and DBP; FW-152 is shown as an oxidative metabolite of the analog TDE.

other polar metabolites(*8*), and this detoxication is increased in DDT-resistant roaches(*140*). It is an oxidation process resident in the microsomal cellular fraction and is probably due to a mixed-function oxidase, which is active if it competes successfully for TPNH with glutathione reductase, otherwise dehydrochlorination prevails over oxidation(*8*). Traces of dicofol are produced from DDT *in vivo* by DDT-resistant strains of the housefly and *Aedes aegypti*, and it is among the five metabolites including DBP produced by the American cockroach *Periplaneta americana*(*113*). The ability of the microsome fraction of housefly homogenates to hydroxylate naphthalene increases in proportion to the

DDT resistance level of the strain(701). The metabolism of TDE by DDT-resistant larvae of *Culex tarsalis*, which is more than twice as fast as by susceptible ones, involves oxidation more than dehydrochlorination; these larvae are also resistant to nondehydrochlorinatable compounds such as PROLAN, *o*-chloro-DDT, and DMC–DDT mixture(647). DDT-resistant larvae of *Drosophila melanogaster* and *D. virilis* produce much dicofol and virtually no DDE, with DBP and FW-152 as secondary metabolites (Fig. 7-2), but the dicofol production does not consistently follow the inheritance of the genetic factor for DDT resistance(774).

In the cone-nose bug, *Triatoma infestans*, the greater tolerance of the mature nymphs as compared with the adults is correlated with a greater detoxication rate, with DDE as the major metabolite(191). However, oxidation to dicofol may be important since this tolerance is increased by a topical application of 3-methylcholanthrene, which stimulates the microsomal hydroxylation, and is decreased by an injection of iproniazid, which inhibits this oxidation process(549). In the cattle ticks *Boophilus microplus* and *B. decoloratus* a metabolism of DDT takes place, but little or no DDE may be detected in susceptible or resistant strains(113); the soft tick, *Ornithodoros coriaceus*, also produces abundant metabolites but no DDE(363).

Among plant-feeding insects, a resistant strain of the pink bollworm, *Pectinophora gossypiella*, absorbed one-third as much DDT as the normal and detoxified three times more of it to DDE(119). By contrast, the intrastrain differences in DDT tolerance of corn earworms (*Heliothis zea*) proved to be inversely proportional to the ease of cuticular penetration (253). In the tobacco budworm, *H. virescens*, one strain with a 20-fold resistance metabolized DDT to DDE and DDA about four times as fast as normal, while another with an eightfold resistance absorbed DDT through the cuticle about half as fast as normal(794). DDT-resistant adults of the spotted root maggot, *Euxesta notata*, absorbed one-half as much as normal and thus detoxified to DDE a greater proportion of the amount absorbed(354).

A DDT-resistant strain of the Japanese cabbageworm (*Pieris rapae erucivora*) produced much more DDE than the susceptible cabbage cutworm *Barathra brassicae*(440). The normally DDT-tolerant tobacco hornworm, *Protoparce sexta*, produces large amounts of DDE(363); so does the DDT-tolerant Mexican bean beetle, *Epilachna varivestris*, by a DDE-dehydrochlorinase requiring glutathione protection and activation (147, 742), and this enzyme has been studied *in vitro*(768). The DDT-tolerant lady beetle, *Coleomegilla maculata*, also rapidly metabolizes DDT to DDE, and is prompt in excreting these compounds(34). Other DDT-tolerant species such as the boll weevil, *Anthonomus grandis*, and the milkweed bug, *Oncopeltus fasciatus*, rapidly detoxify DDT to metabolites other than DDE(*76,225*), whereas *Melanoplus* grasshoppers are refractory to DDT because their cuticle absorbs so little of it in the first place(*742*). The red-banded leafroller, *Argyrotaenia velutinana*, produces DDE from DDT so rapidly that it is DDT tolerant, but dehydrochlorinates TDE to FW-152 so slowly that it is TDE-susceptible(*255,742*). Mature larvae of the European corn borer, *Ostrinia nubilalis*(*469*), resemble those of the Indian-meal moth *Plodia interpunctella*(*363*) in that they produce DDE but not in sufficient quantity to make them significantly DDT tolerant.

B. PROLAN *Resistance*

The insecticide DILAN consists of two DDT analogs, PROLAN and BULAN, that lack the chlorine on the aliphatic nucleus to be dehydrochlorinated (Fig. 7-3). Most laboratory-selected DDT-resistant strains

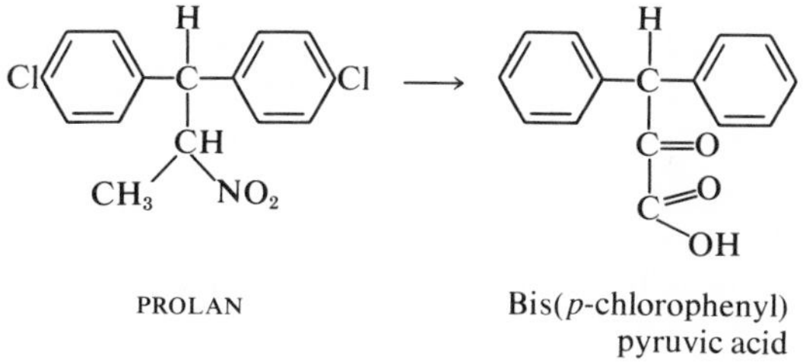

PROLAN   Bis(*p*-chlorophenyl) pyruvic acid

Fig. 7-3. Possible detoxication of PROLAN, a constituent of DILAN, by the oxidative route.

are susceptible to DILAN, with the exception of those resistant to DANP and pyrethrins also. Selection of houseflies with DILAN in due course develops a PROLAN-resistant strain, which is cross-resistant to DDT and contains DDT-dehydrochlorinase(*113*), and in addition has a mechanism making it cross-resistant to *o*-chloro-DDT(*341*), thereby resembling the DMC–DDT-resistant strain. These resistant flies absorb PROLAN more slowly than normal flies; they also can excrete the toxicant in doses that overwhelm the normal flies, first as unchanged PROLAN and later metabolized to an acidic metabolite, possibly bis-(*p*-chlorophenyl) pyruvic acid(*627*). DDT resistance (as well as dieldrin resistance) is also developed in houseflies selected with fluoroacetate, inhibitor of aconitase in the tricarboxylic cycle of carbohydrate metabolism, but these associated resistances eventually fall away as the fluoroacetate resistance reaches its peak(*759*).

C. Cyclodiene Resistance

The mechanism of the strong resistance that insects can develop to the cyclodiene derivatives still remains unknown. Houseflies oxidize aldrin to dieldrin, and similarly epoxidize isodrin and heptachlor (Fig. 7-4), but

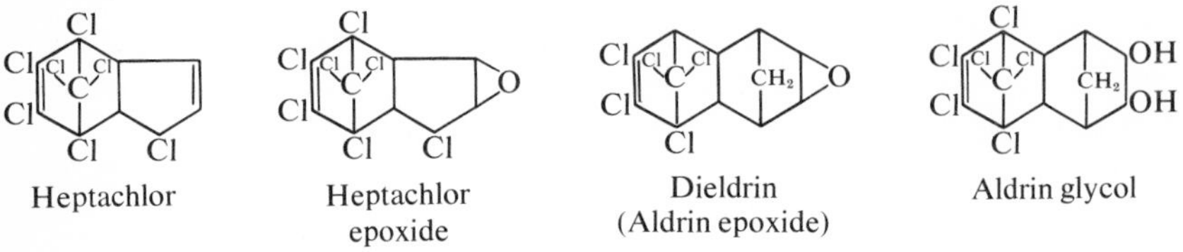

| Heptachlor | Heptachlor epoxide | Dieldrin (Aldrin epoxide) | Aldrin glycol |

Fig. 7-4. Epoxidation of heptachlor, and of aldrin of dieldrin, and further oxidation of dieldrin to aldrin glycol.

there is no evidence that dieldrin or the other epoxides are metabolized further to a significant extent by either susceptible or resistant strains (274,634). Only minute amounts of nontoxic metabolites such as ketonic derivatives of dieldrin have been produced by certain resistant strains(96, 207). On the other hand, the less insecticidal analogs of dieldrin are more readily broken down *in vivo*(98). The fact that these insecticides are not synergized against cyclodiene-resistant flies by compounds such as sesamex, which are known to inhibit the degradation of less toxic cyclodiene derivatives(97, 99, 100), indicates that the resistance cannot involve detoxication.

Some cyclodiene-resistant strains of houseflies absorb less aldrin than some normal strains, but this is not typical. The Malpighian tubules of resistant flies continue to excrete dieldrin long after the susceptible flies have succumbed(824), as if these organs were resistant to the known toxic action of cyclodiene derivatives upon them. In *Stomoxys calcitrans*, no difference could be discerned between resistant and susceptible strains in the absorption, metabolism, or excretion of dieldrin(561). Resistant houseflies or German cockroaches can withstand haemolymph concentrations of chlordane or heptachlor that are many times the normal fatal dose(623,632). The ganglia of resistant flies are more refractory to dieldrin than normal in that the threshold concentration necessary to induce repetitive discharge is greatly elevated(843); yet, when the poisoned nerve cords of resistant German cockroaches are assayed, they are found to contain just as much dieldrin as those of susceptible roaches(672). In dieldrin-resistant houseflies, the rate of uptake of dieldrin into the nerve cords and its subsequent rate of loss were no different from those of normal flies(700). Dieldrin can become bound to proteins and tissue particulate fractions; although dieldrin-resistant *Aedes aegypti* larvae absorbed twice as much dieldrin as normally susceptible larvae, one-third less than the normal was bound to the nuclear fraction(*508*).

In mosquitoes, it has been found that aldrin is converted by cyclodiene-resistant adults of *Culex quinquefasciatus*(*590, 770*) into a metabolite resembling trans-aldrin glycol (Fig. 7-3); normal larvae of *Aedes aegypti* convert aldrin and dieldrin to four water-soluble metabolites that are excreted(*441*), but dieldrin-resistant larvae do not achieve any greater conversion(*274*). Increased lipoid content has accompanied dieldrin resistance in some strains of houseflies(*217*) but not in others, and some strains of mosquitoes(*420*) but not in others. Within a field sample of boll weevils, those individuals that survived a certain dose of endrin contained more total lipoid and stearic acid than those that succumbed(*12, 545*). A cyclodiene-resistant strain of boll weevils absorbed dieldrin no more slowly than a normal strain and could not metabolize it at all(*716*).

D. BHC Resistance

Houseflies and sheep blowflies with cyclodiene resistance are cross-resistant to gamma-BHC (lindane), and houseflies made resistant to gamma-BHC are cyclodiene resistant. Among the BHC analogs, the housefly and blowfly resistance extends to chlorinated adamantane, but not to hexachlorocyclopentadiene, and among the cyclodiene-type compounds it extends to endosulfan but not to KEPONE(*133*). There is reason to believe that in BHC-resistant houseflies the usual cyclodiene resistance is supplemented by a second mechanism not shown in other cyclodiene-resistant species such as *Phaenicia cuprina*, namely an increased detoxication of BHC isomers(*137*). In these BHC-resistant houseflies, a threefold decrease in cuticular absorption may supplement the twofold increase in detoxication(*592*). BHC-resistant strains of the rice stem borer, *Chilo suppressalis*, metabolized gamma-BHC much more rapidly than susceptible strains(*695*).

BHC detoxication products are produced in abundance by houseflies (*84*), and of the 11 metabolites distinguished, the first one to appear is pentachlorocyclohexene (Fig. 7-5), a dehydrochlorinated product that

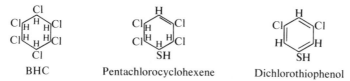

BHC   Pentachlorocyclohexene   Dichlorothiophenol

Fig. 7-5. Metabolities of gamma-hexachlorocyclohexane, the active isomer in BHC ("benzene hexachloride").

is not produced by the DDT-ase enzyme, however (*408*). Dichlorothio-
phenols are the metabolites to which housefly homogenates degrade
gamma-BHC *in vitro*, provided glutathione is added as an activator(*86*).
The enzyme responsible, which is abundant in houseflies but rare in
other insects, has been isolated and found to have a molecular weight of
54,000; but its titer bore no relation to the BHC resistance of the fly, and
it proved to dehydrochlorinate DDT faster than it converted gamma-
BHC to water-soluble products(*382*).

The interstrain differences in absorption and detoxication are insufficient to account for the strong BHC resistance, and these mechanisms are
reinforced by a real difference in the nerve sensitivity, as has already been
observed for cyclodiene resistance. The ganglia and connectives of exposed nerve cords of BHC-resistant flies are highly refractory to gamma-
BHC directly applied to them(*567, 843*). In dieldrin-resistant *Anopheles
gambiae*, the strong cross-resistance to BHC derives presumably from
this mechanism alone since the mosquitoes show no decrease in absorption or increase in detoxication of gamma-BHC(*85*).

E. *Organophosphate Resistance*

The mechanism of organophosphate resistance in houseflies has
proven not to involve any decrease in the sensitivity of the enzyme
cholinesterase or of the isolated nerves and ganglia(*476, 567*); nor does it
derive from a decrease in absorption of the thionophosphate or of its
oxidation to the active phosphate toxicant, e.g., paraoxon and malaoxon
(*497, 593*). Usually the resistance mechanism involves detoxication of the
thionophosphate or its oxidation product (Fig. 7-6), the only notable
exceptions being one diazinon-resistant strain that hydrolyzed diazoxon
no more than the normal(*113*) and two others that were characterized by
lower cuticular absorption as well as a higher detoxication rate(*216, 530*).
The organophosphate-resistant strains of houseflies are characterized by

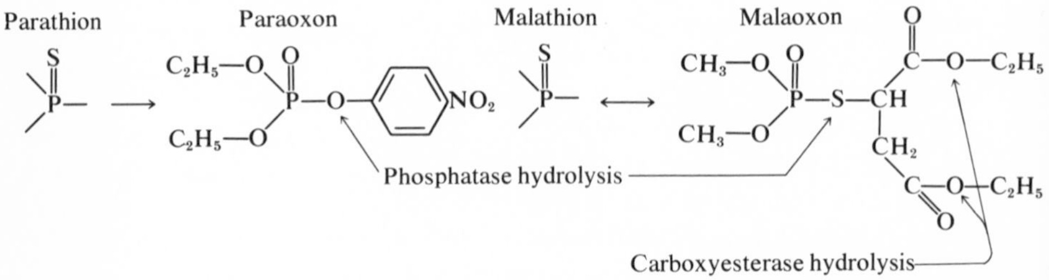

Fig. 7-6. Detoxication of paraoxon and malaoxon by phosphatase, and of malaoxon (or
malathion) by carboxyesterase.

a greatly reduced content of the enzyme aliesterase†(29, 31, 72), and it appears that it has been converted into breakdown enzymes for the appropriate phosphate toxicant(32, 595, 597). The essential change is from an aliesterase with a high turnover number for aliphatic esters and zero turnover for organophosphate compounds and therefore inhibited by them, to a breakdown enzyme with a lower t.o.n. for aliphatic esters but enough turnover for certain oxidized organophosphate compounds to detoxify them(30). Homogenates from malathion-resistant flies could break down malaoxon but not paraoxon or diazoxon, whereas those from parathion-resistant flies could break down all three toxicants and were inhibited by propyl-paraoxon(598).

The parathion-resistant houseflies could also detoxify the unoxidized parathion and diazinon by a phosphatase-type hydrolysis to diethyl phosphorothionate about twice as fast as the normal, the resistance enzyme having shifted in chromatographic characteristics from its normal counterpart(510). This resistance mechanism was different from the "oxonase" type previously mentioned, since it proved not to be easily inhibited by propyl-paraoxon and is essentially a "thionase." The malathion-resistant houseflies could detoxify the unoxidized malathion by a carboxyesterase-type hydrolysis to the monocarboxylic esters of malathion many times faster than the normal, this resistance enzyme being more heat-labile and having shifted its optimum pH from its normal counterpart; moreover, it was inhibited by propyl-paraoxon(509). This carboxyesterase could be a modified aliesterase, since it is the 0,0-dimethyl analog that it is most active in detoxifying, which indicates that its substrates attach at the phosphate as well as the carboxyester sites (507). However, a California organophosphate-resistant strain had a normal aliesterase titer and was able to detoxify methyl parathion and fenitrothion 50% faster than normal, largely by desalkylation(352). Electrophoretic separation of housefly esterases showed a malathion-resistant strain to be richer than normal in enzymes hydrolyzing naphthyl acid phosphate(524).

In *Drosophila melanogaster*, however, the organophosphate-resistant strains were characterized not by decreased but by increased aliesterase content(585). A strain of the mosquito *Culex tarsalis* that was resistant to malathion alone among organophosphate compounds had a normal aliesterase content and a manifold increase in carboxyesterase activity (73, 504). This resistance enzyme was more heat-labile than its normal counterpart, but there was a great deal more of it(505). Malathion-resistant populations of the leafhopper, *Nephotettix cinctipes*, in Japan also showed an increase in carboxyesterase activity, acting only upon

†However, this enzyme should now be termed carboxylesterase following the 1964 recommendations of the International Union of Biochemistry.

malathion(436), but were simultaneously richer in an aliesterase hydrolyzing methyl butyrate(331). Malathion-resistant strains of *Chrysomyia
putoria*(138) and *Tribolium castaneum*(206), which were not cross-
resistant to other organophosphate compounds, degraded more malathion
to carboxyesterase products than normal strains, and their resistance
was eliminated by carboxyesterase inhibitors. Malathion-resistant
German cockroaches suffered a merely transitory inhibition of their
cholinesterase from a malathion dose that almost completely inhibited this
enzyme in normal roaches(496); this was attributed to a protective factor
which could indeed be increased detoxication. In the yellow fever mosquito, *Aedes aegypti*, an increase in larval organophosphate tolerance was
associated not with increased detoxication, but with decreased absorp-
tion(506). However, a definite parathion resistance in *A. nigromaculis* did
involve an increased detoxication, particularly of paraoxon(522). Parathion resistance in the rice stem borer, *Chilo suppressalis*, was due to a
doubling of the ability to hydrolyze parathion and paraoxon(438), chiefly
through desalkylation(437); thin-layer electrophoresis revealed an increased activity of an esterase hydrolyzing β-naphthyl acetate and phenyl
acetate(407).

Some organophosphate-resistant strains of the two-spotted mite,
Tetranychus urticae (*telarius*), are characterized by increased detoxication, others by a less sensitive target enzyme. The Blauvelt strain of
Ithaca, New York, which is not cross-resistant to carbamates, showed a
threefold increase in breakdown of parathion by phosphatase action
(511, 798). This mutant carboxyesterase proved to have a 20-fold
greater affinity for malathion than normal, and its substrate specificity
was that of a β-type aliesterase(512). On the other hand, an organophosphate-resistant strain at Geneva, New York, appeared to contain
less cholinesterase activity than normal(801). A resistant strain developed
at Leverkusen, Germany, by selection with demeton and subsequently
parathion, showed one-third as much cholinesterase activity as normal
(718), and the aliesterase activity was significantly reduced(719). This
mutant-type ChE was less sensitive to paraoxon inhibition and had one-
quarter as much affinity for acetylcholine as the normal enzyme, these
weaknesses being at the esteratic site of the ChE; the hybrids contained
both the susceptible-type and the resistance-type enzyme(798, 799). A
dioxathion-resistant strain of the cattle tick, *Boophilus microplus*, also
proved to have much less ChE than a normal strain, although the hybrids
that were almost fully resistant had almost the normal ChE content(748).
About one-half of the ChE present in the ticks of a resistant colony was
relatively organophosphate insensitive, suggesting that these individuals
were mainly heterozygotes(456).

F. Carbamate Resistance

The resistance mechanism in houseflies made resistant to carbamate insecticides also involves detoxication. A strain resistant to COMPOUND AC-5727 detoxified this isopropylphenyl methylcarbamate 3 times as fast as a normal strain(266), and a strain with tenfold resistance to the isopropoxyphenyl analog propoxur (BAYGON) detoxified this latter insecticide 2.5 times faster than normal(260). Isolan, which is rapidly degraded by normal houseflies, was even more rapidly broken down by flies of an isolan-resistant strain(646). Carbaryl-resistant houseflies have greater activity than normal in hydrolyzing carbaryl to 1-naphthol as the initial metabolite (Fig. 7-7). This hydrolysis is inhibited by sesamex(211),

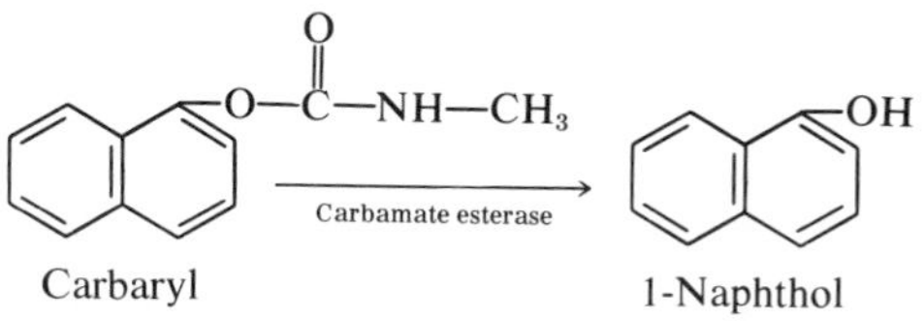

Carbaryl   1-Naphthol

Fig. 7-7. Detoxication of carbaryl by a nonspecific type of esterase.

and the fact that carbamate insecticides are synergized by methylene-dioxyphenyl compounds, such as piperonyl butoxide, suggests that they are detoxified by nonspecific esterases. A propoxur-resistant housefly strain from Japan was more active than normal in detoxifying carbamates *in vitro* by the NADPH system of microsomal oxidation(777).

Neither the AC-5727-selected nor the carbaryl-selected strain of housefly is cross-resistant to allethrin, but the former is cross-resistant to DDT and dieldrin(257), while the latter is not; moreover, the carbaryl-selected strain differs from the other by not being significantly cross-resistant to other carbamate insecticides(380). A strain of *Culex fatigans* made resistant to propoxur was highly cross-resistant to closely related carbamates, but scarcely so to more remotely related carbamates; the resistant larvae metabolised propoxur more than twice as fast as normal (270).

G. Pyrethroid Resistance

The mechanism of tolerance to pyrethroids appears to differ from one strain to the next. Many pyrethrin-tolerant insect populations have arisen from strains resistant to DDT or dieldrin. The DDT-resistant housefly strains that are pyrethrin tolerant have a mechanism making them resist the nondehydrochlorinatable analogs PROLAN and DANP. Selection with pyrethrins has made body lice highly resistant to DDT (*162*) and the same occurs with some housefly strains; yet other housefly strains have become pyrethrin-tolerant without the development of DDT resistance and DDT–dehydrochlorinase(*165*). One strain resistant to both DDT and pyrethrins was found to absorb topically applied pyrethrin I at less than half the normal rate(*229*). The effect of synergists on the comparative toxicity of pyrethrum to a resistant and susceptible housefly strain indicates that the resistance mechanism probably involves detoxication(*729*), but whether hydrolytic or oxidative is unknown. Tissue homogenates from houseflies and the American cockroach to detoxify pyrethrins *in vitro* and are inhibited by piperonyl butoxide, but allethrin is only slowly degraded by housefly preparations and piperonylcyclonene is without effect(*113*).

H. Cyanide Resistance

The resistance to hydrogen cyanide vapor in the California red scale, *Aonidiella aurantii*, was at first considered to be due to the development of the response of closing the spiracles faster and more thoroughly. However it was found that over the entire fumigation period the spiracles of the resistant scales tended to remain open, those of the normal scales to remain closed. The difference involved a more basic physiological change since the rate of oxygen consumption by the resistant strain was scarcely depressed by concentrations of HCN that greatly depressed the respiration of normal scales. Moreover the resistant strain was more susceptible to anoxia than the normal, suggesting that its respiration was mainly dependent on an autoxidizable flavoprotein rather than the CN-sensitive cytochrome oxidase system(*849*).

I. Arsenic Resistance

Since arsenicals are detoxified by conjugation with SH compounds, it was considered that arsenic-resistant cattle ticks might have a higher content of sulfhydryl groups and glutathione. This proved to be the case in larvae and embryos of resistant *Boophilus decoloratus*, but not in embryos of *B. microplus* in Queensland(*113*). The resistance developed to lead arsenate sprays in the codling moth (*Carpocapsa pomonella*) in Colorado proved to have no connection with arsenic *per se*, since the larvae were just as susceptible as normal to oral doses of arsenic and just as resistant to deposits of other types of insecticide. Their difference from the normal larvae was the ability of the newly hatched ones to survive long enough on the apple to find an unsprayed point of entry; it derived from an enhanced resistance to losing weight from starvation and desiccation and was thus an increase in general vigor reflected in their greater resistance to HCN(366). This type of nonspecific resistance has been given the name of "vigor tolerance"(361), and is often developed by strains of insects submitted to any selective stress or in the early stages of selection pressure from insecticides.

7-3 GENETICS OF RESISTANCE

A. Origin of Resistance

It soon became clear that resistance was not a process of the insect adapting or habituating itself to the insecticide during its lifetime. Small doses of sodium arsenate did not make silk-worms any more tolerant of arsenic in the end(139). Small daily doses of DDT, lindane, dieldrin, or diazinon only made houseflies more susceptible to effective doses, in direct proportion to the cumulative nature of the insecticide; nor could wax-moth larvae be thus adapted to nicotine or pyrethrum. Strains of houseflies treated in every generation with nonkilling doses of DDT or BHC have failed to develop resistance, thus proving that it is not post-adaptive in origin, while *Drosophila melanogaster* reared in truly sub-lethal concentrations of DDT for as many as 150 generations have not developed any DDT resistance(483). It was the same with body lice reared on cloth lightly impregnated with DDT, but once the concentration was increased to a level causing more than the control mortality, the DDT tolerance levels proceeded to increase(209).

This resistance develops only if the genetic factors for it, the pre-adaptations, are present in the population; certain strains of the housefly or *Drosophila*, endowed with only a small pool of allelism because they arose from few or single individuals, proved incapable of thus responding to selection(102). A laboratory strain of *Culex fatigans* derived from a single egg-raft failed to develop DDT resistance in 27 generations of selection, whereas a field-collected strain developed strong resistance in 11 generations(442). On the other hand, if the genetic factors for resistance are present, they may be selected out by other means than the insecticide; thus DDT-resistant *Drosophila* strains have been produced by taking those individuals that pupated at the periphery of the medium, DDT-resistant *Musca* by taking those that were the last to pupate, and organophosphate-resistant *Musca* by taking the offspring of females with low aliesterase content(113).

These preadaptations for resistance have proved to be single gene alleles in a strikingly large number of instances(535). It is unlikely that the insecticide induced these alleles, since DDT and lindane at least are known to be nonmutagenic for *Drosophila melanogaster*(*638*), and DDT and dieldrin to be nonmutagenic for *Bracon hebetor*(*297*); it is also unlikely that the mutations of the resistance alleles actually occurred during the regime of selection(*525*). The resistance genes may be produced, however by true mutagens just like any other mutant allele, lethal or visible; the gene *RI* for insecticide resistance in *Drosophila* has thus been produced by irradiating the male parents(*423*), deriving from an intermediate allele that tends to back-mutate(*425*). Thus insecticide resistance is not postadaptive even in the broadest sense; it is truly preadaptive, and the preadaptations are gene alleles.

B. DDT Resistance

When the first resistant house fly strains were crossed with susceptible strains, the lack of segregation in the F_2, which appeared to be intermediate just as the F_1 was, suggested that DDT resistance was due to multiple genes. If, however, the susceptibility levels were assessed not by mortality from topical application but by speed to knockdown on residual deposits, it was found that 75% of the F_2 and 50% of the $F_1 \times R$ backcross offspring were knocked down in a diagnostic time within which none of the R and all of the S parents had succumbed; this indicated that knockdown resistance to DDT was due to a single recessive gene(*326*). Therefore this mutant allele, as studied in an Italian strain(*534*), was called *kdr* (knockdown resistance), and was found to be linked with mutant markers located on the third of the six chromosomes of the housefly (Fig. 7-8). DDT-resistant strains from Florida, Ontario, and the Netherlands each carried a characteristic resistance gene on chromosome 2.

Kill resistance to DDT topically applied also proved to be monofactorial in Illinois and NAIDM strains, once the truly resistant flies were distinguished from the weakly tolerant among the R parental strain(*488*). The resistance allele was incompletely dominant(*463*) so that three phenotypes could be detected in the F_2 in 1:2:1 ratio, and two phenotypes in the $F_1 \times S$ backcross offspring in 1:1 ratio (Fig. 7-9). The homozygotes for this allele were found to contain twice as much DDT-dehydrochlorinase as the heterozygotes, while the susceptible homozygotes lacked it; the resistance character and the detoxifying enzyme were inherited in exact parallel with the kill-resistance allele(*477*). This gene for DDT-dehydrochlorinase activity, which may be called *R-DDT* (*780*) or *Deh*(*371*), has been found to be associated with chromosome 2† in Florida(*372*), Illinois(*463*), and Netherlands(*595*) strains of the house-

†Formerly called linkage group 5; chromosome 3 was formerly linkage group 2, and chromosome 5 was formerly linkage group 3 (*802*).

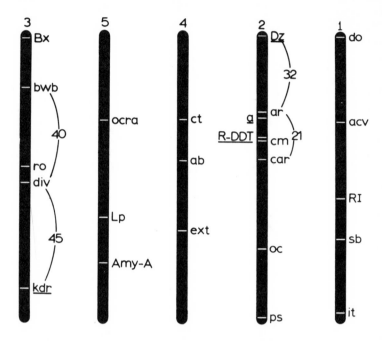

Fig. 7-8. Location of genes for DDT-resistance (*kdr*) for (DDT-knockdown resistance), *Dz* (for Japanese OP resistance), *a* (for European and American OP resistance) and the useful mutant markers on the autosomes of *Musca domestica*.

fly. It is the principal gene for DDT resistance in Japanese strains and is located near the marker *cm* in the middle of this chromosome (*780*). Allelism at the *Deh* (*D-ase*) locus can produce at least three different types of DDT-dehydrochlorinase enzyme differing in activity and substrate specificity (*596*).

A recessive factor on chromosome 3, which is probably a *kdr* allele, has been found in addition to the *Deh* gene in some strains but not in others (*371, 780*). It is responsible for the nerve insensitivity found in certain DDT-resistant strains (*782*). The *kdr-O* allele of this gene for more modest DDT resistance is also responsible for resistance to DILAN and *o*-chloro-DDT and the tolerance to pyrethrins often found in DDT-resistant strains (*643, 644*).† It has become inherited from father to son in an Australian strain (*417*), probably owing to a translocation between chromosome 3 and the Y chromosome. Another factor for moderate DDT resistance has been found on chromosome 5 in a Danish diazinon-resistant strain (*596*); this resistance may be due to a detoxication mechanism since it is inhibited by sesamex (*295*).

†It resembles the dieldrin-resistance gene in houseflies in that it doubles their glucose-6-phosphate dehydrogenase activity (*642a*).

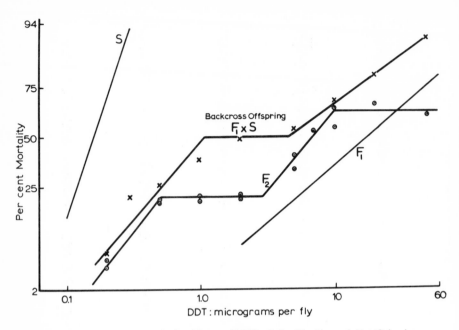

Fig. 7-9. Dosage–mortality relationships to DDT of the F_1, F_2, and $F_1 \times S$ backcross offspring from the cross between a highly-resistant and a susceptible strain of houseflies [after Lovell and Kearns(477)].

DDT resistance in anopheline mosquitoes proved to be predominantly recessive and influenced mainly by a single gene. In *Anopheles sundaicus* and *A. albimanus* the diagnostic dose killing all the S and none of the R parents caused 75% mortality in the F_2 and 50% mortality in the $F_1 \times R$ backcross offspring, indicating a single completely recessive gene. In *A. quadrimaculatus* the DDT-resistance gene was almost completely recessive(*180*), and in *A. stephensi* predominantly recessive(*182*). DDT resistance in the mosquito *Culex tarsalis* was also almost completely recessive(*650*), but in certain strains of *C. fatigans* it was almost completely dominant(*556*), and its inheritance was strongly matroclinous (*181*); its cross-resistance to *o,p'*-DDT was recessive, however(*398*). In *Aedes aegypti*, DDT resistance is clearly due to a single gene allele, the heterozygotes being exactly intermediate in resistance level(*661*); this gene is linked with genes on the second of the three chromosomes (*115, 161*), as is also a modifier gene ensuring that the DDT resistance is not recessive(*431, 832*).

In strains of the cockroach, *Blattella germanica*, a species in which the females are more difficult to kill with DDT (and with other poisons as well) than the males, DDT resistance has proven to be due to a single gene; and its inheritance is independent of the maternal effects originally suspected. Good segregation was shown in the F_2 (1 : 3) and in the $F_1 \times R$ backcross offspring (1 : 1) indicating that the main gene is recessive, and it has been found to be linked with the marker *balloon-wing* on chromosome 2, one of the 12 chromosomes (*158, 159*). DDT resistance in the Australian cattle tick, *Boophilus microplus*, also proved to be due to a single recessive gene (*747*).

Virtually nothing is known about the genetics of DDT resistance in plant-feeding insects. In the spotted root fly, *Euxesta notata*, sufficient segregation has been detected between knockdown times to suggest a single principal gene, for which the heterozygotes are intermediate in DDT resistance (*763*). A California population of the codling moth, *Carpocapsa pomonella*, contained 15% of individuals so sharply separated from the susceptible phenotypes that they could well be heterozygotes for a single DDT-resistance gene (*55*); and this percentage remained stable during years of laboratory culture. With *Euxesta*, laboratory selection with DDT eventually produced a strain comprising only heterozygotes, the homozygotes failing to hatch (*359*). DDT resistance in a Wisconsin population of the cabbage worm, *Pieris rapae*, appeared to be linked with a locus-determining green versus blue larval color, since the mutant-type blue homozygotes were four times more susceptible than the green homozygotes and twice as susceptible as the green heterozygotes (*518*). Resistance to tetradifon, a chlorinated sulfone acaricide somewhat related to DDT, was inherited in *Tetranychus urticae* as if it were due to a single dominant gene making the haploid males as resistant as the diploid females, but the situation was complicated by a distinct matroclinous effect (*854*). Resistance to dicofol in this species was found to be due to a single recessive gene (*851*).

C. Cyclodiene Resistance

This character has nearly always proven to be very definitely determined by a single gene allele, neither dominant nor recessive, the $R \times S$ hybrids being almost exactly intermediate. In *Anopheles* mosquitoes the three phenotypes are so distinct that they be identified by exposing the adults to two diagnostic dosages (Fig. 7-10), thus revealing a 1 : 2 : 1 ratio in the F_2 and 1 : 1 ratio in the offspring of either backcross. Such results were given by dieldrin-resistant strains of *A. gambiae*, *A. albimanus*, *A. quadrimaculatus* (*180, 182*), and three other anophelines, although in one strain of *A. gambiae* (and in one investigation of *A. albimanus*) this monofactorial resistance proved to be fully dominant. The dieldrin-resistance gene was evidently carried by chromosome 3 in *A. quadrimaculatus* (*245*),

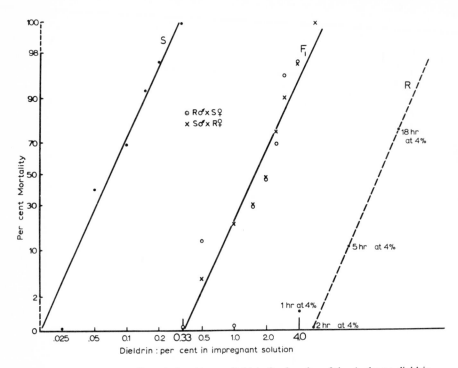

Fig. 7-10. Dosage–mortality relationships to dieldrin for females of the Ambursa dieldrin-resistant and the Lagos susceptible strain of *Anopheles gambiae* and of the F_1 between them, showing the discriminating doses (1-hr exposure) between the three types (after Davidson, 1958).

whereas in *A. albimanus* it was linked with *stripe* on chromosome 2 (*268*). In *Aedes aegypti* the gene for a monofactorial dieldrin resistance without dominance was located on chromosome 2, only a few crossover units from the DDT-resistance gene(*432*); but in *Culex fatigans* a similar dieldrin resistance with the hybrids intermediate(*181, 621*) assorted independently with DDT resistance, the former being on its chromosome 3 and the latter on its chromosome 2(*757*).

Resistance to aldrin, dieldrin, or BHC in the housefly was at first considered to be polyfactorial in origin; the F_1 hybrid was intermediate but no segregation could be detected in the F_2. When a highly resistant Californian strain was crossed with a normal strain pure for dieldrin susceptibility, however, the 1:2:1 ratio for three phenotypes was evident in the F_2, and a 1:1 ratio was shown in the $F_1 \times S$ backcross offspring (Fig. 7-11); since a 1:1 ratio was also shown by the $F_1 \times R$ backcross offspring, it was clear that dieldrin resistance in the housefly was due to a single gene without marked dominance(*269*). When a Sudan strain, long selected with

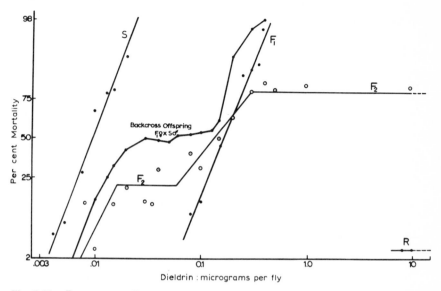

Fig. 7-11. Dosage–mortality relationships to dieldrin for males of the F_1, F_2 and $F_1 \times S$ backcross offspring from the cross between the dieldrin-resistant Super-Pollard strain and a susceptible strain of houseflies [after Georghiou, March, and Printy(269)].

dieldrin, was crossed with a pure susceptible strain, clear segregation was shown in the F_2, the $F_1 \times S$, and the $F_1 \times R$ backcross offspring, indicating that the resistance was due to a single gene without dominance or recessivity(299). In Italian strains a second factor was also involved(536). Although dieldrin resistance in Japanese strains(779) could not be associated with chromosomes 2, 3, or 4, nor in an Oregon strain with chromosomes 2 or 3(643), in a Uruguay strain(600) it was linked with chromosome 4. The resistance to gamma-BHC in a Japanese multiresistant strain(406) proved to be associated with chromosome 3.

In the Australian sheep blowfly, *Phaenicia cuprina*, dieldrin resistance was clearly monofactorial, the three genotypes being readily identified by two diagnostic dosages (which had to be twice as great in the females as in the males); the single resistance allele could be recovered unchanged after six generations of $F_1 \times S$ backcrossing accompanied by selection (705). The gene in South African populations appeared to be identical with that in the Australian(299). Dieldrin resistance in the coprophilous fly, *Chrysomyia putoria*, was also due to a single gene, with the hybrids intermediate(138).

Among insects which attack plants, the onion maggot, *Hylemya antiqua*, and also *H. platura* and *H. brassicae*, have developed a dieldrin resistance that is monofactorial and without dominance(520a, 675, 772), so that the intermediate hybrids produce 1 : 1 ratios in the offspring not only of the $F_1 \times R$ but also of the $F_1 \times S$ backcrosses (Fig. 7-10). Dieldrin resistance in the spotted root fly, *Euxesta notata*, is also due to a single main gene with no dominance, although the segregation is not so distinct (*763*). But a dieldrin-resistant strain in the laboratory came to consist only of heterozygotes, the homozygotes failing to hatch, as occurred with DDT selection(*359*) (Fig. 7-12). The resistance in the boll weevil, *Anthonomus grandis*, is also of the cyclodiene type, being greatest to toxaphene

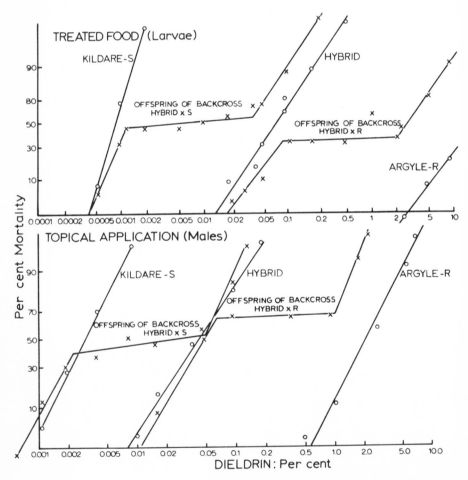

Fig. 7-12. Dosage–mortality relationships to dieldrin for larvae and males of the F_1 and the offspring of its backcross with either the Argyle-R or the Kildare-S strain of *Hylemya brassicae* [after Read and Brown(*675*)].

(*690*), even when the selecting agent is endrin(*93*), and not extending to DDT. The hybrids from an R × S cross were intermediate between the parents, but the dosage–mortality lines overlapped to such an extent (*87*) that it must be concluded either that the R strain was impure or that the resistance could not have been monofactorial.

Cyclodiene resistance in the cockroach *Blattella germanica* was originally concluded to be polyfactorial in origin, but when aldrin was substituted for chlordane as the test insecticide there was clear evidence of segregation for a single gene, located in linkage group VII (*515a*), the hybrids being intermediate(*157*). Similar monofactorial segregation was discovered in the dieldrin resistance of *Cimex lectularius*, *Pediculus corporis*, and *Boophilus microplus*, except that in the louse there was some maternal effect(*300*) and in the tick some overdominance in the heterozygotes(*747*).

D. Organophosphate Resistance

This characteristic has repeatedly proved to be due to a single dominant gene in insects and mites. Crosses between a parathion-resistant house fly strain from Italy and a normal colony gave fully resistant F_1 hybrids, while the mortality at the diagnostic dosage was 25% in the F_2 and 50% in the $F_1 \times S$ backcross offspring. A malathion-resistant strain from Georgia gave similar results. When these two strains were intercrossed, the parathion-R behaved as malathion-S and *vice versa*; but the gene alleles for parathion resistance and malathion resistance, although separate, were allelic with each other(*577*). Diazinon resistance, which is usually the same entity as parathion resistance, proved also to be due to a single dominant allele in a strain from Western Australia(*328*). The diazinon-resistant allele in a Danish strain was called *a* since it simultaneously determined a low aliesterase activity and the appearance of diazinon breakdown-enzyme activity(*594*); another allele that caused malathion resistance also induced a low aliesterase content and the appearance of malathion breakdown activity. These multiple alleles might therefore be termed a_D and a_M, but for the fact that the mutant resistance alleles are peculiar to each strain, thus a_G for the malathion-resistant Savannah G strain, a_F for the diazinon-resistant Italian C strain; their effect is to convert the particular esterase enzyme Ali-E-a into a breakdown enzyme specific to the strain. A subdivision of these alleles, viz., a_{C1} versus a_{C2}, decides whether the Ali-E-a enzyme, determined by the a_{S1} and not the a_{S2} allele in the susceptible strain, is there to be converted(*33*).

The locus of the *a* alleles for low aliesterase was found to lie (Fig. 7-8) very close to the marker *cm* on chromosome 2 of the housefly(*244*), and thus near the *Deh* gene for DDT resistance(*372*). One of the esterase bands distinguished by agar-gel electrophoresis showed negative correlation with organophosphate resistance(*586*), and its inheritance was determined by chromosome 2. The diazinon resistance of a Danish(*596*) and a Japanese(*775*) strain was found to be enhanced by factors on chromosomes 3 and 5. In the Japanese strain, however, the main gene for diazinon resistance, which was termed *Dz*, was located at one end of chromosome 2(*781*). Chromosome 3 exerted slight additional effects on the chromosome-2 parathion resistance of Florida and California strains (*643*); whereas chromosome 5† and not 3 contributed to the diazinon resistance of an Italian–Danish compound strain, which was due mainly to chromosome 2 and detoxication but which was also characterized by a reduced cuticular penetration(*699*).

Malathion resistance in the coprophilous fly, *Chrysomyia putoria*, which involves increased carboxyesterase activity, also proved to be due to a single dominant gene(*138*). The mosquito *Culex tarsalis* shows a remarkable malathion resistance, originating from Fresno, California, so specific that it extends to no other organophosphate compound except malaoxon(*178*). It proved to be due to a single dominant allele causing increased levels of carboxyesterase without any reduction in aliesterase (*504*). This mutant carboxyesterase differs from the normal-type enzyme in being precipitable in dilute acid and in being more heat labile(*505*). On the other hand, an organophosphate tolerance in *Aedes aegypti* involving decreased absorption rather than increased detoxication proved to be due to multiple factors deriving mainly from chromosome 2 but also from chromosome 3(*640*).

Organophosphate resistance in the two-spotted mite, *Tetranychus telarius*, was evidently due to a single dominant gene in a strain from Cranbury, New Jersey; the $F_1 \times S$ backcross offspring showed 50% mortality at a diagnostic dose (a 0.02% malathion dip) that killed none of the $F_1 \times R$ backcross offspring and none of the F_1 hybrids(*762*). In reciprocal $R \times S$ crosses, the male offspring inherited the character of the female parent, since they are haploid (Fig. 7-13); usually the diploid female homozygote for OP resistance shows about twice the LD_{50} (e.g., to TEPP) as the haploid-resistant male(*512*). Similar results were obtained in crossing the Cranbury strain of *T. telarius* with a susceptible strain of *T. cinnabarinus*(*762*).

That parathion resistance inherited as a single dominant gene was also found in a strain of *T. pacificus* from cotton(*14*), sound evidence for

†Termed chromosome 4 by Milani, it is really chromosome 5 of Wagoner(*802*).

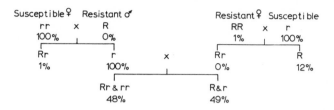

Fig. 7-13. Percentage mortalities for offspring resulting from reciprocal crosses and an $F_1 \times S$ backcross in *Tetranychus telarius* when tested by a diagnostic dose of malathion [after Taylor and Smith(762)].

monofactoriality being obtained in the stepped dosage–mortality relationship for the female offspring of the $F_1 \times S$ backcross (Fig. 7-14). Parathion resistance in two strains of *T. urticae* (one from Lisse, Netherlands, and the other from Leverkusen, Germany, where it had been developed initially by demeton selection) showed the same clear monofactoriality, with its dominance manifested in the F_1 and in the $F_1 \times S$ backcross offspring(336). The parathion resistance in the "Niagara" strain from Gasport, New York, in which the mechanism was an increased detoxication, proved to be due to a single dominant gene(342). Malathion resistance in another New York strain of *T. telarius* (*urticae*), namely the

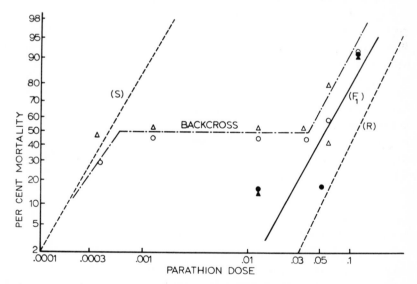

Fig. 7-14. Dosage–mortality relationships to parathion for females of the F_1 and $F_1 \times S$ backcross offspring from the cross between a parathion-resistant and a normal strain of *Tetranychus pacificus* [after Andres and Prout(14)].

Blauvelt strain, in which the mechanism is increased detoxication, was also inherited monofactorially(*515b*). The existence of modifiers for the main gene was suggested by the fact that the homozygotes that recovered with repeated $F_1 \times S$ backcrossing with selection were not so resistant as the original homozygotes(*460*); in *T. pacificus* also there was a suggestion of modifier genes(*14*), but the main gene was linked with the visible mutant *stock*, and not with *white-eye* or *pigmentless* (*853*).

Contrary results were obtained with the Leverkusen strain of *T. urticae* when a different test (the slide-dip method) and a different test insecticide (oxydemetonmethyl instead of parathion) were employed. Repeated $F_1 \times S$ backcrossing with selection eliminated the resistance almost completely, a result taken to indicate polyfactoriality. The resistant types were characterized by a reduced oviposition and hatch and a slower embryonic and larval development(*192*). The semilethal factors responsible for these reductions in biotic potential could be eliminated by a half-dozen generations of inbreeding(*193*); and when the inbred lines were recombined, the new pooled strain gave a stepped dosage–mortality line, indicating that it consisted of a mixture of heterozygotes and homozygotes for a single recessive resistance gene(*194*). The mechanism of resistance in the Leverkusen strain was an organophosphate-insensitive cholinesterase. Supporting evidence that its inheritance was monofactorial(*336*) comes from three other strains characterized by the same mechanism; namely, the Baardse strain from the Netherlands(*702*) and two strains from New Zealand, in which the character was found to be due to a single dominant gene. Nevertheless, strains of *T. urticae* differ greatly in their potential for organophosphate resistance, some achieving it after five generations, others failing to do so after 30 generations of selection(*451*).

Organophosphate resistance developed by the cattle tick, *Boophilus microplus*, to dioxathion was inherited as if due to a single gene allele. Its expression in adults was dominant, but in larvae the dominance was evident only in its cross-resistance to formothion(*748*).

E. Carbamate Resistance

A ten-fold resistance to isolan induced in a Riverside strain of housefly was inherited as a single gene slightly on the dominant side; cross-resistance was shown to ZECTRAN and AC-5727, and the latter carbamate selected for the same gene(*265*). Isolan resistance in an Orlando strain proved to be linked closely with the chromosome-2 marker *stw*(*372*), and carbaryl resistance in an Osaka strain was associated with the same

chromosome(*406*) and, in a Berkeley strain, with chromosome 2 reinforced by chromosome 3(*643*). Chromosome 5 had a considerable influence upon the isolan resistance in the Orlando strain(*643*). Propoxur
resistance in a Berkeley strain was due to a chromosome-2 gene concerned with microsomal detoxication, with influences from chromosomes
3 and 5(*776, 783*). In *Drosophila melanogaster*, carbaryl resistance was
inherited with the chromosome 2 allele *RI* for general insecticide resis-
tance(*424*). By contrast, a tenfold resistance, developed in *Culex fatigans*
to propoxur, was due to multiple genes(*261*).

F. Cyanide Resistance

It was in the California red scale, *Aonidiella aurantii*, that hybrid
crosses of cyanide-resistant and susceptible strains gave the first indication that insecticide resistance could be due to a single gene. The females
divided themselves into three types, resistant (RR), intermediate (Rr),
and susceptible (rr). The males inherited the characters carried by their
female parents, and were of only two types, resistant (R) and susceptible
(r). Thus cyanide resistance was shown to be due to a single gene, without
dominance, that was haploid in the heterogametic males(*189, 850*).

G. Arsenic Resistance

In the codling moth, *Carpocapsa pomonella*, this character, due
essentially to a greater size and vigor in the newly hatched larvae, failed
to show any segregation in the F_2 from the R × S hybrids, or in the backcross offspring. Successive generations from the F_1 hybrid retained their
level of resistance intermediate between the R and S parents, while all the
individuals resulting from either backcross were equally intermediate
between either parent(*365*). This vigor-type resistance is therefore
probably due to multiple factors.

H. Nonspecific Resistance

The initial generations of an insect colony selected with an insecticide
such as DDT at moderate selection pressures often show a nonspecific
low-level tolerance developing to insecticides in general, and this is of the
same order as the modest cross-tolerance shown by a specifically resistant
strain to insecticides in other groups. This change, similar to that resulting
from selection with any stress factor such as heat, and also manifested to
stress factors in general, is called vigor tolerance(*361*). It exhibits a
dosage–mortality line shifted slightly to the right without any decrease in

slope; on the other hand, when specific resistance is really developing the d–m line becomes much shallower as it moves markedly to the right. The shallow regression line results from the wide-ranging heterogeneity resulting from the three greatly differing genotypes consequent on monofactorial inheritance; on the other hand, the steeper slope retained in cases of vigor tolerance is a consequence of the lower variance resulting from a number of genetic factors, each of slight effect (*104*).

A type of resistance extending to a wide variety of insecticides, induced by DDT selection pressure on *Drosophila melanogaster*, proved to be due to multiple genes distributed throughout the three major chromosomes (*170*) and was considered to result from an integration of the gene pool (*430*). In Japanese strains, however, this DDT-induced resistance to many insecticides, including parathion, chlordane, DNOC, and carbaryl, proved to be due to a single gene, called *RI* (*424*), located at map-distance 66 on chromosome 2 (*778*), and enhanced by another gene at locus 50 on chromosome 3 that alone confers nicotine resistance (*601*). These two genes are also found in DDT-selected *Drosophila virilis* (*602*), at the position on its chromosomes 5 and 2, respectively, where they would be expected to be if they were the same in both species (Fig. 7-15).

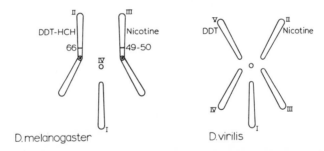

Fig. 7-15. Location of genes for DDT-tolerance and nicotine-tolerance in *Drosophila melanogaster* and *D. virilis* [after Oshima and Hiroyoshi (*602*); Tsukamoto and Ogaki (*778*)].

I. Correlates of Resistance

Some DDT-resistant housefly strains had certain visible characters such as "plexus" and other wing abnormalities, dark abdominal sclerites, or an unusually wide second abdominal sternite. But detailed study of the incidence of morphological characters such as these in nine different housefly strains could establish no correlation between the DDT resistance and any of these visible markers, proving them to have been incidental (*731*, *732*). No consistent differences in size, fecundity, or longevity were found in DDT-resistant strains, but some of them were characterized by a longer larval period (see p. 524).

Other DDT-resistant strains, however, developed faster than normal; it was only within a given strain that there was a positive correlation between LD_{50} and length of the larval period(*185*). Histopathological changes in oogenesis were found to characterize housefly populations in Russia in the year that they were most rapidly developing DDT resistance, but these oogenesis disturbances again reached their maximum in the year after DDT spraying was discontinued(*186*). The development of resistance in *Culex p. molestus* under DDT selection was accompanied by a lengthening of the egg stage and a delay in oviposition until after the second blood meal(*52*). Certain DDT-resistant strains of *Aedes aegypti* were characterized by preferring darker oviposition sites(*830*) or by a longer preoviposition period after the blood meal(*831*); but these recessive characteristics proved to be due to genetic factors other than the DDT-resistance genes.

The chlordane-resistant strain of *Blattella germanica*, discovered at Corpus Christi, Texas, produced smaller broods and showed lower body weights than the normal, but a DDT-resistant strain developed by laboratory selection also showed these differences, probably as a result of the inbreeding(*102*). Similarly, a cyclodiene-resistant strain of the boll weevil developed by laboratory selection with endrin showed a subnormal growth and oviposition rate(*765*), whereas these differences were not shown by the resistant populations in the field(*70*). Aldrin-resistant strains of *Hylemya brassicae* taken from the field into the laboratory were initially characterized by a greatly increased longevity and total egg production as compared with the normal(*675*), but these differences eventually disappeared after prolonged laboratory culture. In Liberia, the first two generations of housefly to be exposed to dieldrin residual sprays showed an increase in egg production(*287*).

The parathion-resistant strain of the mite *Tetranychus urticae*, developed at Leverkusen, was characterized by never entering diapause when transferred to cool temperatures(*692*). This was not an inevitable correlate of parathion resistance, however, since the resistance could be put back into association with the ability to diapause by repeatedly backcrossing it with a normal one while maintaining parathion selection, or conversely the nondiapause characteristic could be introduced into a normal strain by selecting for it(*336*).

J. Chromosome Changes

It was noted that dieldrin-resistant strains of *Anopheles gambiae* from northern Nigeria contained more individuals heterozygous for more chromosomal inversions than a normal strain from coastal Nigeria. Then it was discovered that strains of *A. atroparvus* in which resistance to DDT or to dieldrin had been induced by laboratory selection had come to contain a higher population of inversion heterozygotes, particularly in the females; moreover, the individual heterozygotes proved to be more tolerant than the normal homozygotes. But since similar changes in genotype composition could be induced by the selective effect of stress factors, such as unusual rearing conditions, it was concluded that what had been produced was a nonspecific heterosis vigor(*177*). Strongly DDT-resistant and dieldrin-resistant strains of *A. quadrimaculatus* had no more chromosomal inversions than the normal strains(*501*).

7-4   ELEMENTS FOR UNDERSTANDING THE COURSE OF RESISTANCE

A. Resistance Hazard

Just as insecticides differ in their capacity to induce resistance, so some species of insects are more resistance prone than others. In North America and Europe the housefly has rapidly developed resistance to chlorinated hydrocarbon insecticides where the other barn-inhabiting and synanthropic flies have seldom done so. In India, *Musca domestica vicina* develops DDT resistance(*422*) and dieldrin resistance(*666*) far more strongly than *M.d. nebulo*. A Nebraskan strain of the stable fly failed to develop DDT resistance after 13 generations of laboratory pressure, confirming the general experience with *Stomoxys calcitrans* in the U.S. (*396*); however, Scandinavian populations of this species have developed DDT resistance in the field(*733*). The tropical house mosquito, *Culex fatigans*, develops resistance to DDT or cyclodiene derivatives wherever they are used; yet a large sample of *C. gelidus* from Malaya failed to develop DDT resistance(*767*). Populations of *Aedes aegypti* in West Africa were exceptionally slow to develop DDT resistance(*381*); yet North American populations, when tested in the laboratory, proved to be just as prone to DDT resistance as those that had already achieved it in the Caribbean area(*2*). On the other hand, strains of *Anopheles stephensi*, all derived from the same original collection made in South India and selected with DDT in several laboratories, failed to develop resistance during the very period that strong DDT resistance was developing in field populations of this species around the Persian Gulf.

Among plant-feeding insects, it is truly remarkable that the European corn borer has nowhere yet developed resistance to the DDT applications it has experienced for 15 years. Populations of the apple maggot in Connecticut recently proved to be only slightly less DDT-susceptible in sprayed orchards than in unsprayed ones(459). The California red scale has remained susceptible to parathion for 12 years; when checked by laboratory selection and parathion tolerance rose slightly in the first four generations and then showed no increase in the subsequent 30 generations under parathion pressure(169). Laboratory selection with malathion for 45 generations induced no loss of susceptibility in a West African strain of the body louse(136). Selection with methyl bromide has effected no change in the cadelle *Tenebroides* or in *Tribolium confusum* but has induced tolerance to this and other fumigants in *Sitophilus granarius* (543).

Dieldrin resistance was developed by the malaria mosquito, *Anopheles pseudopunctipennis*, in a few months in an area of Mexico where it had not lost its susceptibility to DDT applications in 13 years; nearly all anophelines are much more liable to cyclodiene resistance than to DDT resistance. Among the root maggots attacking tobacco in Ontario, *Hylemya platura* did not at first develop cyclodiene resistance and therefore became replaced by *H. florilega*, which had developed it(764). Among the capsids attacking cocoa trees in Ghana, *Distantiella theobroma* has developed BHC resistance and is replacing *Sahlbergiella singularis*, which has not developed it; yet in Nigeria *Sahlbergiella* can develop BHC resistance. The gene for cyclodiene resistance in *Anopheles gambiae* is present in West African populations around the Gulf of Guinea(308), but it is apparently absent from East Africa and the south of the continent (Fig. 7-16), with the exception of the Sudan(316), western Kenya, and central Madagascar(151).

B. Development of Resistance

Since resistance is a result of selection, insecticide treatments that cause 100% mortality can no more change the resistance of the population than levels that are completely sublethal. It is when the effect of insecticide deposits falls within the critical range of partial mortality that selection takes place(252), and residual insecticides such as DDT or dieldrin leave deposits that stay in this critical range for months before weathering away. Thus, there are so many instances of DDT resistance and dieldrin resistance and so few cases of resistance to evanescent insecticides such as the pyrethrins and nicotine.

During the initial phase when houseflies are selected with DDT, there is a latent period of a dozen generations or so when the LD_{50} rises only slowly(185), and when an equal tolerance is gained to other insecticides as to DDT. After that the LD_{50} levels proceed to rise abruptly, the resistance being specific to the selecting agent. The pink bollworm under

Fig. 7-16. Distribution limits of *Anopheles gambiae* and areas (black) where it has developed dieldrin-resistance around the Gulf of Guinea [after Hamon and Garrett-Jones(*308*)].

laboratory selection for 20 generations gained its 10-fold DDT resistance very slowly at first(*481*). This exponential growth after a long latent period may be explained partly as an artifact of the topical-application test method, which exaggerates the LC_{50} at the higher levels, and partly as a consequence of the accumulation of resistance alleles, which were initially very rare(*170*). Nevertheless it appears likely that the latent period is necessary for the accumulation of a suitable "residual inheritance" to support the main resistance gene by replacing a number of genes of minor effect with alleles that carry more vigor or are more compatible with the resistance gene(*535*). It would be the preaccumulation of such supporting satellite or ancillary alleles that would subsequently make a DDT-resistant strain develop resistance to BHC(*185*) or to organophosphate compounds(*452*) more quickly than normal.

With dieldrin resistance in most insects, however, no changes seem to be necessary in the residual genotype since the resistance is entirely monofactorial and its development depends simply on the frequency of the principal gene. Thus dieldrin resistance may develop within six months in *Anopheles gambiae* in northern Nigeria(*21*) and within 18 weeks in *Boophilus microplus* in Queensland(*686*). DDT resistance appears to develop first as a polyfactorial system, where a number of minor factors require adjustment, before the specific resistance due to oligogenes rapidly asserts itself. In several species such as anopheline mosquitoes the delay is also due to the main gene being recessive, so that the hybrids are still killed by the regular treatments(*360*). In *Aedes aegypti* there may be a long delay in the development of DDT resistance in certain regions due to the general genome affecting the principal gene, as in West African populations(*381*).

Organophosphate resistance, judging by the experience with houseflies in the laboratory, may require a latent period of scores of generations before the polyfactorial system thus assembled allows principal oligogenes to decide the higher levels of specific resistance; this transfer from polyfactorial to monofactorial systems may also be deduced from the change in slope of the d–m lines during development of organophosphate resistance in *Aedes nigromaculis*(*117*). With houseflies in the field, diazinon resistance or malathion resistance is developed in stages; during the first year, the strain comes to carry resistance heterozygotes but reversion occurs over the winter; in the second year, the resistance homozygotes appear but they have low fecundity; in the third and fourth years, the resistance stabilizes as the homozygotes achieve normal fecundity(*409*).

Carbamate resistance may be very slow in developing. In *Culex fatigans*, the increases induced by larval selection were only 25-fold in 35 generations with propoxur(*270*) and a mere fivefold in 16 generations with COMPOUND AC-5727(*756*). A strain of the body louse selected with carbaryl developed only a fourfold tolerance in 55 generations, but in this case a real 67-fold resistance had developed by the 71st generation(*155*).

The resistance level may finally reach an upper limit, as in a DDT-resistant strain of *Aedes aegypti*, whose final LC_{50} was no greater than the LC_{100} of the normal strain from which it was developed(*195*). DDT resistance in *Anopheles sacharovi* in Greece and in *A. albimanus* in Central America seems to "plateau off" at a level where frequent DDT deposits can still prevent malaria transmission, but this is not so with *A. pharoensis* in Egypt; presumably modifiers often continue to accumulate and enhance the effect of the main gene now in homozygous form.

Often when resistance first appears, or rather becomes noticed, it may be entirely local. DDT resistance in the codling moth often remains restricted to certain orchards for a number of years. HCN resistance in the California red scale first appeared in two localities (Corona and

Orange) and between 1912 and 1938 came to characterize the counties around Los Angeles (657). Cyclodiene resistance in the boll weevil first appeared in 1954 in northeastern Louisiana; by 1959 it had spread into Mississippi, Arkansas, and Oklahoma, with separate areas of resistance in Texas, Alabama, South Carolina, and North Carolina (687). The aldrin resistance of the northern corn rootworm, *Diabrotica longicornis*, in Wisconsin appeared first in a single field eight years after the introduction of aldrin, and two years later it was prevalent throughout two counties (618). Cyclodiene resistance in the western corn rootworm spread from a single spot into seven states in five years (Fig. 7-17); such spread of the

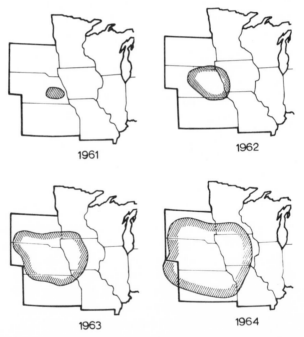

Fig. 7-17. Annual spread of cyclodiene-resistance in the western corn rootworm *Diabrotica virgifera* in South Dakota, Nebraska, Kansas, Minnesota, Iowa, and Missouri. (After Circular GAC 100-203, Geigy Agricultural Chemicals, 1964).

resistance gene allele may also be regarded as a steady recession of the normal susceptibility allele, so that it is insufficient to dilute the resistance at the edges of its distribution.

The higher the selection pressure in terms of per cent mortality, the more rapidly resistance develops, as shown by the effect of DDT selection on *Drosophila melanogaster* (525) and of parathion or malathion selection on *Tetranychus telarius*(*803*). It is possible, however, that at the highest pressures an inadequate number of survivors is left to comprise a gene pool sufficient to integrate into adequate genotypes(*429*); this could happen with DDT resistance, but with dieldrin resistance the quickest way to achieve a finally resistant strain is to select it down to one female resistance homozygote fertilized by a similar male, as was done with *Anopheles gambiae*. Resistance also develops faster when the species has a greater number of generations per annum. In Mediterranean-type climates, the housefly takes about two years to develop DDT resistance and becomes resistant to the dieldrin, chlordane, or BHC introduced as substitutes within an additional year; in more northern climates, DDT resistance may take four to six years to develop(*360*). DDT resistance first appeared in agricultural insects such as the codling moth and cabbage worm in 1951, some five to six years after the introduction of DDT. Cyclodiene resistance became conspicuous in cotton insects between 1955 and 1958, about six years after the introduction of toxaphene and dieldrin. It was the appearance of cyclodiene resistance in insects of public health importance, such as anopheline mosquitoes, that so greatly lengthened the list of resistant species between 1957 and 1960. The numbers of arthropod species that had developed resistant strains in successive years of reviewing were as follows:

1948(*38*)	12	1954(*526*)	25	1963(*111*)	157
1951(*40*)	16	1957(*101*)	76	1965(*262*)	186
		1960(*104*)	137	1969(this review)	224

A deceleration is now detectable in the number of new resistant species; this is attributable to the trend away from the cyclodiene derivatives and toward the organophosphate insecticides.

It may be concluded that the surest way of inducing resistance is to contaminate as wide an area as possible with a residual insecticide which persists in killing a high proportion of the insects for a long period after application. This was the situation with the boll weevil when cyclodiene insecticides were applied over the entire cotton-growing region, and with mosquitoes in coastal Florida, interior California, southern Greece, and lower Egypt, where chlorinated hydrocarbons had been applied as larvicides by aerial spraying, and where these compounds had been even more extensively applied to protect crops such as rice. Application of these insecticides as residual house sprays against the adults reinforces the selective effect already exerted on the larvae. On the other hand, the best way of avoiding or at least postponing resistance is to achieve 100% control in a circumscribed area surrounded by large untouched populations, e.g., antimosquito operations at sites in northern Canada, the substitution of side-dressings and granules instead of broadcast insecticides in root-maggot control. This is done preferably with nonpersistent insecticides, e.g., the nicotine fumigation technique for cabbage-aphid control.

C. Stability of Resistance

It is to be expected that incompletely resistant strains still carrying normal alleles of the resistance gene should revert to susceptibility when the insecticide is removed because it is the wild-type allele that is best adapted to normal insecticide-free conditions. DDT-resistant strains of houseflies taken from the field into laboratory culture eventually lose their resistance(102). On the other hand, a strain of bed bugs from Pittsburg maintained its high DDT resistance for at least five years(494). In laboratory strains of houseflies that are not yet pure for DDT-resistance, the susceptible minus-variants outgrow the resistant plus-variants because they complete their larval and pupal development slightly faster (520); a Danish DDT-resistant strain, however, more than held its own against a normal strain(77). In fact, a resistant strain has as much chance of being superior as does a susceptible strain(60). A strain of *Aedes aegypti* in which DDT resistance had just been induced was very unstable because the plus-variants had a much lower oviposition and hatch rate(1). Once a DDT-resistant strain has become pure for homozygotes it can of course no longer revert, as was found for the strains developed at Urbana, Ill., but this final stage is not reached without considerable loss of material and failure of lines(463). The cyanide-resistant Riverside strain of the California red scale has maintained its resistance for ten years (70 generations) for this reason(850). However, a 25-fold resistance that developed to methyl demeton by *Myzus persicae* reverted to nil in 30 generations of insecticide-free culture, even though all the offspring derived parthenogenetically from a single apterous female(200).

Organophosphate resistance in tetranychid mites is quite stable, although a culture of the Leverkusen demeton-resistant strain reverted because of its low biotic potential(194). Other resistant strains of *Tetranychus urticae* have maintained their resistance(336,789) and shown a normal biotic potential(718). Parathion-resistant strains of *T. urticae* reverted most slowly on hops, which was the crop on which they developed the strongest resistance(691). A parathion-resistant strain eventually came to maintain its numbers in relation to a normal strain, despite a 40% lower oviposition rate(458).

Resistance will revert in the field if the insecticide concerned is withdrawn from use and other insecticides used more sparingly. Thus the resistance of houseflies to DDT and dieldrin reverted in California(526) and Denmark(409) after the use of these chlorinated hydrocarbons had been discontinued in 1951. In parts of New Jersey, however, where organophosphate compounds were extensively used, this type of selection pressure served to increase the resistance to chlorinated hydrocarbons to extremely high levels(239); by 1965 the houseflies in southern California were found to be highly DDT-resistant and cyclodiene-resistant, organophosphate compounds such as diazinon (rather than malathion) having selected for these resistances(262). The resistance to DDT and cyclodiene derivatives developed by *Aedes nigromaculis* and other irrigation-water mosquitoes in central California did not revert subsequent to the replacement of these insecticides as larvicides after 1951, because these insecticides were still used so heavily on crops in the area. Whereas the dieldrin resistance induced in *Anopheles gambiae* in northern Nigeria was nearly always stable, that in *A. culicifacies* in western India did revert when DDT was substituted for dieldrin as the residual treatment for malaria eradication(67).

Unfortunately, populations that have reverted can quickly regain their resistance. By 1956 in Denmark the houseflies had reverted to the point where DDT and chlordane could again be used, but this time they recovered their resistance in six weeks(409). The BHC resistance that had been lost in *Musca domestica vicina* in Egypt returned after a similarly short period(620). The DDT resistance of *Anopheles stephensi* in southern Iran had partially reverted during the three years when dieldrin had been substituted, but it quickly recovered when DDT was reapplied (308). It is therefore clear that although the cessation of an insecticide may result in a decrease of the frequency of the main resistance gene, yet the many minor supporting alleles in the residual inheritance of the resistance genotype are not replaced nearly so quickly. Moreover the substitute insecticides such as DILAN, the organophosphate compounds, and the carbamates may hinder the loss and even favor the further accumulation of resistance gene alleles. Thus, although it occasionally becomes possible to return to the original insecticide for short periods, resistance is essentially a one-way street and "it is unlikely that a genuinely fresh start can ever be made"(409).

D. Test Methods for Resistance

Populations suspected of resistance must be compared with a normal population or laboratory colony by some suitable toxicological method (129). If the method has been standardized for use wherever the species is tested, then the figures obtained for the suspect population alone are sufficient to detect resistance. Standard methods have been set up internationally for a great number of species of public-health importance (*835*). In most of them, the test insecticide has been dissolved in mineral oil (SHELL RISELLA 117) and the solution impregnated into filter paper to a content of 3.6 mg cm^2. These impregnated papers are already prepared for different concentrations of the insecticide; they are used to line the interiors of 5- × 1.75-in plastic tubes. The insects to be tested are exposed to them for 1 hr and then removed to clean containers for 24 hr before the mortality is determined. This type of test is employed for adult mosquitoes, *Phlebotomus* sandflies, *Simulium* blackflies, and tsetse-flies; for fleas, bed bugs, and triatomids, lengths of impregnated paper are inserted into test tubes, and the exposure period is extended to five days for the last two insect types. Tests for body lice are made with dusts impregnated into cloth, while those for mosquito larvae and *Simulium* larvae involve exposure in treated water for 24 hr. Test kits for these various types of resistance tests have been standardized by the World Health Organization (Fig. 7-18). Tests utilizing filter papers impregnated with an acetone

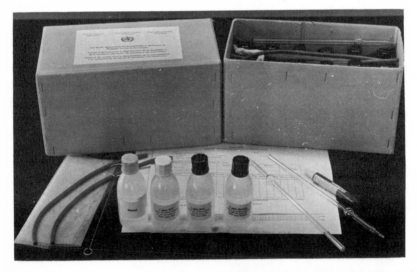

Fig. 7-18. Test kit for determining resistance in mosquito larvae, prepared by the World Health Organization.

solution of DDT have been developed for stored-products insects (*449*).

Standardization is much more difficult for insects of agricultural importance, requiring a variety of methods according to the species (*135, 681*). For the codling moth, eggs about to hatch are confined in individually waxed cells affixed to an apple sprayed with the insecticide (*54*). For aphids and mites, the infested plants may be dipped in an emulsion made from an acetone or xylene solution of the insecticide(*15,740*). In the tests for the two-spotted mite, the host plant may be sprayed in a Potter tower(*655*), or the mites may be affixed to cellophane tape and dipped in the insecticide emulsion(*797*). For leaf-chewing caterpillars such as the beet armyworm and the red-banded leafroller(*276*), the leaves are dipped and the insects placed on them in vials. *Lygus* bugs have been tested, like houseflies, by exposing them for 1 hr in shell-vials coated with a mineral oil solution of DDT(*16*). *Erythroneura* leafhoppers have been tested, like cockroaches and houseflies, in flasks treated with an acetone solution of the insecticide(*56*). For large insects such as the boll weevil, topical application of the insecticide is employed as a standard test (*210,689*); this method is also useful for houseflies and ticks, the solution being applied by means of a syringe, microloop, or capillary buret(*835*). As the solvent for chlorinated hydrocarbons, mineral oil is preferable to acetone because its droplet is capable of spreading on the insect; for organo-phosphate compounds, dioctyl phthalate has been employed; but a suit-able solvent for both groups of insecticides is methyl ethyl ketone, or a mixture of acetone and olive oil if the insects show control mortality. Several test methods are available for root maggots in the adult and larval stages(*674*), and conversion factors are available between the Potter-tower, impregnated-paper, and topical-application test methods(*772*). Corn rootworms are tested by topical application to the adult stage(*48*). Many of these methods have been standardized by the Food and Agriculture Organization(*236a*) and the Entomological Society of America(*93a, 210*).

7-5 COUNTERMEASURES FOR RESISTANCE

A. Changing to New Insecticides

The usual countermeasure for resistance is to switch to a new insecti-cide. DDT resistance in the housefly was countered by substituting lindane or dieldrin, and the inevitable cyclodiene resistance demanded a switch to malathion or diazinon, which remained effective until organo-phosphate resistance developed. DDT resistance in the codling moth has been met by the substitution of organophosphate insecticides or carbaryl. DDT-resistant cotton bollworms (*Heliothis zea*) in the Rio Grande valley of Texas remained susceptible not only to methyl parathion and carbaryl but also to endrin and toxaphene(*482*); at College Station in that state the cotton bollworm had become tolerant to endrin and toxaphene–DDT mixture, while the tobacco budworm (*H. virescens*) had developed cyclodiene resistance as well as DDT resistance(*6*). DDT-resistant pink bollworms (*Pectinophora gossypiella*) remained susceptible to azin-phosmethyl and carbaryl(*144*). Cyclodiene-resistant tobacco hornworms

(*Protoparce sexta*) also remained susceptible to carbaryl, with very low cross-resistance to endosulfan(*664*).

The cross-resistance pattern of cyclodiene-resistant root maggots (*Hylemya antiqua, brassicae*, and *liturata*) has been extensively studied by tests on the adult flies. Developed mainly by aldrin applications, the cross-resistance is equally great to dieldrin and heptachlor, moderate to endrin, gamma-BHC, and isobenzan(*321,323*), and least to endosulfan (*323*); there was no cross-resistance to DDT or the organophosphate compounds(*321,580*). Thus, cyclodiene resistance in root maggots has dictated the replacement of aldrin or heptachlor with organophosphate compounds such as diazinon and DURSBAN. Conversely, organophosphate resistance in red mites has been countered successively with the chlorinated acaricides ovex, dicofol, and tetradifon.

Occasionally advantage may be taken of the fact that there are two separate types of resistance among the chlorinated hydrocarbons; for the control of *Anopheles sundaicus* in Java, DDT resistance on the north coast has been satisfactorily combatted with dieldrin, and dieldrin resistance of this mosquito on the south coast has been satisfactorily met by the use of DDT(*308*). In agriculture, however, DDT resistance is almost invariably countered by means of organophosphate compounds in order to avoid the residue problems of cyclodiene insecticides.

Occasionally DDT resistance fails to reach a level where it cannot be countered by heavier doses of DDT; normally it is a bad practice because it leads to more intense resistance, but for DDT resistance in certain species of *Anopheles*, the most practical answer has been the shortening of spray cycles while retaining the use of DDT. Cyclodiene resistance in *A. pharoensis* and *A. gambiae* has been countered by gamma-BHC, which, having a rapid fumigant action, is effective against the adults despite their reduced susceptibility levels. Organophosphate resistance in houseflies has often been countered by the use of baits that ensure a stiff dose, and thus trichlorfon has remained effective. Certain organophosphate compounds suffer only lightly from cross-resistance, such as dimethoate against houseflies†, fenthion against *Aedes nigromaculis* and houseflies(*846*), and phosphoroamidothioates against tetranychid mites. Hopes that the residual fumigant technique would not select for resistance have been abandoned because of the recent finding that houseflies can become specifically resistant to the vapor of dichlorvos (*411*).

---

†Although dimethoate resistance up to 120 times the normal appeared on some Danish forms in 1966(*414*).

B. Analogs Not Detoxified by Resistant Strains

DDT analogs that lack the terminal chlorination on the central aliphatic chain, such as the mixture of PROLAN and BULAN, known commercially as DILAN, or DDT molecules to which have been added substituents conferring steric hindrance against their breakdown, such as *o*-chloro-DDT (*341*), have both been initially effective against houseflies (Fig. 7-19).

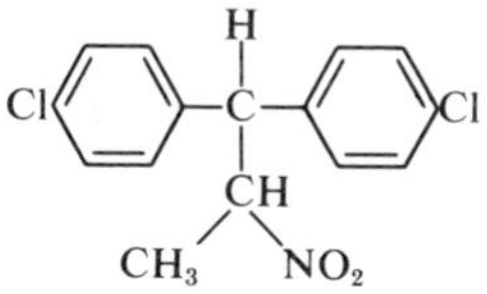

Prolan 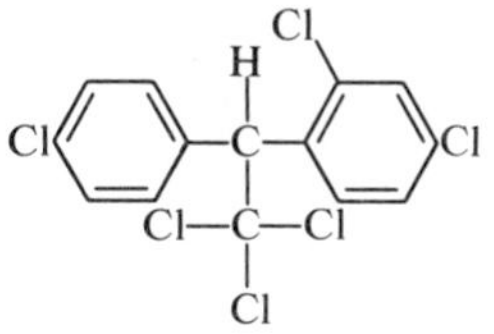   *o*-Chloro-DDT 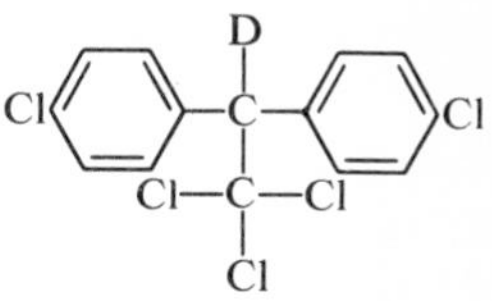   Deutero-DDT

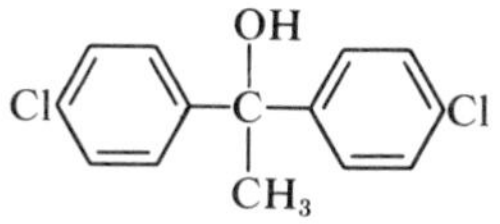

DMC 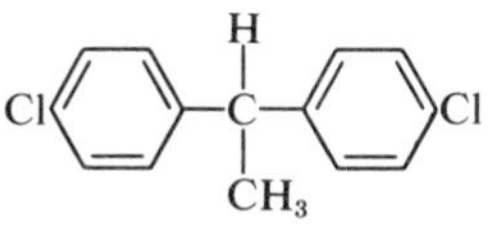   K-3926

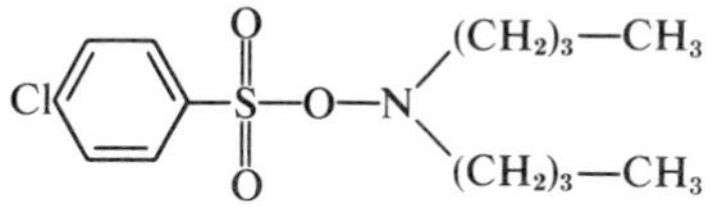

WARF-Antiresistant

Fig. 7-19. Analogs of DDT liable to be not detoxified by DDT-resistant strains (upper line), and (lower line) analogs effective as DDT-synergists against DDT-resistant strains.

Subsequently, however, DILAN resistance develops (*412,627*) and it carries a cross-resistance to *o*-chloro-DDT. This resistance is not associated with any reduction in cuticular absorption or any increase in detoxication of *o*-chloro-DDT; a Rutgers strain could dehydrochlorinate it to *o*-chloro-DDE about one-quarter as fast as it detoxified DDT to DDE; but it gained thereby only a threefold tolerance to *o*-chloro-DDT (*630*). The isotopic effect in reducing detoxicability has been exploited in deutero-DDT, in which deuterium has been substituted for the central aliphatic hydrogen. DDT-resistant *Aedes aegypti*, which can degrade *o*-chloro-DDT, cannot dehydrochlorinate deutero-DDT and develops resistance to deutero-DDT only very slowly (*642*). Although effective also against DDT-resistant *Culex fatigans*, deutero-DDT is dehydrochlorinated and thus successfully resisted by DDT-resistant houseflies *(53,547)*.

Among the organophosphate compounds, the carbomethoxy analog of malathion cannot be detoxified by malathion-resistant *Culex tarsalis* and is thus an effective remedial larvicide*(179,648)*. Houseflies made resistant by malathion pressure are at first susceptible to diazinon and parathion, and *vice versa*; this differential, however, soon disappears with extensive use of organophosphate insecticides in the field. Neither malathion-resistant nor parathion-resistant houseflies show any material resistance to the O,O-di-isopropyl analogs of malathion or parathion *(647)*, which are potent inhibitors of the aliesterase from which the breakdown enzymes are derived. In organophosphate-resistant greenhouse mites, however, the detoxifying esterases and the resistance mechanisms in general are so versatile that all the different alkyl substituents in the phosphate and carboxy ester positions of malathion are equally resisted*(800)*.

C. Synergists for Resistant Strains

One countermeasure would be to add to the insecticide a compound that inhibits its enzymic detoxication. DMC inhibits the dehydrochlorinase of houseflies*(631)* and mosquitoes*(427)*, and thus initially it is an effective DDT synergist against DDT-resistant strains. Unfortunately, chlorinated compounds such as DMC and K-3926 (Fig. 7-19) are competitive inhibitors of the dehydrochlorinase, so that houseflies can develop resistance to the synergist-DDT mixture*(498)* by coming to contain more enzyme than the absorbed DMC can inhibit*(546)*. The DDT synergist N,N-dibutyl-*p*-chlorobenzenesulfonamide (called WARF ANTIRESISTANT) has also evoked resistance to the synergized mixture by *Aedes aegypti(641)* and by houseflies, which simultaneously develop cross-tolerance to organophosphate compounds and carbamates but not to the cyclodienes*(238)*.

Pyrethrin synergists such as piperonyl butoxide and other methylenedioxyphenyl compounds serve to synergize many of the modern insecticides against houseflies that have become resistant to them. They are thus effective for DDT*(628)*, for carbamates*(529)*, and for diazinon trichlorfon, and other phosphates*(625,667)*, but they are antagonistic for malathion*(668)*; in explanation, they are known to inhibit not only esterase enzymes but also the process of microsomal hydroxylation. A number of organophosphate analogs are effective inhibitors for the detoxifying enzymes involved in certain organophosphate resistances.

Thus, propyl-paraoxon can synergize malathion and parathion by inhibiting the housefly phosphatase(*598*), and the ethyl and isopropyl phosphorotetrathioates could synergize malathion by inhibiting the carboxyesterase(*649*). Other carboxyesterase inhibitors are malathion synergists against malathion-resistant strains, such as EPN for *Culex tarsalis*(*504*) and naled and dichlorvos for the rice leafhopper, *Nephotettix*(*436*). Aliesterase inhibitors such as tributyl phosphorothioate and phosphorothioate also synergize malathion against malathion-resistant houseflies and *Culex tarsalis*(*645*), the same effect is noted in *Tetranychus telarius*(*340*). A curious synergism occurs between toxaphene and DDT against cyclodiene-resistant *Anthonomus*; toxaphene enhances the cuticular absorption of DDT, which is normally not very insecticidal for the weevil(*470*).

D. Mixtures and Alternations

A mixture of two insecticides of independent action has been suggested as a countermeasure for resistance, e.g., BHC–DDT mixtures against malaria mosquitoes, the gamma-BHC to kill the DDT-resistant survivors, and *vice versa*. Calculations based on the selection pressure indicate, however, that resistance should develop to the two components faster than if either component were used alone. When a mixture of malathion and chlordane was applied against the German cockroach for 15 generations, the resistance developed to both compounds was as great as the single resistances induced in the same time by one or other insecticide employed alone(*120*). Similar results were obtained when the insecticides were not mixed but alternated in each generation against the roaches, or with mixtures or alternations of malathion and DDT against houseflies and bed bugs. The speed with which the boll weevil developed resistance to endrin was reduced by nearly one-half when azinphosmethyl was alternated with it(*292*). Selection of the weevil with a mixture of toxaphene and DDT for 42 generations induced only a slight tolerance to the mixture but imparted an equal cross-tolerance to organophosphate compounds and a real cross-resistance to carbamate insecticides.

The experience has been different, however, with the European red mite in the field. In Ohio, the organophosphate resistance that had developed in most of the state was avoided in four orchards where different acaricides were used in successive years, or sulfur-containing compounds had been alternated with the organophosphate insecticides (*175*). In Pennsylvania, mixtures or alternations of tetradifon and dicofol remained effective after three years when either compound used alone had already encountered resistance(*27*). In England, however, the alternation of dicofol with dimethoate did not prevent the development of organophosphate resistance(*188*).

E. Negatively Correlated Insecticides

If specific resistance to one insecticide carries an enhanced susceptibility to another insecticide, the cross-resistance characteristics of the two compounds being thus negatively correlated, each should thus constitute a countermeasure for resistance to the other(*24, 108*). Cyclodiene resistant boll weevils have been found that have become more susceptible to malathion, DDT resistant houseflies more susceptible to malathion, and cyclodiene resistant houseflies more susceptible to cyclopropane and CO_2 anaesthesia. DDT-resistant houseflies were found to be unusually susceptible to a technical grade of diisopropyl tetrachloroethyl phosphate (DTP), and DTP selection was reported to restore the DDT susceptibility; but special studies with a new preparation could not confirm this(*527*). Cetyl bromoacetate (CBA) was found to knock down DDT-resistant, chlordane-resistant, and organophosphate-resistant strains of houseflies faster than normal strains(*23*); but in terms of kill, it proved no more insecticidal to resistant houseflies either in the laboratory or in the field(*108*). Cetyl fluoroacetate (CFA) and cetyl fluoride (CF) showed negative correlation with DDT on housefly larvae and so did CF on larvae of *Anopheles atroparvus*(*25*).

Selection of a dieldrin-resistant strain of *A. albimanus* with compound AC-5727 resulted in its reverting toward susceptibility, which did not happen when this strain was maintained without selection(*267*), although another dieldrin-resistant strain did(*415b*). Two DDT-resistant strains of body lice were more susceptible than normal to carbaryl(*163*), but selection of a third with DDT did not make it any more susceptible to this carbamate(*154*).† In *Drosophila melanogaster* it was found(*583*) that the *RI* resistance allele itself conferred unusual susceptibility to PTU (phenylthiourea, i.e., the substance known as phenylthiocarbamide in the genetical taste test). Thus, selection with PTU restored the DDT susceptibility by selecting for the wild-type, PTU-tolerant allele; and a mixture of PTU and DDT could kill any of the genotypes(*584*). Unfortunately, PTU does not show this negative correlation with DDT for houseflies. Nonetheless, compounds of this type, to which insects are made the more susceptible by the pleiotropic effect of the usual resistance alleles, offer the best prospects for reliable remedial insecticides to counter directly the resistance problem.

---

†Negative correlation with OP resistance in *Tetranychus urticae* has recently been found in the carbamate products formetanate(*738a*) and chlorophenamidine(*194a*).

REFERENCES

1. Z. H. Abedi and A. W. A. Brown, *Can. J. Genetics Cytol.*, **2**, 252 (1960).
2. Z. H. Abedi and A. W. A. Brown, *Mosquito News*, **21**, 1 (1961).
3. Z. H. Abedi and J. R. Duffy, and A. W. A. Brown, *J. Econ. Entomol.*, **56**, 511 (1963).
4. M. Acosta, *Rev. Med. (Peru)*, **30**, 53 (1960).
5. C. H. Adams and W. H. Cross, *J. Econ. Entomol.*, **60**, 1016 (1967).
6. P. L. Adkisson and S. J. Nemec, *Texas Agr. Exptl. Sta.*, **B-1048**, 4 pp. (1966).
7. V. E. Adler and C. C. Blickenstaff, *J. Econ. Entomol.*, **57**, 299 (1964).
8. M. Agosin, D. Michaeli R. Miskus, S. Nagasawa, and W. M. Hoskins, *J. Econ. Entomol.*, **54**, 340 (1961).
9. M. Agosin, B. C. Fine, N. Scaramelli, J. Ilivicky, and L. Aravena, *Comp. Biochem. Physiol.*, **18**, 101 (1966).
10. W. W. Allen, A. K. Ota, and R. D. Gehring, *J. Econ. Entomol.*, **57**, 187 (1964).
11. H. E. Aller and P. C. Lippold, *J. Econ. Entomol.*, **56**, 721 (1963).
12. W. E. Allison, Ph. D. Dissertation, Texas A. & M. (1963).
13. R. M. Altman, Report U.S. Armed Forces Pest Control Board, Washington, Oct. 31, pp. 7–8 (1958).
14. L. A. Andres and T. Prout, *J. Econ. Entomol.*, **53**, 626 (1960).
15. L. A. Andres and H. T. Reynolds, *J. Econ. Entomol.*, **51**, 285 (1958).
16. L. A. Andres and V. E. Burton, R. F. Smith, and J. E. Swift, *J. Econ. Entomol.*, **48**, 509 (1955).
17. G. W. Ankersmit, *Bull. Entomol. Res.*, **44**, 421 (1953).
18. E. A. Anthon, *J. Econ. Entomol.*, **48**, 56 (1955).
19. D. W. Anthony, *J. Econ. Entomol*, **52**, 782 (1959).
20. Army Environmental Health Agency, Insecticide Resistance Records of U.S. Armed Forces Pest Control Board (1962).
21. J. A. Armstrong, V. Ramakrishna, and C. D. Ramsdale, *Ann. Trop. Med. Parasit.*, **52**, 247 (1958).
22. S. Ashina, K. Yasutomi, Y. Inouye, K. Ogata, and Y. Noguchi, *Japan J. Sanit. Zool.*, **17**, 243 (1966).
23. K. R. S. Ascher, *Bull. World Health Organ.*, **18**, 675 (1958).
24. K. R. S. Ascher, Drug. Res. (Arzneim.-Forsch.), **10**, 450 (1960).
25. K. R. S. Ascher and E. D. Bergmann, *Entomol. Exp. Appl.*, **5**, 88 (1962).
26. K. R. S. Ascher and A. S. Tahori, *Nature* **179**, 324 (1957).
27. D. Asquith, *J. Econ. Entomol.*, **54**, 439 (1961).
28. D. Asquith, *J. Econ. Entomol.*, **57**, 905 (1964).
29. K. van Asperen, *J. Insect Physiol.*, **8**, 401 (1962).
30. K. van Asperen, *Entomol. Exp. Appl.*, **7**, 205 (1964).
31. K. van Asperen and F. J. Oppenoorth, *Entomol. Exp. Appl.*, **2**, 48 (1959).
32. K. van Asperen and F. J. Oppenoorth, *Entomol. Exp. Appl.*, **3**, 68 (1960).
33. K. van Asperen, M. van Mazijk, and F. J. Oppenoorth, *Entomol. Exp. Appl.*, **8**, 163 (1965).
34. Y. F. Atallah and W. C. Nettles, *J. Econ. Entomol.*, **59**, 560 (1966).
35. E. L. Atkins and L. D. Anderson, *J. Econ. Entomol.*, **55**, 791 (1962).
36. C. N. Ault, *Records Med. Vet. (Buenos Aires)*, **30**, 174 (1948).
37. S. A. Azeez, *World Health Organ. Inform. Circ. Insecticide Resistance*, **55**, 8 (1966).
38. F. H. Babers, U.S. Dept. Agr., Bur. Entomology and Plant Quarantine E-776, 31 pp. (1949).
39. F. H. Babers, *J. Econ. Entomol.*, **46**, 869 (1953).

40. F. H. Babers and J. J. Pratt, U.S. Dept. Agr., Bur. Entomology and Plant Quarantine E-818, 45 pp. (1951).
41. R. W. Babione, Mim. Doc. WHO/Mal/482, pp. 6–14 (1965).
42. O. G. Bacon, W. D. Riley, and G. Zweig, *J. Econ. Entomol.*, **57**, 225 (1964).
43. A. S. Badawi, WHO Assign. Rept. SEA/MAE/52 (1965).
44. M. L. Baerecke, *Z. Pflanzenkrankh Pflanzenschutz*, **69**, 453 (1962).
45. L. O. Baker, experiments at Montana State College, Bozeman; information from H. E. Gray (1964).
46. J. A. F. Baker, and R. D. Shaw, *J. S. African Vet. Med. Assoc.*, **36**, 3 (1965).
47. H. J. Ball, *J. Econ. Entomol.*, **61**, 496, (1968).
48. H. J. Ball and G. T. Weekman, *J. Econ. Entomol.*, **55**, 439 (1962) and **56**, 553 (1963).
49. G. H. Ballantyne and R. A. Harrison, *Entomol. Exp. Appl.*, **10**, 231 (1967).
50. F. L. Banham and D. G. Finlayson, *J. Entomol. Soc. (Brit. Colombia)*, **64**, 17 (1967).
51. A. Barkai, Mim. Doc. WHO/Vector Control/58 (1964).
52. A. Barkai and P. Rosen, *Riv. Parasitol.*, **25**, 217 (1964).
53. R. J. Barker, *J. Econ. Entomol.*, **53**, 35 (1960).
54. M. M. Barnes, *J. Econ. Entomol.*, **51**, 547 (1958).
55. M. M. Barnes and H. R. Moffitt, *J. Econ. Entomol.*, **56**, 722 (1963).
56. M. M. Barnes, R. A. Flock, and R. D. Garmus, *J. Econ. Entomol.*, **47**, 238 (1954).
57. A. R. Barr, *Proc. Calif. Mosquito Contr. Assoc.*, **30**, 88 (1962).
58. S. Basheir and K. A. Lord, *Chem. Ind.*, **7**, 1598 (1965).
59. R. L. Beard, *Conn. Agr. Exp. Sta. New Haven Bull.*, **631**, 22 pp. (1960).
60. R. L. Beard, *J. Econ. Entomol.*, **58**, 584 (1965).
61. J. A. Begg, *Can. Entomologist*, **93**, 1022 (1961).
62. G. D. Belios and G. Fameliaris, *Bull. World Health Organ.*, **15**, A15 (1956).
63. G. W. Bennet and W. T. Spink, *J. Econ. Entomol.*, **61**, 426 (1968).
64. R. S. Berger and R. G. Young, *J. Econ. Entomol.*, **55**, 533 (1962).
65. F. Beye, *Verhandl. Naturforsch. Ges. Basel*, **71**, 283 (1960).
66. S. C. Bhatia and R. B. Deobhankar, *Indian J. Entomol.*, **24**, 36 (1962).
67. S. C. Bhatia and R. B. Deobhankar, *Indian J. Malaria*, **17**, 339 (1963).
68. S. R. Bhombore, R. G. Roy, and F. Samson, *Bull. Natl. Soc. India Malaria*, **11**, 31 (1963).
69. S. R. Bhombore, R. G. Roy, and S. Franco, *Bull. Natl. Soc. India Malaria*, **11**, 35 (1963).
70. R. V. Bielarski, J. S. Roussel, and D. F. Clower, *J. Econ. Entomol.*, **50**, 481 (1957).
71. J. H. Bigger, *J. Econ. Entomol.*, **56**, 118 (1963).
72. W. S. Bigley and F. W. Plapp, *Ann. Entomol. Soc. Am.*, **53**, 360 (1960).
73. W. S. Bigley and F. W. Plapp, *J. Insect Physiol.*, **8**, 545 (1962).
74. A. Bishop, *Biol. Rev. Cambridge Phil. Soc.*, **34**, 445 (1959).
75. B. D. Blair, C. A. Triplehorn, and G. W. Ware, *J. Econ. Entomol.*, **56**, 894 (1963).
76. M. S. Blum, N. W. Earle, and J. S. Roussel, *J. Econ. Entomol.*, **52**, 17 (1959).
77. O. Boggild and J. Keiding, *Oikos*, **9**, 1 (1958).
78. A. Bojanowska and Z. Wojciak, *Przeglad Epidemiol.*, **14**, 67 (1960).
79. C. C. Bowling, *J. Econ. Entomol.*, **61**, 1027 (1968).
80. A. M. Boyce, C. A. Persing, and C. S. Barnhart, *J. Econ. Entomol.*, **35**, 790 (1942).
81. C. E. Boyd and D. E. Ferguson, *J. Econ. Entomol.*, **57**, 430 (1964).
82. C. E. Boyd, S. B. Vinson, and D. E. Ferguson, *Copeia*, 1963, 426.
83. C. M. Boyle, *Nature*, **188**, 517 (1960).
84. F. R. Bradbury, *J. Sci. Food Agr.*, **8**, 90 (1957).
85. F. R. Bradbury, and H. Standen, *Nature*, **178**, 1053 (1956).
86. F. R. Bradbury and H. Standen, *Nature*, **183**, 983 (1959).

87. C. B. Bragassa and J. R. Brazzel, *J. Econ. Entomol.*, **54**, 311 (1961).
88. L. Bravenboer, *Mededel. Directeur Tuinbouyl*, **18**, 672 (1955).
89. L. Bravenboer, *Proefsta. Naaldwijk.*, 1963, p. 172 (1964).
90. J. R. Brazzel, *J. Econ. Entomol.*, **56**, 571 (1963).
91. J. R. Brazzel, *J. Econ. Entomol.*, **57**, 455 (1964).
92. J. R. Brazzel, *Mississippi Farm Res.*, **28** (8), 8, also **27** (4), 1 (1965).
93. J. R. Brazzel and D. A. Lindquist, *J. Econ. Entomol.*, **53**, 551 (1960).
93a. J. R. Brazzel et al., *Bull. Entomol. Soc. Am.* **16**, 147 (1970).
94. C. H. Brett and R. W. Brubaker, *J. Econ. Entomol.*, **48**, 343 (1955).
95. R. G. Bridges, *Nature*, **184**, 1337 (1959).
96. G. T. Brooks, *Nature*, **186**, 96 (1960).
97. G. T. Brooks and A. Harrison, *Nature*, **198**, 1169 (1963).
98. G. T. Brooks and A. Harrison, *Biochem. J.*, **87**, 5 (1963).
99. G. T. Brooks and A. Harrison, *Biochem. Pharmacol.*, **13**, 827 (1964).
100. G. T. Brooks and A. Harrison, *J. Insect Physiol.*, **10**, 633 (1964).
101. A. W. A. Brown, *Advan. Pest Control. Res.*, **2**, 351 (1958).
102. A. W. A. Brown, *World Health Organ. Monogr. Series No.* **38**, 240 pp. (1958).
103. A. W. A. Brown, *Bull. World Health Organ.*, **19**, 1053 (1958).
104. A. W. A. Brown, *Misc. Publ. Entomol. Soc. Am.*, **1**, 20 (1959).
105. A. W. A. Brown, *Ann. Rev. Entomol.*, **5**, 301 (1960).
106. A. W. A. Brown, *Bull. Entomol. Soc. Am.*, **7**, 6 (1961).
107. A. W. A. Brown, *Ind. Trop. Health (Harvard)*, **4**, 197 (1961).
108. A. W. A. Brown, *Pest Control*, **29** (9), 24 (1961).
109. A. W. A. Brown, et al., *Mosquito News*, **22**, 205 (1962).
110. A. W. A. Brown, *Bull. World Health Organ.*, **29** Suppl., 41 (1963).
111. A. W. A. Brown, *Farm Chemicals* **126**, Nos. 10, 11, 12 and **127** No. 1 (1963).
112. A. W. A. Brown, *Bull. World Health Organ.*, **30**, 97 (1964).
113. A. W. A. Brown, *Handbook of Physiology—Environment*, American Physiology Society, 1964, Chap. 48.
114. A. W. A. Brown and Z. H. Abedi, *Mosquito News*, **20**, 118 (1960).
115. A. W. A. Brown and Z. H. Abedi, *Can. J. Genet. Cytol.*, **4**, 319 (1962).
116. A. W. A. Brown and E. Wambera, *Pest Control*, **29** (6), 62 (1961); also *EP Technical Rep. 9*, Direct. Biol. Sci. Res. Defence Res. Board Canada 4 pp. (1965).
117. A. W. A. Brown, L. L. Lewallen, and P. A. Gillies, *Mosquito News*, **23**, 341 (1963).
118. L. B. Browne and R. W. Kerr, *Entomol. Exp. Appl.*, **10**, 337 (1967).
119. D. L. Bull and P. L. Adkisson, *J. Econ. Entomol.*, **56**, 641 (1963).
120. G. S. Burden, C. S. Lofgren, and C. N. Smith, *J. Econ. Entomol.*, **53**, 1138 (1960).
121. C. C. Burkhardt, *Proc. N. Central Branch Entomol. Soc. Am.*, **18**, 82 (1963).
122. G. F. Burnett and L. H. Ash, *Bull. World Health Organ.*, **24**, 547 (1961).
123. E. C. Burns and B. H. Wilson, *J. Econ. Entomol.*, **56**, 718 (1963).
124. G. J. Burton, *Mosquito News*, **24**, 200 (1964).
125. E. C. Burts, *Abstr. Pacific Branch Entomol. Soc. Am.*, **2** (1) (1959).
126. J. R. Busvine, *Nature*, **17**, 118 (1953).
127. J. R. Busvine, *Nature*, **174**, 783 (1954).
128. J. R. Busvine, *Trans. Roy. Soc. Trop. Med. Hyg.*, **51**, 11 (1957).
129. J. R. Busvine, *Techniques for Testing Insecticides*, Commonwealth Institute of Entomology, London, 1957.
130. J. R. Busvine, *Bull. World Health Organ.*, **19**, 1041 (1958).
131. J. R. Busvine, Working Paper 5.14.5, 13th. Exptl. Commission on Insecticides WHO (1962).
132. J. R. Busvine, Proc. XIth Internat. Congr. Entomol., **3**, 220 (1962).

133. J. R. Busvine, *Bull. Entomol. Res.*, **55**, 271 (1964).

134. J. R. Busvine, *Bull. World Health Organ.*, **31**, 645 (1964).

135. J. R. Busvine, *World Rev. Pest Control*, **7**, 27 (1968).

136. J. R. Busvine and G. J. Shanahan, *Entomol. Exp. Appl.*, **4**, 1 (1961).

137. J. R. Busvine and M. G. Townsend, *Bull. Entomol. Res.*, **53**, 763 (1963).

138. J. R. Busvine, J. D. Bell, and A. M. Guneidy, *Bull. Entomol. Res.*, **54**, 589 (1963).

139. F. L. Campbell, *J. Econ. Entomol.*, **19**, 516 (1962).

140. W. R. Campbell and D. G. Cochran, *Bull. Entomol. Soc. Am.*, **11**, 157 (1965).

141. Canada Dept. Agr., Report on Research, p. 40, Ottawa, 1957.

142. B. R. Champ and J. Cribb, *J. Stored Products Res.*, **1**, 9 (1965).

143. B. R. Champ and R. C. H. Shepherd, *Queensland J. Agr. Sci.*, **22**, 69, 461, 511 (1965).

144. A. J. Chapman and L. B. Coffin, *J. Econ. Entomol.*, **57**, 148 (1964).

145. R. K. Chapman, *Proc. N. Central Branch, Entomol. Soc. Am.*, **9**, 88 (1954).

146. R. K. Chapman, *Misc. Publ. Entomol. Soc. Am.*, **2**, 27 (1960).

147. A. N. Chattoraj and C. W. Kearns, *Bull. Entomol. Soc. Am.*, **4**, 95 (1958).

148. R. P. Chaudhuri and R. C. Naithani, *Bull. Entomol. Res.*, **55**, 405 (1964).

149. G. Chauvet, *Bull. Soc. Pathol. Exotique.*, **55**, 1156 (1962).

150. G. Chauvet and J. Coz, *Publ. Inst. Rech. Sci.*, Madagascar, Tananarive (1960).

151. G. Chauvet and G. Davidson, Mim. Doc. WHO/Vector Control/66.224 (1966).

152. C. Y. Chow, *Indian J. Malaria*, **12**, 34 (1958).

153. C. Y. Chow, *WHO Inform. Circ. Instecticide Resistance*, **54**, 6 (1965).

154. P. H. Clark and M. M. Cole, *J. Econ. Entomol.*, **57**, 205 (1964).

155. P. H. Clark and M. M. Cole, *J. Econ. Entomol.*, **60**, 398 (1967).

156. T. H. Coaker, D. J. Mowat, and G. A. Wheatley, *Nature*, **200**, 664 (1963).

157. D. G. Cochran, 12th Intern. Congr. Entomol. London, Programme p. 34.

158. D. G. Cochran and M. H. Ross, *J. Econ. Entomol.*, **55**, 88 (1962).

159. D. G. Cochran and M. H. Ross, *Bull. World Health Organ.*, **27**, 257 (1962).

160. D. G. Cochran and M. H. Ross, *J. Econ. Entomol.*, **57**, 485 (1964).

161. W. Z. Coker, *Ann. Trop. Med. Parasitol.*, **60**, 347 (1966).

162. M. M. Cole and P. H. Clark, *J. Econ. Entomol.*, **54**, 649 (1961).

163. M. M. Cole and P. H. Clark, *J. Econ. Entomol.*, **55**, 98 (1962).

164. E. Collyer and A. H. M. Kirby, *Ann. Rept. East Malling Res. Sta. Kent*, **1958**, 131 (1958).

165. R. Comes, I. Frada, and D. Macri, *Riv. Parassit.*, **23**, 151 (1962).

166. C. C. Compton and C. W. Kearns, *J. Econ. Entomol.*, **30**, 512 (1937).

167. R. R. Correa, *Arch. Hig. (San Paulo)*, **27**, 341 (1962).

168. J. Coulon, J. Lhoste and J. Missioner, Abst. 6th Intern. Congr. Plant Prot. p. 578 (1967).

168a. J. Coz, G. Davidson, G. Chauvet, and J. Hamon, Cahiers ORSTOM, *Ser. Ent. Med.*, **6**, 207 (1968).

169. A. W. Cressman, *J. Econ. Entomol.*, **56**, 884 (1963).

170. J. F. Crow, *Ann. Rev. Entomol.*, **2**, 227 (1957).

171. J. H. Cuthbert, *Nature*, **198**, 807 (1963).

172. F. P. Cuthbert and W. J. Reid, U.S. Dept. Agr. *ARS-33-72 mim.* (1962).

173. L. K. Cutkomp, A. G. Peterson, and P. E. Hunter, *J. Econ. Entomol.*, **51**, 828 (1958).

174. C. R. Cutright, *J. Econ. Entomol.*, **48**, 304 (1955).

175. C. R. Cutright, *J. Econ. Entomol.*, **52**, 432 (1959).

176. R. Cwilich and K. R. S. Ascher, *Israel J. Agr. Res.*, **11**, 135 (1961).

177. G. D'Alessandro, C. Bruno-Smiraglia, and M. Mariani, *Riv. Parassit.*, **23**, 227 (1962).

178. D. I. Darrow and F. W. Plapp, *J. Econ. Entomol.*, **53**, 777 (1960).

179. W. C. Dauterman and F. Matsumura, *Science*, **138**, 694 (1962).

180. G. Davidson, *Bull. World Health Organ.*, **28**, 25 & 29, 177 (1963).

181. G. Davidson, *Ann. Trop. Med. Parasitol.*, **58**, 180 (1964).
182. G. Davidson and C. E. Jackson, *Nature*, **190**, 364 (1961).
183. M. Davies, J. Keiding, and C. G. von Hofsten, *Nature*, **182**, 1816 (1958).
184. D. W. Davis and G. L. Nielsen, *Bull. Entomol. Soc. Am.*, **4**, 93 (1958).
185. G. C. Decker and W. N. Bruce, *Am. J. Trop. Med. Hyg.*, **1**, 395 (1952).
185a. J. Dekker, *World Rev. Pest Control*, **8**, 79 (1969).
186. V. P. Derbeneva-Ukhova, V. A. Lineva, and V. P. Drobozina, *Bull. World Health Organ.*, **34**, 939 (1966).
187. V. P. Derbenova-Ukhova and V. A. Lineva, Mim. Doc. WHO/Vector Control/ 164 (1965).
188. G. H. L. Dicker, *Ann. Rept. East Malling Res. Sta. Kent*, 1963, p. 33 (1964).
189. R. C. Dickson, *Hilgardia*, **13**, 515 (1941).
190. R. C. Dickson and D. L. Lindgren, *Calif. Citrograph*, **32**, 524 (1947).
191. M. L. Dinamarca, M. Agosin, and A. Neghme, *Exp. Parasitol.*, **12**, 61 (1962).
192. V. Dittrich, *Z. Angew Entomol.*, **48**, 34 (1961).
193. V. Dittrich, *Advan. Acarology (Itacha, N.Y.)*, **1**, 238 (1963).
194. V. Dittrich, *J. Econ. Entomol.*, **56**, 182 (1963).
194a. V. Dittrich, *J. Econ. Entomol.*, **62**, 44 (1969).
195. J. M. Doby, J. Corbeau, and B. Rault, *Bull. Soc. Pathol. Exotique*, **54**, 1345 (1961).
196. G. Dosse, *Hoefchen Briefe (English Ed.)*, **5**, 238 (1952).
197. R. S. Downing, *Proc. Entomol. Soc. (Brit. Colombia)*, **51**, 10 (1954).
198. J. A. Dunn, *Nature*, **199**, 1207 (1963).
199. J. A. Dunn, *Entomol. Exp. Appl.*, **6**, 304 (1963).
200. J. A. Dunn and D. P. Kempton, *Entomol. Exp. Appl.*, **9**, 67 (1966).
201. M. Duport, Mim. Doc. WHO/Vector Control/119 (1965).
202. M. Duport, I. Combiesco, A. Atanasiu, G. Constantinesco, and M. Scarlat, *Arch. Roumaines Pathol. Exp. Microbiol.*, **23**, 1045 (1964).
203. J. P. Duret, *Bol. Ofic. Sanit. Panam.*, **51**, 285 (1961).
204. R. Dutoit, Communication to WHO March 1, 1968.
205. C. E. Dyte, and D. G. Blackman, *J. Stored Products Res.*, **2**, 211 (1967).
206. C. E. Dyte and D. G. Rowlands, *J. Stored Products Res.*, **4**, 157 (1968).
207. N. W. Earle, *J. Agr. Food Chem.*, **11**, 281 (1963).
208. E. F. Eastin, R. D. Palmer, and C. O. Grogan, *Weeds*, **12**, 49 and 64 (1964).
209. G. E. Eddy, M. M. Cole, M. D. Couch, and A. Selhime, *Public Health Rept. (U.S.)*, **70**, 1035 (1955).
210. W. G. Eden et al., *Bull. Entomol. Soc. Am.*, **14**, 31 (1968).
211. M. E. Eldefrawi and W. M. Hoskins, *J. Econ. Ent.* **54**, 401 (1961).
212. O. Enan, R. Miskus, and R. Craig, *J. Econ. Entom.*, **57**, 364 (1964).
213. P. F. Entwhistle, *Proceedings of the Conference on Mirid Pests of Cacao*, W. African Cacao Research Institute, Ibadan, 1964, p. 9.
214. B. R. Evans, J. E. Porter, and G. Kozuchi, *Mosquito News*, **20**, 116 and **21**, 4 (1961).
215. W. H. Ewart, F. A. Gunther, J. H. Barkley, and H. S. Elmer, *J. Econ. Entomol.*, **45**, 578 (1952).
216. A. W. Farnham, K. A. Lord, and R. M. Sawicki, *J. Insect Physiol.*, **11**, 1475 (1965).
217. P. G. Fast, *Memoir Entomol. Soc. Can.* No. 37, 50 pp. (1964).
218. J. M. Feigin, *Ann. Entomol. Soc. Am.*, **56**, 878 (1963).
219. J. C. Felton, *Shell Public Health Agr. News (SPAN)* **3**, 33 (1960).
220. D. E. Ferguson, *Agr. Chem.*, **18** (9), 32 (1963).
221. D. E. Ferguson and C. E. Boyd, *Copeia*, **4**, 706 (1964).
222. D. E. Ferguson, R. L. Callahan and W. D. Cotton, *J. Mississippi Acad. Sci.*, **11**, 229 (1965).

223. D. E. Ferguson, D. D. Culley, and W. D. Cotton, *J. Mississippi Acad. Sci.*, **11**, 235 (1965).

224. D. E. Ferguson, D. D. Culley, W. D. Cotton, and R. P. Dodds, *BioScience* **14**, 43 (1964).

225. W. C. Ferguson and C. W. Kearns, *J. Econ. Entomol.* **42**, 810 (1949).

226. F. H. Finch, *Pastoral Review (Melbourne)*, **58** 537 (1948).

227. B. C. Fine, *Nature*, **191**, 884 (1961).

228. B. C. Fine, *Pyrethrum Post*, **7** (2), 18 (1963).

229. B. C. Fine, P. J. Godin, and E. M. Thain, *Nature*, **199**, 927 (1963).

230. D. G. Finlayson and M. D. Noble, *Pesticide Research Report*, NCPUA, Ottawa, 1964, p. 214.

231. D. G. Finlayson, H. H. Crowell, A. J. Howitt, D. R. Scott, and A. J. Walz, *J. Econ. Entomol.*, **52**, 851 (1959).

232. E. H. Fisher, *Wisconsin Univ. Agr. Exp. Sta. Bull.*, **520**, 35 pp. (1957).

233. R. W. Fisher, *Can. J. Plant Sci.*, **40**, 580 (1960).

234. J. Fjelddalen and T. Daviknes, *Gartneryket*, **1952** No. 13.

235. W. P. Flint, *J. Econ. Entomol.*, **16**, 209 (1923).

236. Food and Agriculture Organization, *1st Rept. Working Party on Resistance of Pests to Pesticides*, (1967).

236a. Food and Agriculture Organization, *FAO Plant Prot. Bull.*, **17**, 76 (1969) and succeeding numbers.

237. W. H. Foott, N. A. Patterson, and F. T. Lord, Insecticide *Newsletter*, Can. Dept. Agr., **8** (1), 2 (1959).

238. A. J. Forgash, *J. Econ. Entomol.*, **60**, 1750 (1967).

239. A. J. Forgash and E. J. Hansens, *J. Econ. Entomol.*, **52**, 733 (1959).

240. H. Y. Forsythe, *J. Econ. Entomol.*, **58**, 811 (1965).

241. I. Fox, *Bull. World Health Organ.*, **24**, 489 (1961).

242. A. D. Flynn and H. F. Schoof, *Mosquito News*, **25**, 411 (1965).

243. A. D. Flynn, H. F. Schoof, H. B. Morlan, and J. E. Porter, *Mosquito News*, **24**, 118 (1964).

244. M. G. Franco and F. J. Oppenoorth, *Entomol. Exp. Appl.*, **5**, 119 (1962).

245. W. L. French and J. B. Kitzmiller, *Mosquito News*, **24**, 32 (1964).

246. R. E. Furr, E. P. Lloyd, and M. E. Merke, Mississippi Agr. Exptl. Sta. Inform. Sheet 560 (1957).

247. J. B. Gahan, C. N. Smith, and B. M. Glancey, *Mosquito News*, **26**, 330 (1966).

248. J. C. Gaines and P. L. Adkisson, *Bull. Entomol. Soc. Am.*, **10**, 176 (1964).

249. R. C. Gaines, *Agr. Chem.*, **12** (4), 41 (1957).

250. P. Garman, *J. Econ. Entomol.*, **43**, 53 (1950).

251. R. Garms, *Z. Tropenmed. Parasitol.*, **2**, 353 (1960).

252. C. Garrett-Jones and G. Gramiccia, *Bull. World Health Organ.*, **11**, 865 (1954).

253. R. T. Gast, *J. Econ. Entomol.*, **54**, 1203 (1961).

254. R. Gasser, *Proc. Intern. Congr. Plant Protection 4th*, **1**, 643 (1959).

255. P. E. Gatterdam, R. K. De, F. E. Guthrie, and T. G. Bowery, *J. Econ. Entomol.*, **57**, 258 (1964).

256. J. A. George and A. W. A. Brown, *J. Econ. Entomol.*, **60**, 974 (1967).

257. G. P. Georghiou, *J. Econ. Entomol.*, **55**, 494 (1962).

258. G. P. Georghiou, *J. Econ. Entomol.*, **55**, 768 (1962).

259. G. P. Georghiou, *J. Econ. Entomol.*, **56**, 655 (1963).

260. G. P. Georghiou, *Nature*, **207**, 883 (1965).

261. G. P. Georghiou, *Advan. Pest Control Res.*, **6**, 171 (1965).

262. G. P. Georghiou, *Proc. Calif. Mosquito Contr. Assoc.*, **33**, 34 (1965).

263. G. P. Georghiou, *J. Econ. Entomol.*, **60**, 1338 (1967).

263a. G. P. Georghiou, *World Rev. Pest Control*, **8**, 86 (1969).

264. G. P. Georghiou and W. R. Bowen, *J. Econ. Entomol.*, **59**, 204 (1966).

265. G. P. Georghiou and M. J. Garber, *Bull. World Health Organ.*, **32**, 181 (1965).

266. G. P. Georghiou and R. L. Metcalf, *J. Econ. Entomol.*, **54**, 231 (1961).

267. G. P. Georghiou and R. L. Metcalf, *Science*, **140**, 301 (1963).

268. G. P. Georghiou, F. E. Gidden, and J. W. Cameron, *Ann. Entomol. Soc. Am.*, **60**, 323 (1967).

269. G. P. Georghiou, R. B. March, and G. E. Printy, *Bull. World Health Organ.*, **29**, 155 (1963).

270. G. P. Georghiou, R. L. Metcalf, and F. E. Gidden, *Bull. World Health Organ.*, **35**, 691 (1966).

271. G. P. Georghiou, R. L. Metcalf, and R. B. March, *J. Econ. Entomol.*, **54**, 132 (1961).

271a. G. P. Georghiou, P. A. Gillies, and D. J. Womeldorf, *Calif. Vector Views*, **16**, 115 (1969).

271b. S. G. Georgopoulos, *Nature*, **194**, 148 (1962).

272. P. D. Gerhardt and G. P. Wene, *J. Econ. Entomol.*, **52**, 760 (1959).

273. J. L. Gerold and J. J. Laarman, *Nature*, **204**, 500 (1964).

274. P. Gerolt, *J. Econ. Entomol.*, **58**, 849 (1965).

275. B. M. Glancey, C. S. Lofgren, and T. W. Miller, *Mosquito News*, **26**, 439 (1966).

276. E. H. Glass, *J. Econ. Entomol.*, **50**, 674 (1957).

277. E. H. Glass, *Misc. Publ. Entomol. Soc. Am.*, **2**, 17 (1960).

278. E. H. Glass and B. Fiori, *J. Econ. Entomol.*, **48**, 598 (1955).

279. J. S. Glover, *Public Health (Johannesburg)*, **66**, 209 (1966).

280. C. Gokberk, *WHO Inform. Circ. Insecticide Resistance*, **56**, 30 (1966).

281. G. E. Gould, *J. Econ. Entomol.*, **54**, 475 (1961).

282. G. E. Gould and L. L. McCrasky, *J. Econ. Entomol.*, **47**, 190 (1954).

283. H. J. Gould, *Proc. 3rd Brit. Insecticide Fungicide Conf.*, Ref. No. 11-3 (1965).

284. J. M. G. Gradidge, *Pest Technol.*, **2**, 229 (1960).

285. C. D. Grant and A. W. A. Brown, *Can. Entomol.*, **99**, 1040 (1967).

286. N. Gratz, *Bull. World Health Organ.*, **24**, 668 (1961).

287. N. Gratz, *Acta Trop.*, (Basel) **23**, 108 (1966).

288. J. B. Graves and O. Mackensen, *J. Econ. Entomol.*, **58**, 990 (1965).

289. J. B. Graves, D. F. Clower, and J. R. Bradley, *J. Econ. Entomol.*, **60**, 887 (1967).

290. J. B. Graves, J. S. Roussel, and J. R. Phillips, *J. Econ. Entomol.*, **56**, 442 (1963).

291. J. B. Graves, T. R. Everett, and R. D. Hendrick, *J. Econ. Entomol.*, **60**, in press (1967).

292. J. B. Graves, J. S. Roussel, J. Gibbens, and D. Patton, *J. Econ. Entomol.*, **60**, 47 (1967).

293. J. M. Grayson, *Bull. World Health Organ.*, **24**, 563 (1961).

294. J. M. Grayson, *J. Econ. Entomol.*, **58**, 956 (1965).

295. A. Grigolo and F. J. Oppenoorth, *Genetica*, **37**, 159 (1966).

296. C. O. Grogan, E. F. Eastin, and R. D. Palmer, *Crop Sci.*, **3**, 451 (1963).

297. D. S. Grosch and L. R. Valcovic, *J. Econ. Entomol.*, **60**, 1177 (1967).

298. H. Gruchet, *Bull. Soc. Pathol. Exotique.*, **54**, 1358 (1961).

299. A. M. Guneidy and J. R. Busvine, *Bull. Entomol. Res.*, **55**, 499 (1964).

300. A. M. Guneidy and J. R. Busvine, *Bull. Entomol. Res.*, **55**, 509 (1964).

301. F. E. Guthrie, R. L. Rabb, and D. A. Mount, *J. Econ. Entomol.*, **56**, 7 (1963).

302. G. Guyer and A. Wells, *Quart. Bull. Mich. Agr. Exptl. Sta.*, **41**, 614 (1959).

303. D. W. Hamilton, *J. Econ. Entomol.*, **49**, 866 (1956).

304. D. W. Hamilton, *Am. Fruit Grower*, **79** (4), 48 (1959).

305. D. W. Hamilton and J. E. Fahey, *J. Econ. Entomol.*, **47**, 861 (1954).

306. E. W. Hamilton, *J. Econ. Entomol.*, **58**, 296 (1965).

307. K. C. Hamilton and H. Tucker, *Weeds*, **12**, 220 (1964).

308. J. Hamon and C. Garrett-Jones, *Bull. World Health Organ.*, **28**, 1 (1963).

309. J. Hamon and J. Mouchet, *Med. Trop.*, **21**, 565 (1961).

310. J. Hamon, S. Sales, P. Venard, J. Coz, and J. Brengues, *Med. Trop.*, **28**, 222 (1968).

311. E. O. Hamstead, *J. Econ. Entomol.*, **50**, 109 (1957).

312. C. O. Hansen, J. A. Naegele, and H. E. Everett, *Advan. Acarol. (Ithaca, N.Y.)*, **1**, 257 (1963).

313. D. G. Harcourt, *Can. J. Agr. Sci.*, **36**, 430 (1956).

314. D. G. Harcourt and L. M. Cass, *J. Econ. Entomol.*, **52**, 221 (1959).

315. P. R. Harding, *Plant Diseases Reptr.*, **48** (1), 43 (1964).

316. A. M. Haridi, *WHO Inf. Circ. Insecticide Resistance*, **58**, 10 (1966).

317. F. C. Harmston, *Communication to Pacific Slope Branch, Entomol. Soc. Am.*, June 21, 1967.

318. F. H. Harries and E. Burts, *J. Econ. Entomol.*, **52**, 530 (1959).

319. F. H. Harries and E. Burts, *J. Econ. Entomol.*, **58**, 172 (1965).

320. C. R. Harris, *Proc. N. Central Branch Entomol. Soc. Am.*, **20**, 159 (1965).

321. C. R. Harris and J. L. Hitchon, *J. Econ. Entomol.*, **59**, 650 (1966).

322. C. R. Harris, G. F. Manson, and J. H. Mazurek, *J. Econ. Entomol.*, **55**, 777 (1962).

323. C. R. Harris, J. H. Mazurek, and H. J. Svec, *J. Econ. Entomol.*, **57**, 702 (1964).

324. C. R. Harris, H. J. Svec, and J. H. Mazurek, *J. Econ. Entomol.*, **56**, 563 (1963).

325. R. L. Harris, E. D. Frazar, and O. H. Graham, *J. Econ. Entomol.*, **59**, 387 (1966).

326. C. M. Harrison, *Nature*, **167**, 855 (1951).

327. I. R. Harrison, *Vet. Records*, **80**, 205 (1967).

328. R. J. Hart, *Nature*, **195**, 1123 (1962).

329. T. L. Harvey and D. E. Howell, *J. Invert. Pathol.*, **7**, 92 (1965).

330. W. O. Haufe, *Entomology Newsletter* (Research Branch, Canada Dept. Agr.) **44** (11), 2 (1966).

331. M. Hayashi and M. Hayakawa, *Japan J. Appl. Entomol. Zool.*, **6**, 250 (1962).

332. N. C. Hayslip, W. G. Genung, E. G. Kelsheimer, and J. W. Wilson, *Florida Univ. Agr. Exp. Sta. (Gainseville) Bull.*, **534**, 9 (1953).

333. E. I. Hazard, C. S. Lofgren, D. B. Woodard, H. R. Ford, and B. M. Glancey, *Science*, **145**, 500 (1964).

334. R. A. Hedeen, *Mosquito News*, **23**, 100 (1963).

335. E. Heidenreich, *Proc. 11th Intern. Congr. Entomol.*, **2**, 611 (1962).

336. W. Helle, *Tijdschr. Plantenziektan*, **63**, 155 (1962).

337. W. Helle, *Tijdschr. Plantenziekten*, **65**, 107 and **67**, 28 (1963).

338. W. Helle, *Advan. Acarol.*, **2**, 71 (1965).

339. L. S. Henderson, communication to U.S. Armed Forces Pest Control Board, Oct. 25, 1960.

340. T. J. Henneberry and W. R. Smith, *J. Econ. Entomol.*, **58**, 312 (1965).

341. D. J. Hennessy et al., *Nature*, **190**, 341 (1961).

342. D. H. C. Herne and A. W. A. Brown, *J. Econ. Entomol.*, **62**, 205 (1969).

343. G. E. R. Hervey and K. G. Swenson, *J. Econ. Entomol.*, **47**, 564 (1954).

344. M. van den Heuvel and D. G. Cochran, *J. Econ. Entomol.*, **58**, 872 (1965).

345. A. Hikichi, *Proc. Entomol. Soc. Ontario*, **92**, 182 (1962).

346. O. A. Hills, E. A. Taylor, and A. C. Valcarce, *J. Econ. Entomol.*, **49**, 94 (1956).

347. S. Hirakoso, *Japan. J. Exp. Med.*, **32**, 207 (1962).

348. T. Hiroyoshi, *Genetics*, **46**, 1373 (1961).

349. L. F. Hitchcock and I. M. Mackerras, *J. Counc. Sci. Ind. Res. Australia*, **20**, 43 (1947).

350. J. M. Hodgson, *Weeds*, **12**, 167 (1964).

351. R. N. Hofmaster, *J. Econ. Entomol.*, **49**, 530 (1956).
352. R. M. Hollingworth, R. L. Metcalf, and T. R. Fukuto, *J. Agr. Food Chem.*, **15**, 25c (1967).
353. K. Honma, A. Toshima, and H. Furihata, *Japan. J. Appl. Entomol. Zool.*, **5**, 225 (1961).
354. G. H. S. Hooper, *J. Econ. Entomol.*, **58**, 608 (1965).
355. G. H. S. Hooper, *Mosquito News*, **26**, 552 (1966).
356. G. H. S. Hooper, *J. Econ. Entomol.*, **61**, 490 (1968).
357. G. H. S. Hooper, *J. Econ. Entomol.*, **61**, 858 (1968).
358. G. H. S. Hooper and A. W. A. Brown, *Entomol. Exp. Appl.*, **8**, 263 (1965).
359. G. H. S. Hooper and A. W. A. Brown, *J. Econ. Entomol.*, **58**, 824 (1965).
360. W. M. Hoskins, *Intern. Rev. Trop. Med.*, **2**, 119 (1963).
361. W. M. Hoskins and H. T. Gordon, *Ann. Rev. Entomol.*, **1**, 89 (1956).
362. W. M. Hoskins and S. Nagasawa, *Botyu-Kagaku*, **26**, 115 (1961).
363. W. M. Hoskins and J. M. Witt, *Proc. 10th Intern. Congr. Entomol.*, **2**, 151 (1958).
364. W. S. Hough, *J. Econ. Entomol.*, **21**, 325 (1928).
365. W. S. Hough, *J. Agr. Res.*, **38**, 245 (1929).
366. W. S. Hough, *J. Agr. Res.*, **48**, 533 (1934).
367. W. S. Hough, *Virginia Agr. Exp. Sta. Tech. Bull.*, **91**, 32 pp. (1943).
368. W. L. Howe, E. E. Ortman, and B. W. George, *Proc. N. Central Branch Entomol. Soc. Am.*, **18**, 83 (1963).
369. A. J. Howitt, *J. Econ. Entomol.*, **51**, 883 (1958).
370. A. J. Howitt and S. G. Cole, *J. Econ. Entomol.*, **52**, 963 (1959).
371. R. F. Hoyer and F. W. Plapp, *J. Econ. Entomol.*, **59**, 495 (1966).
372. R. F. Hoyer, F. W. Plapp, and R. D. Orchard, *Entomol. Exp. Appl.*, **8**, 65 (1965).
373. S. C. Hoyt, *J. Econ. Entomol.*, **59**, 1278 (1966).
374. S. C. Hoyt and F. H. Harries, *J. Econ. Entomol.*, **54**, 12 (1961).
375. J. Hrdy, J. Hurkova and J. Zeleny, *Proc. 13th Intern. Congr. Entomol.*, Moscow (1968).
376. J. Hurkova and J. Hrdy, *Agrochemia*, **6**, 66 (1966).
377. M. Husseiny, Proc. Reg. Symp. Ins. Res. Insecticides, Middle East Science Coop. Office UNESCO, Cairo, pp. 175–177 (1960).
378. N. W. Hussey, *Proc. 3rd Brit. Insecticide Fungicide Conf.*, Ref. No. II-2 (1965).
379. C. H. Hwang, K. H. Paik, and C. M. Kahn, *Korean Cent. J. Med.*, **9**, 161 (1965).
380. H. Ikemoto, *Botyu-Kagaku*, **29**, 68 (1964).
381. E. E. Inwang, M. A. Q. Khan, and A. W. A. Brown, *Bull. World Health Organ.*, **36**, 409 (1967).
382. M. Ishida and P. A. Dahm, *J. Econ. Entomol.*, **58**, 383 and 602 (1965).
383. T. Ishii and M. Sherman, *J. Econ. Entomol.*, **58**, 48 (1965).
384. E. E. Ivy and A. L. Scales, *J. Econ. Entomol.*, **47**, 981 (1954).
385. R. Iyengar, Mim. Doc. WHO/Mal/526 (1965).
386. S. Jatanasen and J. Mouchet, Mim. Doc. WHO/Vector Control/66.222 (1966).
387. L. R. Jeppson, *Misc. Publ. Entomol. Soc. Am.*, **2**, 13 (1960).
388. L. R. Jeppson, *Advan. Acarol. (Ithaca, N.Y.)*, **1**, 276 (1963).
389. L. R. Jeppson and M. J. Jesser, *J. Econ. Entomol.*, **55**, 78 (1962).
390. L. R. Jeppson, M. J. Jesser, and M. R. Complin, *J. Econ. Entomol.*, **51**, 232 (1958).
391. L. R. Jeppson, M. J. Jesser, and M. R. Complin, *J. Econ. Entomol.*, **54**, 439 (1964).
392. L. R. Jeppson, M. R. Complin, and M. J. Jesser, *J. Econ. Entomol.*, **55**, 17 (1962).
393. R. N. Jefferson and F. S. Morishita, *J. Econ. Entomol.*, **49**, 151 (1956).
394. C. Johansen, *J. Econ. Entomol.*, **58**, 857 (1965).

395. H. G. Johnston, *Misc. Publ. Entomol. Soc. Am.*, **2**, 41 (1960).

396. C. M. Jones, *Proc. N. Central Branch, Entomol. Soc. Am.*, **17**, 133 (1962).

397. W. Kalow and W. B. Saunders, 231 pp. Philadelphia, 1962.

398. R. L. Kalra, *Indian J. Exptl. Biol.*, **5**, 187 (1967).

399. R. L. Kalra and G. C. Joshi, *Indian J. Malaria*, **16**, 327 (1962).

400. R. L. Kalra and B. S. Krishnamurthy, *Bull. Indian Soc. Nalan Communicable Diseases*, **2**, 223 (1965).

401. R. L. Kalra and R. Pal, *Bull. Natl. Soc. India Malaria*, **4**, 123 (1959).

402. R. L. Kalra, A. S. Perry and J. W. Miles, *Bull. World Health Organ.*, **37**, 651 (1967).

403. O. M. Kamel and A. M. Gad, *J. Egypt. Public Health Assoc.*, **41**, 263 (1966).

404. O. M. Kamel, *J. Egypt. Public Health Assoc.*, **43**, 223 (1968).

405. N. D. P. Karani and P. B. Menon, *Armed Forces Med. J. India*, **1**, 22 (1960).

406. T. Kasai and Z. Ogita, *Botyu-Kagaku*, **30**, 12 (1965).

407. T. Kasai and Z. Ogita, *SABCO Journal*. **1**, 130 (1965).

408. C. W. Kearns, in *Origins of Resistance to Toxic Agents*, Academic, New York, 1955, pp. 148–159.

409. J. Keiding, *Bull. World Health Organ.*, **29** Suppl., 51 (1963).

410. J. Keiding, *Ann. Rep. Gov. Pest Infest. Lab.*, Springforbi, Denmark, 1963, p. 44 (1964).

411. J. Keiding, Mim. Doc. WHO/Vector Control/160 (1965).

412. J. Keiding, *Ann. Rep. Gov. Pest. Infest. Lab.*, Springforbi, Denmark, 1955–1956, p. 52 (1958).

413. J. Keiding, *Mededel. Landbouwhogeschool Opzoekingsstat. Staat. Gent.* **30**, 1362 (1965).

414. J. Keiding, *Ann. Rep. Gov. Pest Infest. Lab.*, Springforbi, Denmark, 1966, p. 40 (1967).

415. J. C. Keller, P. H. Clark, and C. S. Lofgren, *Pest Control*, **24** (9), 12 (1956).

415a. C. E. Kennett, *J. Econ. Entomol.*, **63**, 1999 (1970).

415b. W. J. Keppler, W. Klassen, and J. B. Kitzmiller, *Mosquito News*, **25**, 415 (1965).

416. J. A. Kerr, S. de Camargo, and Z. H. Abedi, *Mosquito News*, **24**, 276 (1964).

417. R. W. Kerr, *Australian J. Biol. Sci.*, **14**, 605 (1961).

418. R. W. Kerr, *J. Australian Inst. Agr. Sci.*, **30**, 33 (1964).

419. T. W. Kerr and C. E. Olney, *J. Econ. Entomol.*, **52**, 519 (1959).

420. M. A. Q. Khan and A. W. A. Brown, *J. Econ. Entomol.*, **59**, 1512 (1966).

421. M. A. Q. Khan and L. C. Terriere, *J. Econ. Entomol.*, **61**, 732 (1968).

422. N. H. Khan and J. A. Ansari, *Botyu-Kagaku*, **29**, 15 (1964).

423. H. Kikkawa, *Botyu-Kagaku*, **29**, 37 (1964).

424. H. Kikkawa, *Botyu-Kagaku*, **29**, 42 (1964).

425. H. Kikkawa, *Botyu-Kagaku*, **32**, 101 (1967).

426. J. W. Kilpatrick and H. F. Schoof, *J. Econ. Entomol.*, **51**, 18 (1958).

427. T. Kimura and A. W. A. Brown, *J. Econ. Entomol.*, **57**, 710 (1964).

428. T. Kimura, J. R. Duffy, and A. W. A. Brown, *Bull. World Health Organ.*, **32** 557 (1965).

429. J. C. King, *J. Econ. Entomol.*, **47**, 387 (1954).

430. J. C. King, *Am. Naturalist*, **89**, 39 (1955).

431. W. Klassen, *Mosquito News*, **26**, 309 (1966).

432. W. Klassen and A. W. A. Brown, *Can. J. Genetics Cytol.*, **6**, 61 (1964).

433. W. Klassen and F. Matsumura, *Nature*, **209**, 1155 (1966).

434. J. M. Klein and R. Michel, *Bull. Soc. Pathol. Exotique*, **52**, 295 (1959).

435. E. F. Knipling, *Agr. Chem.*, **9** (6), 46 (1954).

436. K. Kojima, T. Ishizuka, and S. Kitakata, *Botyu-Kagaku*, **28**, 17 (1963).
437. K. Kojima, T. Ishizuka, A. Shiino, and S. Kitakata, *Japan. J. Appl. Zool.*, **7**, 63 (1963).
438. K. Kojima, T. Ishizuka, and S. Kitakata, *Botyu-Kagaku*, **28**, 55 (1963).
439. K. Kojima, S. Kitakata, and A. Shine, *Botyu-Kagaku*, **28**, 13 (1963).
440. K. Kojima, Y. Nagae, T. Ishizuka, and A. Shiino, *Botyu-Kagaku*, **23**, 12 (1958).
441. F. Korte, G. Ludwig, and J. Vogel, *Liebig's Ann. Chem.*, **656**, 135 (1962).
442. T. Koshi, R. S. Dixit, and S. L. Perti, *Indian J. Malaria*, **17**, 23 (1963).
443. J. B. Kring, *Conn. Agr. Exp. Sta. New Haven Circ.*, **193**, 23 pp. (1955).
444. J. B. Kring, *J. Econ. Entomol.*, **49**, 557 (1956).
445. J. B. Kring, *J. Econ. Entomol.*, **51**, 823 (1958).
446. B. S. Krishnamurthy and G. C. Joshi, *Indian J. Malaria*, **16**, 137 (1962).
447. B. S. Krishnamurthy and N. N. Singh, *Indian J. Malaria*, **16**, 375 (1962).
448. R. Kuhn and I. Löw, in *Origins of Resistance to Toxic Agents*, Academic, New York. 1955, p. 122.
449. V. Kumar and F. O. Morrison, *Phytoprotection*, **44**, 101 (1963).
450. V. Kumar and F. O. Morrison, *J. Econ. Entomol.*, **60**, 1430 (1967).
451. P. A. van der Laan, *Mededel. Landbouwhogeschool Opzoekingsstat. Staat. Gent.*, **27**, 730 (1962).
452. G. C. Labrecque and H. G. Wilson, *J. Econ. Entomol.*, **53**, 320 (1960).
453. M. Lahav, M. Shilo, and S. Sarig, *Bamidgeh*, **14**, 67 (1967).
454. A. Lakocy, *Biulet. Inst. Ochr. Roslin (Poznan)*, **27**, 129 and 145 (1964).
455. F. A. Lari, *Pakistan J. Health*, **12**, 1 (1962).
456. R. M. Lee and P. Batham, *Entomol. Exp. Appl.*, **9**, 13 (1966).
457. T. Legg and G. J. Shanahan, *Australian Vet. J.*, **30**, 95 (1954).
458. R. Lehr and F. F. Smith, *J. Econ. Entomol.*, **50**, 634 (1957).
459. D. E. Leonard, *J. Econ. Entomol.*, **54**, 880 (1961).
460. L. L. Lewallen, *Mosquito News*, **21**, 310 (1961).
461. J. Lhoste, *Phytiat-Phytopharm.*, **9**, 161 (1960).
462. J. Lhoste, *Cahiers Ingen. Agronom.*, No 163, pp. 21–38 (1962).
463. E. T. Lichtwardt, *J. Heredity*, **47**, 11 (1956).
464. E. T. Lichtwardt, *Entomol. Exp. Appl.*, **7**, 296 (1964).
465. S. E. Lienk, *J. Econ. Entomol.*, **61**, 1130 (1968).
466. S. E. Lienk, R. W. Dean, and P. J. Chapman, *J. Econ. Entomol.*, **45**, 1082 (1952).
467. N. Lieutand, *WHO Inform. Circ. Insecticide Resistance*, **27**, 2 (1961).
468. D. L. Lindgren and L. E. Vincent, *J. Econ. Entomol.*, **58**, 551 (1965).
469. D. A. Lindquist and P. A. Dahm, *J. Econ. Entomol.*, **49**, 579 (1956).
470. D. A. Lindquist, J. R. Brazzel, and T. B. Davish, *J. Econ. Entomol.*, **54**, 299 (1961).
471. H. Lipke and J. Chalkley, *Bull. World Health Organ.*, **30**, 57 (1964).
472. H. Lipke and C. W. Kearns, *J. Biol. Chem.*, **234**, 2123 (1959).
473. H. Lipke and C. W. Kearns, *Advan. Pest Control Res.*, **3**, 253 (1960).
474. N. C. Lloyd, *Agr. Gaz. N.S. Wales*, **71**, 35 (1960).
475. V. K. Lohmeyer, *J. Dep. Agr. S. Australia*, **61**, 127 (1957).
476. K. A. Lord, F. M. Molloy, and C. Potter, *Bull. Entomol. Res.*, **54**, 189 (1963).
477. J. B. Lovell and C. W. Kearns, *J. Econ. Entomol.*, **52**, 931 (1959).
478. J. I. Lowe, *Trans. Am. Fisheries Soc.*, **93**, 396 (1964).
479. W. L. Lowry and R. S. Berger, *J. Econ. Entomol.*, **58**, 590 (1965).
480. W. L. Lowry and C. H. Tsao, *J. Econ. Entomol.*, **54**, 1209 (1961).
481. W. L. Lowry, M. T. Ouye, and R. S. Berger, *J. Econ. Entomol.*, **58**, 781 (1965).
482. W. L. Lowry, R. L. McGarr, O. T. Robertson, R. S. Berger, and H. M. Graham, *J. Econ. Entomol.*, **58**, 732 (1965); **59**, 479 (1966).

483. H. Lüers and V. Bochnig, *Zool. Beitr.*, **8**, 15 (1963).

484. M. Lund, *Nature*, **203**, 778 (1964).

485. G. Lupasco, M. Duport, I. Combiesco, and A. Atanasiu, *Arch. Roumainas Pathol. Exp. Microbiol.*, **22**, 749 (1963).

486. H. F. Madsen and L. A. Falcon, *J. Econ. Entomol.*, **53**, 1083 (1960).

487. H. F. Madsen and S. C. Hoyt, *J. Econ. Entomol.*, **51**, 422 (1958).

488. D. A. Maelzer and R. L. Kirk, *Australian J. Biol. Sci.*, **6**, 244 (1953).

489. D. A. Maelzer and V. K. Lohmeyer, *Dept. Agr. S. Australia, Tech. Bull.*, **30**, (1960).

490. A. A. Maher et al., *Agr. Res. Rev. (Cairo)*, **40**, 51 (1962).

491. Malaria Eradication Div., Mim. WHO/Mal/266 (1960).

492. Malaria Eradication Div., Mim. Doc. WHO/Mal/325 (1961).

493. Malaria Eradication Div., *WHO/Inf. Circ. Insecticide Resistance*, **54**, 21 (1965).

494. A. Mallis and A. C. Miller, *J. Econ. Entomol.*, **57**, 608 (1964).

495. A. Mallis, W. C. Easterlin and R. J. Astor, *Pest Control*, **34** (6), 22 (1965).

496. A. Mansingh, *J. Econ. Entomol.*, **58**, 580 (1965).

497. R. B. March, *Misc. Publ. Entomol. Soc. Am.*, **1**, 13 (1959).

498. R. B. March, R. L. Metcalf, and L. L. Lewallen, *J. Econ. Entomol.*, **45**, 851 (1952).

499. A. Martinez-Palacios, *CNEP Bol. (Dept. Health Mex.)*, **3** (1), 38 (1959).

500. A. Martinez-Palacios and J. de Zulueta, *Nature*, **203**, 940 (1964).

501. G. F. Mason and A. W. A. Brown, *Bull. World Health Organ.*, **28**, 77 (1963).

502. W. Mathis, A. D. Flynn and H. F. Schoof, *J. Econ. Entomol.*, **60**, 1407 (1967).

503. W. Mathis, H. B. Morlan and H. F. Schoof, *Mosquito News*, **25**, 196 (1965).

504. F. Matsumura and A. W. A. Brown, *J. Econ. Entomol.*, **54**, 1176 (1961).

505. F. Matsumura and A. W. A. Brown, *J. Econ. Entomol.*, **56**, 381 (1963).

506. F. Matsumura and A. W. A. Brown, *Mosquito News*, **23**, 26 (1963).

507. F. Matsumura and W. C. Dauterman, *Nature*, **202**, 1356 (1964).

508. F. Matsumura and M. Hayashi, *Mosquito News*, **26**, 190 (1966).

509. F. Matsumura and C. J. Hogendijk, *Entomol. Exp. Appl.*, **7**, 179 (1964).

510. F. Matsumura and C. J. Hogendijk, *J. Agr. Food Chem.*, **12**, 447 (1964).

511. F. Matsumura and G. Voss, *J. Econ. Entomol.*, **57**, 911 (1964).

512. F. Matsumura and G. Voss, *J. Insect Physiol.*, **11**, 147 (1965).

513. K. Matsuo, *Endemic Diseases Bull. Nagasaki Univ.*, **5**, 123 (1963).

514. R. J. McClanahan, C. R. Harris, and L. A. Miller, *Ann. Rep. Entomol. Soc. Ontario*, **89**, 55 (1958).

515. I. C. McDonald and D. G. Cochran, *J. Econ. Entomol.*, **61**, 670 (1968).

515a. I. C. McDonald, M. H. Ross, and D. G. Cochran, *Bull. World Health Organ.*, **40**, 745 (1969).

515b. W. D. McEnroe, personal communication, in **511**.

516. W. C. McDuffie, *Misc. Publ. Entomol. Soc. Am.*, **2**, 49 (1960).

517. F. L. McEwen and R. K. Chapman, *J. Econ. Entomol.*, **45**, 717 (1952).

518. F. L. McEwen and C. M. Splitstoesser, *J. Econ. Entomol.*, **57**, 197 (1964).

519. F. L. McEwen et al., *J. Econ. Entomol.*, **60**, 1261 (1967).

520. R. E. McKenzie and W. M. Hoskins, *J. Econ. Entomol.*, **47**, 984 (1954).

520a. D. G. R. McLeod, C. R. Harris, and G. R. Driscoll, *J. Econ. Entomol.*, **62**, 427 (1969).

521. A. L. Melander, *J. Econ. Entomol.*, **7**, 167 (1914).

522. D. C. Mengle and L. L. Lewallen, *Bull. Entomol. Soc. Am.*, **10**, 165 (1964).

523. H. F. Menke, *J. Econ. Entomol.*, **47**, 704 (1954).

524. D. B. Menzel, R. Craig, and W. M. Hoskins, *J. Insect Physiol.*, **9**, 479 (1963).

525. D. J. Merrell and J. C. Underhill, *J. Econ. Entomol.*, **49**, 300 (1956).

526. R. L. Metcalf, *Physiol. Rev.*, **35**, 197 (1955).

527. R. L. Metcalf and T. R. Fukuto, *Bull. World Health Organ.*, **24**, 670 (1961).
528. R. L. Metcalf and G. P. Georghiou, *Bull. World Health Organ.*, **27**, 251 (1962).
529. R. L. Metcalf, T. R. Fukuto, and M. Y. Winton, *J. Econ. Entomol.*, **53**, 828 (1960).
530. R. L. Metcalf, T. R. Fukuto and M. Y. Winton, *Bull. World Health Organ.*, **29**, 219 (1963).
531. A. E. Michelbacher, O. H. Fullmer, C. C. Cassil, and C. S. Davis, *J. Econ. Entomol.*, **47**, 366 (1954).
532. D. W. Micks, *Bull. World Health Organ.*, **22**, 519 (1960).
533. D. W. Micks, W. M. Cox, and J. C. McNeill, *Mosquito News*, **21**, 229 (1961).
534. R. Milani, *Riv. Parassit.*, **15**, 513 (1954).
535. R. Milani, *Riv. Parassit.*, **17**, 233 (1956); **18**, 43 (1956).
536. R. Milani, *Bull. World Health Organ.*, **29** Suppl., 77 (1963).
537. E. S. Millar, *WHO Inf. Circ. Insecticide Resistance*, **56**, 7 (1966).
538. S. Miller and A. S. Perry, *Agr. Food Chem.*, **12**, 167 (1964).
539. S. S. Miyake, C. W. Kearns, and H. Lipke, *J. Econ. Entomol.*, **50**, 359 (1957).
540. B. N. Mohan, *Indian J. Malaria*, **14**, 175 (1960).
541. B. N. Mohan, *Indian J. Malaria*, **16**, 277 (1962).
542. H. A. U. Monro, *Pest Control*, **32** (7), 11 (1964).
543. H. A. U. Monro, A. J. Musgrave, and E. Upitis, *Ann. Appl. Biol.*, **49**, 373 (1961).
544. B. Moore, R. O. Drummond, and H. M. Brundrett, *J. Econ. Entomol.*, **52**, 980 (1959).
545. R. F. Moore, A. R. Hopkins, H. M. Taft, and L. L. Anderson, *J. Econ. Entomol.*, **60**, 64 (1967).
546. H. H. Moorefield and C. W. Kearns, *J. Econ. Entomol.*, **48**, 403 (1955).
547. H. H. Moorefield, M. H. J. Widen, and D. J. Hennessy, *Contrib. Boyce Thompson Inst.*, **21**, 481 (1962).
548. P. C. Morel, *Rev. Elevage Vet. Pays. Trop.*, **20**, 451 (1967).
549. A. Morello, *Nature*, **203**, 785 (1964).
550. C. V. G. Morgan and N. H. Anderson, *Canad. Entomologist*, **90**, 92 (1958).
551. D. S. Morris and R. Van Baer, *J. Dept. Agr. Vict.*, **57**, 619 (1959).
552. R. F. Morris, *Can. Entomol.*, **95**, 81 (1963).
553. F. O. Morrison, *Ann. Rep. Entomol. Soc. Ont.*, **1950**, 41 (1951).
554. J. Mouchet, *Mim. Doc. WHO/Vector Control/190*, pp. 91–108 (1965).
555. J. Mouchet, *WHO Inform. Circ. Insecticide Resistance*, **55**, 11 (1966).
556. J. Mouchet, *WHO Inform. Circ. Insecticide Resistance*, **56**, 26, 27 (1966).
557. J. Mouchet, *Cahiers ORSTOM* (Paris), *Ser. Med. Entomol.*, **6**, 225 (1968).
558. J. Mouchet and G. Chastel, *Med. Trop.*, **26**, 505 (1966).
559. J. Mouchet and J. Laigret, *Med. Trop.*, **27**, 685 (1967).
560. G. A. Mount, *J. Econ. Entomol.*, **58**, 794 (1965).
561. G. A. Mount, C. S. Lofgren, M. C. Bowman, and F. Acree, *J. Econ. Entomol.*, **59**, 1352 (1966).
562. R. C. Muirhead-Thomson, *Bull. World Health Organ.*, **22**, 721 (1960).
563. M. S. Mulla, *J. Econ. Entomol.*, **55**, 130 (1962).
564. M. S. Mulla, L. W. Isaak, and H. Axelrod, *Mosquito News*, **20**, 256 (1960).
565. F. Munger, J. E. Gilmore, and A. W. Cressman, *J. Econ. Entomol.*, **53**, 384 (1960).
566. A. C. Myburgh and A. J. Heyns, *Farming S. Africa*, **33**, No. 5 (1957).
567. T. Narahashi, *Japan. J. Med. Sci. Biol.*, **17**, 46 (1964).
568. D. J. H. Neate, *Trop. Agr.*, **26**, 93 (1957).
569. P. H. Needham, *Ref. Rothamsted Exp. Sta.*, 1964, p. 164 (1965).
570. J. M. Neely, *Mosquito News*, **25**, 344 (1965).
571. I. Neri, K. R. S. Asher, and E. Mosna, *Indian J. Malaria*, **16**, 41 (1962).

572. E. J. Newcomer, *Proc. Wash. State Hort. Assoc.*, **47**, 67 (1951).

573. E. J. Newcomer and F. P. Dean, *J. Econ. Entomol.*, **46**, 894 (1953).

574. H. D. Newsom, *Proc. Conf. Milit. Entomol.*, *Washington* Oct. 23, p. 80 (1961).

575. L. D. Newsom, *Agr. Chem.*, **12** (1), 48 (1957).

576. W. L. Newton, *Bull. World Health Organ.*, **29**, 539 (1963).

577. V. D. Nguy and J. R. Busvine, *Bull. World Health Organ.*, **22**, 531 (1960).

578. H. D. Niemczyk, *J. Econ. Entomol.*, **58**, 163 (1965).

579. H. D. Niemczyk and C. R. Harris, *J. Econ. Entomol.*, **55**, 560 (1962).

580. H. D. Niemczyk and G. Prins, *J. Econ. Entomol.*, **58**, 1074 (1965).

581. D. M. Norris, *Bull. Entomol. Soc. Am.*, **3**, 40 (1957).

582. V. Notananda, *Minuten-Nadeln*, **1**, 7 (1960).

583. Z. Ogita, *Botyu-Kagaku*, **23**, 188 (1958).

584. Z. Ogita, *Nature*, **182**, 1529 (1958).

585. Z. Ogita, *Botyu-Kagaku*, **26**, 93 (1961).

586. Z. Ogita and T. Kasai, *Japan. J. Genetics*, **40**, 4, 173, 185 (1965).

587. T. A. Omardeen, *Nature*, **190**, 559 (1961).

588. T. A. Omardeen, *Bull. World Health Organ.*, **24**, 495 (1961).

589. J. A. Onsager and J. C. Maitler, *J. Econ. Entomol.*, **59**, 1120 (1966).

590. E. S. Oonnithan and R. Miskus, *J. Econ. Entomol.*, **57**, 425 (1964).

591. A. J. Ophof and D. W. Langeveld, *WHO Working Papers*, INS/IT/WP/68.20 (1968).

592. F. J. Oppenoorth, *Arch. Neerl. Zool.*, **12**, 1 (1956).

593. F. J. Oppenoorth, *Nature*, **181**, 425 (1958).

594. F. J. Oppenoorth, *Entomol. Exp. Appl.*, **2**, 304 (1959).

595. F. J. Oppenoorth, *Ann. Rev. Entomol.*, **10**, 185 (1965).

596. F. J. Oppenoorth, *Mededel. Landbouwhogeschool Opzoekingsstat Staat Gent*, **30**, 1390 (1965).

597. F. J. Oppenoorth and K. van Asperen, *Science*, **132**, 298 (1960).

598. F. J. Oppenoorth and K. van Asperen, *Entomol. Exp. Appl.*, **4**, 311 (1961).

599. F. J. Oppenoorth and N. W. H. Houx, *Entomol. Exp. Appl.*, **11**, 81 (1968).

600. F. J. Oppenoorth and G. E. Nasrat, *Entomol. Exp. Appl.*, **9**, 223 (1966).

601. C. Oshima, *Botyu-Kagaku*, **19**, 93 (1954).

602. C. Oshima, and T. Hiroyoshi, *Botyu-Kagaku*, **21**, 65 (1956).

603. J. D. Ouzts and R. E. Hutchins, *J. Econ. Entomol.*, **55**, 1011 (1962).

604. K. Ozaki, *Botyu-Kagaku*, **27**, 81 (1962).

605. G. W. Ozburn and F. O. Morrison, *Nature*, **196**, 1009 (1962).

606. G. W. Ozburn and F. O. Morrison, *Can. J. Zool.*, **42**, 519 (1964).

607. G. W. Ozburn and F. O. Morrison, *Can. J. Zool.*, **43** 709 (1965).

608. R. Pal, *Indian J. Malaria*, **12**, 383 (1958).

609. R. Pal and R. L. Kalra, Mim. Doc. WHO/Vector Control/120 (1965).

610. R. Pal and R. L. Kalra, Mim. Doc. WHO/Vector Control/122 (1965).

611. C. E. Palm, *Advan. Chem. Ser.*, **1**, 218 (1949).

611a. Pan American Health Organization, pre-publication report (1970).

612. C. R. Parencia and C. B. Cowan, *J. Econ. Entomol.*, **53**, 52 (1960).

613. C. R. Parencia, C. B. Cowan, and J. W. Davis, *J. Econ. Entomol.*, **53**, 1051 (1960).

614. E. A. Parkin and C. J. Lloyd, *J. Sci. Food Agr.*, **11**, 471 (1960).

615. E. A. Parkin, R. Forster, C. E. Dyte, and D. G. Blackman, *Pest Infestation Research*, *Ann. Rept. Pest Inf. Lab. Agr. Res. Council U.K.* (1962). p. 37.

616. T. B. Patel, S. C. Bhatia, and R. B. Deobhankar, *Bull. World Health Organ.*, **23**, 301 (1960).

617. K. K. Patel and J. W. Apple, *J. Econ. Entomol.*, **59**, 522 (1966).

618. R. S. Patterson, C. S. Lofgren and M. D. Boston, *J. Econ. Entomol.*, **60**, 1673 (1967).
619. R. L. Peffly, *Bull. World Health Organ.*, **20**, 757 (1959).
620. R. L. Peffly and A. A. Shawarby, *Am. J. Trop. Med. Hyg.*, **5**, 183 (1956).
621. J. T. Pennell and W. M. Hoskins, *Bull. World Health Organ.*, **31**, 669 (1963).
622. N. E. Pennington, *Mosquito News*, **28**, 193 (1968).
623. A. S. Perry, *Misc. Publ. Entomol. Soc. Am.*, **2**, 119 (1960).
624. A. S. Perry, *Bull. World Health Organ.*, **22**, 743 (1960).
625. A. S. Perry, in *The Physiology of Insecta*, Vol. III, Academic, New York, 1964, pp. 285–378.
626. A. S. Perry and A. J. Buckner, *Am. J. Trop. Med. Hyg.*, **7**, 620 (1958).
627. A. S. Perry and A. J. Buckner, *J. Econ. Entomol.*, **52**, 997 (1959).
628. A. S. Perry and W. M. Hoskins, *Science*, **111**, 600 (1950).
629. A. S. Perry and W. M. Sacktor, *Ann. Entomol. Soc. Am.*, **48**, 329 (1958).
630. A. S. Perry, D. J. Hennessy, and J. W. Miles, *J. Econ. Entomol.*, **60**, 568 (1967).
631. A. S. Perry, A. M. Mattson, and A. J. Buckner, *Biol. Bull.*, **104**, 426 (1953).
632. A. S. Perry, A. M. Mattson, and A. J. Buckner, *J. Econ. Entomol.*, **51**, 346 (1958).
633. A. S. Perry, S. Miller, and A. J. Buckner, *J. Arg. Food Chem.*, **11**, 457 (1963).
634. A. S. Perry, G. W. Pearce, and A. J. Buckner, *J. Econ. Entomol.*, **57**, 867 (1964).
635. Pest Infestation Laboratory, *Report on Pest Infestation Research*, U.K. Agr. Res. Council (1964).
636. D. G. Peterson, *Pesticides Progress, Dept. Agr. Can.*, **4** (3), 58 (1966).
637. A. G. Peterson, M. S. Silberman, and A. B. Meade, *J. Econ. Entomol.*, **56**, 580 (1963).
638. D. P. Pielou, *Can. J. Zool.*, **30**, 375 (1952).
639. D. P. Pielou and R. F. Glasser, *Science*, **115**, 117 (1952).
640. M. K. K. Pillai and A. W. A. Brown, *J. Econ. Entomol.*, **58**, 255 (1965).
641. M. K. K. Pillai, Z. H. Abedi, and A. W. A. Brown, *Mosquito News*, **23**, 112 (1963).
642. M. K. K. Pillai, D. J. Hennessy, and A. W. A. Brown, *Mosquito News*, **23**, 118 (1963).
642a. F. W. Plapp, *J. Econ. Entomol.*, **63**, 1768 (1970).
643. F. W. Plapp and R. F. Hoyer, *J. Econ. Entomol.*, **60**, 768 (1967).
644. F. W. Plapp and R. F. Hoyer, *J. Econ. Entomol.*, **61**, 761 (1968).
645. F. W. Plapp and H. H. C. Tong, *J. Econ. Entomol.*, **59**, 11 (1966).
646. F. W. Plapp, G. A. Chapman, and W. S. Bigley, *J. Econ. Entomol.*, **57**, 692 (1964).
647. F. W. Plapp, G. A. Chapman, and J. W. Morgan, *J. Econ. Entomol.*, **58**, 1064 (1965).
648. F. W. Plapp, R. D. Orchard, and J. W. Morgan, *J. Econ. Entomol.*, **58**, 953 (1965).
649. F. W. Plapp, W. S. Bigley, G. A. Chapman, and G. W. Eddy, *J. Econ. Entomol.*, **56**, 643 (1963).
650. F. W. Plapp, D. E. Borgard, D. I. Darrow, and G. W. Eddy, *Mosquito News*, **21**, 315 (1961).
651. M. Privora, *Proc. 12th Intern. Congr. Entomol., London*, p. 835 (1965).
652. M. Privora and E. Radova, *J. Epidemiol. Microbiol. Immunol. (Praha)*, **6**, 265 (1962).
653. R. L. Post, *Proc. N. Central Branch Entomol. Soc. Am.*, **9**, 94 (1954).
654. R. D. Powell, G. J. Brewer, A. S. Alving, and J. W. Millar, *Bull. World Health Organ.*, **30**, 29 (1964).
655. C. Potter, *Ann. Appl. Biol.*, **28**, 142 (1941).
656. W. L. Putman and D. C. Herne, *Can. Entomologist*, **91**, 567 (1959).
656a. K. D. Quarterman, pre-publication report, AID-USPHS (1969).
657. H. J. Quayle, *Hilgardia*, **11**, 183 (1938).
658. H. J. Quayle, *J. Econ. Entomol.*, **36**, 493 (1943).
659. R. J. Quinton, Ph.D. Thesis, Cornell University (1955).
660. A. H. Qureshi, E. J. Bond, and H. A. U. Monro, *J. Econ. Entomol.*, **58**, 324 (1965).

661. M. Qutubuddin, *Bull. World Health Organ.*, **19**, 1109 (1959).

662. M. Qutubuddin, *Am. J. Trop. Med. Hyg.*, **10**, 773 (1961).

663. Do Van Quy, *Rapp. Ann. Inst. Pasteur Vietnam 1965*, p. 21 (1965).

664. R. L. Rabb and F. E. Guthrie, *J. Econ. Entomol.*, **57**, 995 (1964).

665. R. G. Rachou, M. A. Souza, M. M. Lima, and J. M. P. Memoria, *Rev. Brasil. Malaria*, **12**, 47 (1960).

666. S. J. Rahman, *Angew. Parasit.*, **7**, 115 (1966).

667. L. Rai and C. C. Roan, *J. Econ. Entomol.*, **49**, 591 (1956).

668. L. Rai and C. C. Roan, *J. Econ. Entomol.*, **52**, 218 (1959).

669. S. P. Ramakrishnan, B. S. Krishnamurthy, S. N. Ray, and N. N. Singh, *WHO Inform. Circ. Insecticide Resistance*, **40**, 7 (1963).

670. A. P. Randall, *Can. Entomologist*, **97**, 1281 (1965).

671. V. S. Rao and N. L. Sitaraman, *Bull. Indian Soc. Malar. Comm. Dis.*, **1**, 127 (1964).

672. J. W. Ray, *Nature*, **197**, 1226 (1963).

673. D. C. Read, *Proc. Entomol. Soc. Ontario*, **95**, 128 (1965).

674. D. C. Read, *J. Econ. Entomol.*, **58**, 719 (1965).

675. D. C. Read and A. W. A. Brown, *Can. J. Genet. Cytol.*, **8**, 71 (1966).

676. E. T. Reid, *Central African J. Med.*, **6**, 528 (1960).

677. W. J. Reid and F. P. Cuthbert, *J. Econ. Entomol.*, **49**, 879 (1956).

678. M. Reiff, *Rev. Suisse Zool.*, **63**, 317 (1956).

679. M. Reiff, *Rev. Suisse Zool.*, **65**, 411 (1958).

680. Research & Development Committee, AMCA, *Mosquito News*, **22**, 205 (1962).

681. H. T. Reynolds, *Misc. Pub. Entomol. Soc. Am.*, **2**, 103 (1960).

682. J. G. Robertson, *Can. J. Zool.*, **35**, 629 (1957).

683. G. C. Rock, C. H. Hill, and J. M. Grayson, *J. Econ. Entomol.*, **54**, 88 (1961).

684. J. L. Rodriguez and L. A. Riehl, *J. Econ. Entomol.*, **56**, 509 (1963).

685. L. H. Rolston, R. Mayer, and Y. H. Bang, *Arkansas Farm Res.*, Nov.–Dec., 8 (1965).

686. W. J. Roulston, *Australian J. Agr. Res.*, **15**, 490 (1964).

687. J. S. Roussel, *Misc. Publ. Entomol. Soc. Am.*, **2**, 45 (1960).

688. J. S. Roussel and D. Clower, *Louisiana Agr. Exp. Sta. Circ. 41* (1955).

689. J. S. Roussel and D. Clower, *J. Econ. Entomol.*, **50**, 463 (1957).

690. J. S. Roussel, M. S. Blum, and N. W. Earle, *J. Econ. Entomol.*, **52**, 403 (1959).

691. F. Saba, *Z. Angew. Entomol.*, **48**, 265 (1961).

692. F. Saba, *Entomol. Exp. Appl.*, **4**, 264 (1961).

693. F. Saba, *Anz. Schaedlingskunde*, **35**, 141 (1962).

694. G. Sacca and A. Scirocchi, Mim. Doc. WHO/Vector Control/192 (1966).

695. T. Saito, K. Kojima, and O. Morikawa, *11th Pacific Science Congr., Tokyo*, Symp. 44 (1966).

696. K. G. Samnotra, *Bull. Natl. Soc. India Malaria*, **9**, 417 (1961).

697. K. G. Samnotra, *Bull. Indian Soc. Malaria Communicable Diseases*, **1**, 57 (1964).

698. F. F. Sanchez and M. Sherman, *J. Econ. Entomol.*, **59**, 272 (1966).

699. R. M. Sawicki, M. G. Franco, and R. Milani, *Bull. World Health Organ.*, **35**, 893 (1966).

700. C. H. Schaefer and T. P. Sun, *J. Econ. Entomol.*, **60**, 1580 (1967).

701. R. D. Schonbrod, W. W. Philleo, and L. C. Terriere, *J. Econ. Entomol.*, **58**, 74 (1965).

702. G. G. M. Schulten, *Genetica*, **37**, 207 (1966).

703. M. W. Service and G. Davidson, *Nature*, **203**, 209 (1964).

704. C. Sevintuna and A. J. Musgrave, *Can. Entomologist*, **93**, 545 (1961).

705. G. J. Shanahan, *Proc. 11th Intern. Congr. Entomol.*, **1**, 432 (1961).

706. G. J. Shanahan, personal communication (1969).

707. C. H. Shanks, *J. Econ. Entomol.*, **60**, 968 (1967).
708. M. I. D. Sharma and G. C. Joshi, *Bull. World Health Organ.*, **25**, 270 (1961).
709. M. I. D. Sharma and K. G. Samnotra, *Bull. Nat. Soc. India Malaria*, **10**, 1 (1962).
710. M. I. D. Sharma, B. N. Mohan, and N. N. Singh, *Indian J. Malaria*, **16**, 1 (1962).
711. R. D. Shaw, *Bull. Entomol. Res.*, **56**, 389 (1966).
712. R. D. Shaw and H. A. Malcolm, *Vet. Records*, **76**, 210 (1964).
712a. R. D. Shaw, M. Cook, and R. E. Carson, *J. Econ. Entomol.*, **61**, 1590 (1968).
713. D. R. Shepherd, personal communication, Jan. 7, 1965.
714. F. H. Sherck, *J. Econ. Entomol.*, **53**, 84 (1960).
715. H. Shiino, *Shokubutsu-Boeki (Plant Protection)*, **15**, 206 (1961).
716. O. E. Shipp and J. R. Brazzel, *J. Econ. Entomol.*, **57**, 174 (1964).
717. J. P. Sleesman, *Market Growers J.*, **84** (4), 6 (1955).
718. H. R. Smissaert, *Science*, **143**, 129 (1964).
719. H. R. Smissaert, *Nature*, **205**, 158 (1965).
720. A. G. Smith, *Orchardist New Zeal.*, **34**, 315 (1963).
721. A. J. Smith, *Farming S. Africa*, **20**, 753 (1946).
722. A. N. Smith, *Bull. World Health Organ.*, **21**, 240 (1959).
723. F. F. Smith, *Misc. Pub. Entomol. Soc. Am.*, **2**, 5 (1960).
724. F. F. Smith and R. A. Fulton, *J. Econ. Entomol.*, **44**, 229 (1951).
725. L. C. Smith, *J. Dept. Agr. S. Australia*, **59**, 12 (1955).
726. L. C. Smith and V. K. Lohmeyer, *J. Dep. Agr. S. Australia*, **60**, 185 (1956).
727. W. W. Smith and W. C. Yearian, *J. Kansas Entomol. Soc.*, **37**, 63 (1964).
728. T. Smyth and C. C. Roys, *Biol. Bull.*, **108**, 66 (1955).
729. F. M. Snyder and L. E. Chadwick, *Entomol. Exp. Appl.*, **7**, 229 (1964).
730. M. Soerono, G. Davidson, and D. A. Muir, *Bull. World Health Organ.*, **32**, 161 (1965).
731. R. R. Sokal and T. Hiroyoshi, *J. Econ. Entomol.*, **52**, 1077 (1959).
732. R. R. Sokal and P. E. Hunter, *Ann. Entomol. Soc. Am.*, **48**, 499 (1955).
733. L. Somme, *J. Econ. Entomol.*, **51**, 599 (1958).
734. T. T. Son, *WHO Inform. Circ. Insecticide Resistance*, **44**, 5 (1964).
735. R. D. Speirs, L. M. Redlinger, and H. P. Bolles, *J. Econ. Entomol.*, **60**, 1373 (1967).
736. R. D. Speirs, H. P. Bolles, J. H. Lang, and L. M. Redlinger, *Bull. Entomol. Soc. Am.*, **10**, 176 (1964).
737. E. M. Stafford and F. L. Jensen, *Calif. Agr.*, **7** (4), 6 (1953).
738. R. R. Stapp, *J. Econ. Entomol.*, **57**, 327 (1964).
738a. W. R. Steinhausen, *Z. Angew. Entomol.*, **55**, 107 (1968).
739. V. M. Stern and H. T. Reynolds, *Calif. Agr.*, **11** (4), 14 (1957).
740. V. M. Stern and H. T. Reynolds, *J. Econ. Entomol.*, **51**, 312 (1958).
741. J. Sternburg and C. W. Kearns, *Ann. Entomol. Soc. Am.*, **43**, 444 (1950).
742. J. Sternburg and C. W. Kearns, *J. Econ. Entomol.*, **45**, 497 (1952).
743. J. Sternburg, C. W. Kearns, and H. H. Moorefield, *J. Agr. Food Chem.*, **2**, 1125 (1954).
744. J. Sternburg, E. B. Vinson, and C. W. Kearns, *J. Econ. Entomol.*, **46**, 513 (1953).
745. W. A. Stevenson, W. Kaufmann, and L. W. Sheets, *J. Econ. Entomol.*, **50**, 279 (1957).
746. W. W. A. Stewart, *Pesticide Research Report*, NCPUA, Ottawa, 1964, pp. 191–192.
747. B. F. Stone, *Australian J. Agr. Res.*, **13**, 984 & 1008 (1962).
748. B. F. Stone, *Ann. Rep. Div. Entomol. C.S.I.R.O. Australia*, (1964–5), p. 61.
749. W. A. K. Stubbings, *Fruit Res. Sta. Stellenbosch Entomol. Circ. No. 18* (1948).
750. F. M. Summers, *J. Econ. Entomol.*, **42**, 22 (1949).
751. D. J. Sutherland and L. E. Hagmann, *Proc. N.J. Mosquito Extermination Assoc.*, **30**, 297 (1963).
752. T. Suzuki and R. Kano, *Japan. J. Exp. Med.*, **32**, 309 (1962).

753. T. Suzuki and K. Mizutani, *Japan. J. Exp. Med.*, **32**, 397 (1962).
754. T. Suzuki, S. Hirakoso, and H. Matsunaga, *Japan. J. Exp. Med.*, **31**, 351 (1961).
755. T. Suzuki, Y. Ito, and S. Harada, *Japan. J. Exp. Med.*, **33**, 41 (1963).
756. T. Tadano and A. W. A. Brown, *Bull. World Health Organ.*, **35**, 189 (1966).
757. T. Tadano and A. W. A. Brown, *Bull. World Health Organ.*, **36**, 101 (1967).
758. A. S. Tahori, *J. Econ. Entomol.*, **56**, 67 (1963).
759. A. S. Tahori, *J. Econ. Entomol.*, **59**, 462 (1966).
760. G. Taksdal, *Acta. Agr. Scand.*, **16**, 129 (1966).
761. R. G. Tapley, *E. African Agr. J.*, **23**, 82 (1957).
762. E. A. Taylor and F. F. Smith, *J. Econ. Entomol.*, **49**, 858 (1956).
763. J. N. Telford and A. W. A. Brown, *Can. Entomologist*, **96**, 625 (1964).
764. J. N. Telford and A. W. A. Brown, *Can. Entomologist*, **96**, 758 (1964).
765. J. G. Thomas and J. R. Brazzel, *J. Econ. Entomol.*, **54**, 417 (1961).
766. T. C. A. Thomas, *Report of Entomologist*, Ministry of Health, Sierra Leone (1963).
767. V. Thomas, *Indian J. Malaria*, **16**, 203 (1962).
768. A. S. Tombes and A. J. Forgash, *J. Insect Physiol.*, **7**, 216 (1961).
769. C. Tomizawa and A. A. Maher, *Abstr. 6th Intern. Congr. Plant Protection*, **53** (1967).
770. A. D. Tomlin, *J. Econ. Entomol.*, **61**, 855 (1968).
771. W. A. Tompkins, *Trans. Kentucky Acad. Sci.*, **14**, 43 (1953).
772. J. G. Towgood and A. W. A. Brown, *Can. J. Genetics Cytol.*, **4**, 160 (1962).
773. H. Trapido, *Am. J. Trop. Med. Hyg.*, **1**, 853 (1952).
774. M. Tsukamoto, *Botyu-Kagaku*, **26**, 74 (1961).
775. M. Tsukamoto, *Japan. J. Sanit. Zool.*, **13**, 179 (1962).
776. M. Tsukamoto, *Bull. Entomol. Soc. Am.*, **12**, 292 (1966).
777. M. Tsukamoto and J. E. Casida, *Nature*, **213**, 49 (1967).
778. M. Tsukamoto and M. Ogaki, *Botyu-Kagaku*, **19**, 25 (1954).
779. M. Tsukamoto and R. Suzuki, *Inst. Toxic. Inform. Service (Utrecht)*, **6**, 128 (1963).
780. M. Tsukamoto and R. Suzuki, *Botyu-Kagaku*, **29**, 76 (1964).
781. M. Tsukamoto and R. Suzuki, *Botyu-Kagaku*, **31**, 1 (1966).
782. M. Tsukamoto, T. Narahashi, and T. Yamasaki, *Botyu-Kagaku*, **30**, 128 (1965).
783. M. Tsukamoto, S. P. Shrivastava, and J. E. Casida, *J. Econ. Entomal.*, **61**, 50 (1968).
784. N. Turner, *J. Econ. Entomol.*, **46**, 369 (1953).
785. U.S. Dept. of Agri., *summary prepared for Entomological Society of America Committee on an Insecticide Resistance Survey* (1964).
786. G. Uilenberg, *Rev. Elevage*, **16**, 137 (1963).
787. E. M. Ungureanu, Mim. Doc. WHO/Mal/446 (1964).
788. E. M. Ungureanu and J. Haworth, Mim. Doc. WHO/Mal/482 (1965).
789. G. Unterstenhofer, *Hoefchen Briefe (English Ed.)*, **14**, 1 (1961).
790. M. VandeVrie, *Insecticide Newsletter, Can. Dept. Agr.*, **8** (6), 16 (1959).
791. P. Vermes, *Proc. 6th Intern. Congr. Plant Protection*, **18** (1967).
792. A. Viladebo, *Abstr. 6th Intern. Congr. Plant Protection*, **586** (1967).
793. L. E. Vincent and D. L. Lindgren, *J. Econ. Entomol.*, **60**, 1763 (1967).
794. S. B. Vinson and J. R. Brazzel, *Intern. J. Econ. Entomol.*, **59**, 600 (1966).
795. S. B. Vinson, C. E. Boyd, and D. E. Ferguson, *Science*, **139**, 217 (1963).
796. S. B. Vinson, C. E. Boyd, and D. E. Ferguson, *Herpetologica*, **19**, 77 (1963).
797. G. Voss, *Anz. Schaedlingskunde*, **34**, 76 (1961).
798. G. Voss and F. Matsumura, *Nature*, **202**, 319 (1964).
799. G. Voss and F. Matsumura, *Can. J. Biochem.*, **43**, 63 (1965).
800. G. Voss, W. C. Dauterman, and F. Matsumura, *J. Econ. Entomol.*, **57**, 808 (1964).
801. A. C. Wagenknecht, *Insect Toxicol. Inform. Serv.*, **4**, 148 (1961).

802. D. E. Wagoner, *Genetics*, **57**, 729 (1967).

803. D. L. Watson and J. A. Naegele, *J. Econ. Entomol.*, **53**, 80 (1959).

804. D. L. Watson, C. O. Hansen and J. A. Naegele, *Advan. Acarol. (Ithaca, N.Y.)*, **1**, 248 (1963).

805. B. C. Walton, M. M. Winn, and J. E. Williams, *Am. J. Trop. Med. Hyg.*, **7**, 618 (1958).

806. N. Ward, reported by W. J. Fischang in *Pest Control*, **28** (3), 38 (1960).

807. M. J. Way, *N.A.A.S. Quarterly Review (U.K.)* No. 46, pp. 60–67 (1959).

808. J. E. Webb, *J. Econ. Entomol.*, **54**, 805 (1961).

809. R. E. Webb and F. Horsfall, *Science*, **56**, 1762 (1967).

810. R. L. Webster, *J. Econ. Entomol.*, **26**, 1016 (1933).

811. E. A. Weiant, *Ann. Entomol. Soc. Am.*, **48**, 489 (1955).

812. G. P. Wene, D. M. Tuttle, and L. W. Sheets, *J. Econ. Entomol.*, **53**, 78 (1960).

813. A. S. West, *WHO Inform. Circ. Insecticide Resistance*, **61**, 14 (1967).

814. G. A. Wheatley, *Conn. Agr. Exp. Sta. New Haven Bull.*, **590**, 15 pp. (1955).

815. G. B. Whitehead, *Nature*, **184**, 378 (1959).

816. G. B. Whitehead, *J. Entomol. Soc. S. Africa*, **25**, 121 (1962).

817. G. B. Whitehead and J. A. F. Baker, *Bull. Entomol. Res.*, **51**, 755 (1961).

818. A. B. M. Whitnall, *J. Entomol. Soc. S. Africa*, **21**, 239 (1958).

819. A. B. M. Whitnall and B. Bradford, *Bull. Entomol. Res.*, **38**, 353 (1947).

820. A. B. M. Whitnall and B. Bradford, *Bull. Entomol. Res.*, **40**, 207 (1959).

821. R. Wiesmann, *Mitt. Schweiz. Entomol. Ges.*, **20**, 484 (1947).

822. R. Wiesmann, *Mitt. Biol. Bundesanstalt Land Forstwirtsch. Berlin-Dahelm*, **83**, 17 (1955).

823. R. Wiesmann, *J. Insect Physiol.*, **1**, 187 (1957).

824. F. P. W. Winteringham and P. S. Hewlett, *Chem. Ind.*, **6** (35), 1512 (1964).

825. F. P. W. Winteringham and A. Harrison, *Nature*, **184**, 608 (1959).

826. R. S. Woglum, *J. Econ. Entomol.*, **18**, 593 (1925).

827. D. O. Wolfenbarger, *J. Econ. Entomol.*, **51**, 357 (1958).

828. D. O. Wolfenbarger, *J. Econ. Entomol.*, **53**, 403 (1960).

829. D. J. Womeldorf, P. A. Gillies, and W. H. Wilder, *Proc. Calif. Mosquito Control Assoc.*, **34**, 77 (1966).

830. R. J. Wood, *Bull. Entomol. Res.*, **52**, 541 (1961).

831. R. J. Wood, *Bull. Entomol. Res.*, **53**, 287 (1962).

832. R. J. Wood, *Bull. World Health Organ.*, **32**, 563 (1965).

833. H. C. Woodville, *Plant Pathol.*, **14**, 138 (1965).

834. World Health Organization, *11th Rept. Exp. Committee Malar.*, *Tech. Rept. Series*, **291**, 46 pp. (1964).

835. World Health Organization, *17th Rept. Exp. Committee Insecticides*, *Tech. Rept. Ser.*, **443**, 279 pp. (1970).

836. World Health Organization, *Inform. Circ. Insecticide Resistance*, **58–59**, 13 (1966).

837. World Health Organization, *WHO Chronicle*, **20**, 297 (1966).

838. World Health Organization, *Inform. Circ. Insecticide Resistance*, **60**, 11 (1966).

839. J. W. Wright and R. Pal, *Bull. World Health Organ.*, **33**, 485 (1966).

840. N. Wu and D. Djou, *Acta. Entomol. Sinica*, **13**, 172 (1964).

841. I. J. Wyatt, *Proc. 3rd British Insecticide Fungicide Conf.*, Ref. No. II-5 (1965).

842. R. P. Yadav, H. L. Anderson, and W. H. Long, *J. Econ. Entomol.*, **58**, 1122 (1965).

843. T. Yamasaki and T. Narahashi, *Botyu-Kagaku*, **23**, 146 (1958).

844. T. Yamasaki and T. Narahashi, *Japan. J. Appl. Entomol. Zool.*, **6**, 293 (1962).

845. K. Yasutomi, *Japan. J. Med. Sci. Biol.*, **15**, 29 (1961).

846. K. Yasutomi, *Japan. J. Sanit. Zool.*, **17**, 71 (1966).

847. Y. Yu, S. J. Hsu, and S. C. Lu, *Acta Entomol. Sinica*, **12**, 163 (1963).

848. H. R. Yust, R. A. Fulton, and H. D. Nelson, *J. Econ. Entomol.*, **44**, 833 (1951).

849. H. R. Yust and F. F. Shelden, *Ann. Entomol. Soc. Am.*, **45**, 220 (1952).

850. H. R. Yust, H. D. Nelson, and R. L. Busbey, *J. Econ. Entomol.*, **36**, 744 (1943).

851. I. V. Zilbermints et al., *Selsk. Biologia (Moskva)*, **3**, 125 (1968).

852. G. Zoebelein, *Anz. Schäedlingskunde*, **36**, 1 (1963).

853. A. Q. van Zon and W. Helle, *Entomol. Ext. Appl.*, **10**, 69 (1967).

854. A. Q. van Zon, W. P. J. Overmeer, and W. Helle, *Entomol. Exp. Appl.*, **7**, 270 (1964).

855. J. de Zulueta, *Bull. World Health Organ.*, **20**, 797 (1959).

856. J. de Zulueta, *Riv. Malariol.*, **41**, 169 (1962).

857. J. de Zulueta, *Riv. Malariol.*, **43**, 29 (1964).

AUTHOR INDEX

Numbers in parentheses are reference numbers and indicate that an author's work is referred to although his name is not cited in the text. Numbers in italics show the page on which the complete reference is listed.

X Y Z

SUBJECT INDEX

A

ABATE, residue, 348
Absorption, cuticular, 490
AC-5727, 501
  tolerance, 521
  genetics, 514
Acaricides, 4, 98
  biological activity, *115*
  chemistry, *114*
  cross tolerance, *475*
  mode of action, *114*
  resistance ratios, 474
  structure activity, *115*
  tolerance, 528, 531
  toxicity, *115*
Acarina, 114
  resistance, 462
Acceptable daily intake, 445
1,2-Acenaphthoquinone, *9*
3-(α-Acetonylbenzyl)-4-hydroxy, coumarin, *132*
1-Acetoxy mercuri-2-hydroxyethane, *21*
Acetyl choline, 113, 500
ACh, 105
Acid lead arsenate, tolerance, 463
Aconitase, 495
Acris crepitans, resistance, 459
Acris gryllus, resistance, 460
Acrolein, *66*
Acrylonitrile, *122*
Activated carbon, 393
Active ingredient, analysis, *273*
Aculus cornutus, resistance, *471*
Acute oral toxicity, *427*
2-Acylindandiones, *138*
Adenine diphosphoglucose, 257, 262
Adenosine triphosphate, 25

ADI – acceptable daily intake, *447*
Adsorbents, 338
Adsorption chromatography, 247
Aedes aegypti
  aldrin, 210
  control, 83
  DDT, 202
  isobenzan, 213
  resistance, 460, *484*, 485, 488, 490, 492, 493, 496, 497, 500, 518, 521, 524, 529, 530
  genetics, 506, 508
  reproduction, 517
Aedes albimanus, 485
Aedes albopictus, *484*
Aedes atropalpus, 485
Aedes cantans, *484*
Aedes cantator, *484*
Aedes detritus, 489
Aedes dorsalis, *484*, 485
Aedes fijiensis, 485
Aedes melanimon, *484*, 485
Aedes nigromaculis, *484*, 485
  resistance, 521, 525, 528
Aedes pseudoscutellaris, 485
Aedes sollicitans, *484*, 485
Aedes taeniorhynchus, *484*, 485
Aedes vexans, 485
Aedes viltatus, *484*
Aeneolamia varia, *470*
Aerobacter aerogenes, 204
Aesclepias syriaca, 70
African red tick, 482
Agmatine conjugation, 263
Agranulocytosis, 444
Agricultural insecticides, tolerance, 463
Agricultural insects, resistance, *462*, 468, *469, 470*

W

X

Y